Lecture Notes in Computer Science 9497

Commenced Publication in 1973
Founding and Former Series Editors:
Gerhard Goos, Juris Hartmanis, and Jan van Leeuwen

More information about this series at http://www.springer.com/series/7410

Liqun Chen · Shin'ichiro Matsuo (Eds.)

Security Standardisation Research

Second International Conference, SSR 2015
Tokyo, Japan, December 15–16, 2015
Proceedings

 Springer

Editors
Liqun Chen
Hewlett Packard Laboratories
Bristol
UK

Shin'ichiro Matsuo
NICT
Tokyo
Japan

ISSN 0302-9743   ISSN 1611-3349   (electronic)
Lecture Notes in Computer Science
ISBN 978-3-319-27151-4   ISBN 978-3-319-27152-1   (eBook)
DOI 10.1007/978-3-319-27152-1

Library of Congress Control Number: 2015955372

LNCS Sublibrary: SL4 – Security and Cryptology

Springer Cham Heidelberg New York Dordrecht London

Printed on acid-free paper

Springer International Publishing AG Switzerland is part of Springer Science+Business Media
(www.springer.com)

Preface

The Second International Conference on Research in Security Standardisation was hosted by the Internet Initiative of Japan, in Tokyo, Japan, during December 15–16, 2015. This event was the second in what is planned to become a series of conferences focusing on the theory, technology, and applications of security standards.

SSR 2015 built on the successful SSR 2014 conference, held at Royal Holloway, University of London, UK, in December 2014. The proceedings of SSR 2014, containing 14 papers, were published in volume 8893 of the *Lecture Notes in Computer Science*.

The conference program consisted of two invited talks, 13 contributed papers, and a panel session. We would like to express our special thanks to the distinguished keynote speakers, Kenny Paterson and Pindar Wong, who gave very enlightening talks. Special thanks are due also to the panel organizer, Randall Easter, and the panel members.

Out of 18 submissions from 10 countries, 13 papers were selected, presented at the conference, and are included in these proceedings. The accepted papers cover a range of topics in the field of security standardisation research, including Bitcoin and payment, protocol and API, analysis of cryptographic algorithms, privacy, and trust and formal analysis.

The success of this event depended critically on the hard work of many people, whose help we gratefully acknowledge. First, we heartily thank the Program Committee and the additional reviewers, listed on the following pages, for their careful and thorough reviews. Each paper was reviewed by at least three people, and most by four. A significant amount time was spent discussing the papers. Thanks must also go to the hard-working shepherds for their guidance and helpful advice on improving a number of papers. We also thank the general co-chairs for their excellent organization of the conference.

We sincerely thank the authors of all submitted papers. We further thank the authors of accepted papers for revising papers according to the various reviewer suggestions and for returning the source files in good time. The revised versions were not checked by the Program Committee, and thus authors bear final responsibility for their contents.

Thanks are due to the staff at Springer for their help with producing the proceedings. We must further thank the developers and maintainers of the EasyChair software, which greatly helped simplify the submission and review process.

December 2015

Liqun Chen
Shin'ichiro Matsuo

Security Standardisation Research 2015

Tokyo, Japan
December 15–16, 2015

General Chairs

Yuji Suga	Internet Initiative Japan, Japan
Hajime Watanabe	National Institute of Advanced Industrial Science and Technology, Japan

Program Chairs

Liqun Chen	Hewlett-Packard Laboratories, UK
Shin'ichiro Matsuo	NICT, Japan

Steering Committee

Liqun Chen	Hewlett-Packard Laboratories, UK
Shin'ichiro Matsuo	NICT, Japan
Chris Mitchell	Royal Holloway, University of London, UK
Bart Preneel	Katholieke Universiteit Leuven, Belgium
Sihan Qing	Peking University, China

Program Committee

David Chadwick	University of Kent, UK
Lily Chen	NIST, USA
Liqun Chen	Hewlett-Packard Laboratories, UK
Takeshi Chikazawa	IPA, Japan
Cas Cremers	University of Oxford, UK
Andreas Fuchsberger	Microsoft, Germany
Phillip H. Griffin	Griffin Information Security Consulting, USA
Feng Hao	Newcastle University, UK
Jens Hermans	KU Leuven - ESAT/COSIC and iMinds, Belgium
Dirk Kuhlmann	HP, UK
Eva Kuiper	Hewlett-Packard, Canada
Pil Joong Lee	Postech, Republic of Korea
Peter Lipp	IT-Security, Austria
Joseph Liu	Monash University, Australia
Javier Lopez	University of Malaga, Spain
Shin'Ichiro Matsuo	NICT, Japan
Catherine Meadows	NRL, USA
Jinghua Min	China Electronic Cyberspace Great Wall Co., Ltd., China

Chris Mitchell	Royal Holloway, University of London, UK
Atsuko Miyaji	School of Information Science, Japan Advanced Institute of Science and Technology, Japan
Kenny Paterson	Royal Holloway, University of London, UK
Angelika Plate	HelpAG, UAE
Kai Rannenberg	Goethe University Frankfurt, Germany
Christoph Ruland	University of Siegen, Germany
Mark Ryan	University of Birmingham, UK
Gautham Sekar	The Indian Statistical Institute, India
Ben Smyth	Huawei, France
Jacques Traore	Orange Labs, France
Vijay Varadharajan	Macquarie University, Australia
Claire Vishik	Intel Corporation, UK
Debby Wallner	National Security Agency, USA
Michael Ward	MasterCard, UK
Yanjiang Yang	Institute for Infocomm Research, Singapore

Additional Reviewers

Batten, Ian
Chen, Jiageng
Costello, Craig
Franklin, Joshua
Hegen, Marvin
Kim, Geonwoo
Künnemann, Robert
Lee, Jinwoo

Mancini, Loretta
Moody, Dustin
Omote, Kazumasa
Pape, Sebastian
Schantin, Andreas
Shin, Jinsuh
Slamanig, Daniel

Contents

Bitcoin and Payment

Authenticated Key Exchange over Bitcoin 3
  Patrick McCorry, Siamak F. Shahandashti, Dylan Clarke,
  and Feng Hao

Tap-Tap and Pay (TTP): Preventing the Mafia Attack in NFC Payment 21
  Maryam Mehrnezhad, Feng Hao, and Siamak F. Shahandashti

Protocol and API

Robust Authenticated Key Exchange Using Passwords and Identity-Based
Signatures 43
  Jung Yeon Hwang, Seung-Hyun Kim, Daeseon Choi, Seung-Hun Jin,
  and Boyeon Song

Non-repudiation Services for the MMS Protocol of IEC 61850 70
  Karl Christoph Ruland and Jochen Sassmannshausen

Analysis of the PKCS#11 API Using the Maude-NPA Tool 86
  Antonio González-Burgueño, Sonia Santiago, Santiago Escobar,
  Catherine Meadows, and José Meseguer

Analysis on Cryptographic Algorithm

How to Manipulate Curve Standards: A White Paper for the Black Hat
http://bada55.cr.yp.to 109
  Daniel J. Bernstein, Tung Chou, Chitchanok Chuengsatiansup,
  Andreas Hülsing, Eran Lambooij, Tanja Lange, Ruben Niederhagen,
  and Christine van Vredendaal

Security of the SM2 Signature Scheme Against Generalized Key
Substitution Attacks 140
  Zhenfeng Zhang, Kang Yang, Jiang Zhang, and Cheng Chen

Side Channel Cryptanalysis of Streebog 154
  Gautham Sekar

Privacy

Improving Air Interface User Privacy in Mobile Telephony 165
  Mohammed Shafiul Alam Khan and Chris J. Mitchell

Generating Unlinkable IPv6 Addresses 185
    Mwawi Nyirenda Kayuni, Mohammed Shafiul Alam Khan, Wanpeng Li,
    Chris J. Mitchell, and Po-Wah Yau

Trust and Formal Analysis

A Practical Trust Framework: Assurance Levels Repackaged Through
Analysis of Business Scenarios and Related Risks.................... 203
    Masatoshi Hokino, Yuri Fujiki, Sakura Onda, Takeaki Kaneko,
    Natsuhiko Sakimura, and Hiroyuki Sato

First Results of a Formal Analysis of the Network Time Security
Specification... 218
    Kristof Teichel, Dieter Sibold, and Stefan Milius

Formal Support for Standardizing Protocols with State................. 246
    Joshua D. Guttman, Moses D. Liskov, John D. Ramsdell,
    and Paul D. Rowe

Author Index 267

Bitcoin and Payment

Authenticated Key Exchange over Bitcoin

Patrick McCorry$^{(\boxtimes)}$, Siamak F. Shahandashti, Dylan Clarke, and Feng Hao

School of Computing Science, Newcastle University, Newcastle upon Tyne, UK
{patrick.mccorry,siamak.shahandashti,dylan.clarke,
feng.hao}@ncl.ac.uk

Abstract. Bitcoin is designed to protect user anonymity (or pseudo-nymity) in a financial transaction, and has been increasingly adopted by major e-commerce websites such as Dell, PayPal and Expedia. While the anonymity of Bitcoin transactions has been extensively studied, little attention has been paid to the security of post-transaction correspon-dence. In a commercial application, the merchant and the user often need to engage in follow-up correspondence after a Bitcoin transaction is completed, e.g., to acknowledge the receipt of payment, to confirm the billing address, to arrange the product delivery, to discuss refund and so on. Currently, such follow-up correspondence is typically done in plaintext via email with no guarantee on confidentiality. Obviously, leakage of sensitive data from the correspondence (e.g., billing address) can trivially compromise the anonymity of Bitcoin users. In this paper, we initiate the first study on how to realise end-to-end secure commu-nication between Bitcoin users in a post-transaction scenario without requiring any trusted third party or additional authentication creden-tials. This is an important new area that has not been covered by any IEEE or ISO/IEC security standard, as none of the existing PKI-based or password-based AKE schemes are suitable for the purpose. Instead, our idea is to leverage the Bitcoin's append-only ledger as an additional layer of authentication between previously confirmed transactions. This naturally leads to a new category of AKE protocols that bootstrap trust entirely from the block chain. We call this new category "Bitcoin-based AKE" and present two concrete protocols: one is non-interactive with no forward secrecy, while the other is interactive with additional guarantee of forward secrecy. Finally, we present proof-of-concept prototypes for both protocols with experimental results to demonstrate their practical feasibility.

Keywords: Authenticated key exchange · Bitcoin · Diffie-Hellman · YAK

1 Introduction

Bitcoin [22] is an online currency whose value is not endorsed by any central reserve, but is based on the perception of its users [15]. In recent years it has surged in value, reaching a peak of \$1147 per bitcoin in December 2013.

© Springer International Publishing Switzerland 2015
L. Chen and S. Matsuo (Eds.): SSR 2015, LNCS 9497, pp. 3–20, 2015.
DOI: 10.1007/978-3-319-27152-1_1

The currency is supported by a decentralised network of users whose collective computational power provides a guarantee of integrity for an append-only ledger. Any attempt to change the ledger's history (a history-revision attack [4]) would require an adversary with at least, in theory 51 % of the networks computational resources to be successful[1]. Several central banks have evaluated the value of digital currencies and their potential impact on society [15,25,29].

Bitcoin is increasingly being accepted by many e-commerce websites as a form of payment. For example, Dell, one of the largest computer retailers in the world, now allows customers to use Bitcoin to pay for online purchases on the Dell website [9]. Recently, PayPal [5] and Expedia [24] have also endorsed support for using Bitcoin. Similarly, many community-driven organisations allow anonymous donations using Bitcoin. Examples include the TOR project [27], Mozilla Foundation [21] and the Calyx Institute [18],

While Bitcoin is designed to support anonymity (or pseudonymity) in a transaction, little attention has been paid to the anonymity in the post-payment scenario. As with any on-line payment system, the payer and the payee may need to engage in follow-up correspondence after the payment has been made, e.g., to acknowledge the receipt, to confirm billing information, to amend discrepancies in the order if there are any and to agree on the product delivery or pick-up. Such correspondence can involve privacy-sensitive information, which, if leaked to a third party, may trivially reveal the identity of the user involved in the earlier transaction (e.g., information about product delivery may contain the home address).

Currently, the primary mechanism to support follow-up correspondence after a Bitcoin transaction is through email. The Dell website requires shoppers to provide their email address when making a Bitcoin payment to facilitate post-payment correspondence. The Calyx Institute, a non-profit research organization dedicated to providing "privacy by design for everyone", also recommends using e-mails for follow-up correspondence after a donation is made in Bitcoin. On its website, the instruction is given as the following [18]:

> "Note that if you make a donation by Bitcoin, we have no way to connect the donation with your email address. If you would like us to confirm receipt of the donation (and send a thank you email!), you'll need to send an email with the details of the transaction. Otherwise, you have our thanks for your support in advance".

However, emails are merely a communication medium and have no built-in guarantees of security. First of all, there is no guarantee that the sender of the email must be the same person who made the Bitcoin payment. The details of the transaction cannot serve as a means of authentication, since they are publicly available on the Bitcoin network. Furthermore, today's emails are usually not encrypted. The content of an email can be easily read by third parties (e.g., ISPs) during the transit over the Internet. The leakage of privacy-sensitive information

---

[1] An adversary may not require 51 % of computational power in reality [3,4,10].

in email can seriously threaten the anonymity of the user who has made an "anonymous" payment in Bitcoin previously.

So far the importance of protecting post-payment communication has been largely neglected in both the Bitcoin and the security research communities. To the best of our knowledge, no solution is available to address this practical problem in the real world. This is a gap in the field, which we aim to bridge in our work.

One trivial solution is to apply existing Authenticated Key Exchange (AKE) protocols to establish a secure end-to-end (E2E) communication channel between Bitcoin users. Two general approaches for realising secure E2E communication in cryptography include using 1) PKI-based AKE (e.g., HMQV), and 2) Password-based AKE (e.g., EKE and SPEKE). The former approach would require Bitcoin users to be part of a global PKI system, with each user holding a public key certificate. This is not realistic in current Bitcoin applications. The second approach requires Bitcoin users to have a pre-shared secret password. However, securely distributing pairwise shared passwords over the internet is not an easy task. Furthermore, passwords are a weak form of authentication and they may be easily guessed or stolen (e.g. by shoulder-surfing). A solution that can provide a stronger form of authentication without involving any passwords will be desirable.

Following the decentralised and anonymity-driven nature of the Bitcoin network [17], we propose new AKE protocols to support secure post-payment communication between Bitcoin users, without requiring any PKI or pre-shared passwords. Our solutions leverage the transaction-specific secrets in the confirmed Bitcoin payments published on the public blockchain to bootstrap trust in establishing an end-to-end secure communication channel. Given each party's transaction history and our AKE protocols, both parties are guaranteed to be speaking to the other party who was involved in the transactions, without revealing their real identities.

Contributions. Our contributions in this paper are summarised below.

- We propose two authenticated key exchange protocols – one interactive and the other non-interactive – using transaction-specific secrets and without the support of a trusted third party to establish end-to-end secure communication. These are new types of AKE protocols, since they bootstrap trust from Bitcoin's public ledger instead of a PKI or shared passwords.
- We provide proof-of-concept implementations for both protocols in the Bitcoin Core client with performance measurements. Our experiments suggest that these protocols are feasible for practical use in real-world Bitcoin applications.

Organization. The rest of the paper is organised as follows. Section 2 explains the background of Bitcoin and the ECDSA signature that is used for authenticating Bitcoin transactions. Section 3 proposes two protocols to allow post-payment secure communication between users based on their transaction history. One protocol is non-interactive with no forward secrecy, while the other is interactive with the additional guarantee of forward secrecy. Security proofs for both

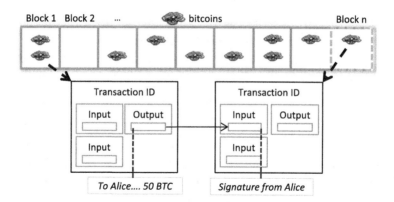

Fig. 1. Transactions stored on the Blockchain based on [19]

protocols are provided in Sect. 4. Section 5 describes the proof-of-concept implementations for both protocols and reports the performance measurements. Finally, Sect. 6 concludes the paper.

2 Background

In this section, we will provide brief background information about the Bitcoin protocol, the transaction signature and the underlying Elliptic Curve Digital Signature Algorithm (ECDSA). This information will be needed for understanding the two protocols presented in this paper.

2.1 Bitcoin

Bitcoin is a digital currency that allows a network of nodes to access a public ledger and to agree upon append-only changes periodically. We will outline the three main mechanisms in the Bitcoin protocol which include Bitcoin addresses, transactions and the Blockchain. Together, they allow users to identify each other pseudonymously, transfer bitcoins and record the transaction in the public ledger.

Each user is responsible for generating their Bitcoin address, which is simply the hash of an ECDSA public key. The corresponding private key is required to spend bitcoins. This approach for user identification is considered appropriate as the probability that two users generate the same public key is negligible due to the high number of possible ECDSA public keys.

A common belief in the community is that Bitcoin offers *pseudonymity* that can help disguise their real-world identity due to the random nature of ECDSA public keys. This belief is bolstered as users are recommended to create a new Bitcoin address per transaction to increase the difficulty of tracking their transactions. However, it should be noted that Bitcoin was not designed with *anonymity*

Algorithm 1. ECDSA Signature Generation algorithm [11]

Input: Domain parameters $D = (q, P, n, \text{Curve})$, private key d, message m.
Output: Signature (r, s).

1: Select $k \in_R [1, n-1]$.
2: Compute $kP = (x_1, y_1)$ where $x_1 \in_R [0, q-1]$
3: Compute $r = x_1 \bmod n$. If $r = 0$, then go to Step 1.
4: Compute $e = H(m)$.
5: Compute $s = k^{-1}(e + dr) \bmod n$. If $s = 0$, then go to Step 1.
6: Return (r, s).

---

in mind [23] and studies have shown with limited success that it is possible to link Bitcoin addresses to real-world identities [3,23,26].

Transactions are created by users to send bitcoins. All transactions are sent to the network and its correctness is verified by other peers before it is accepted into the public ledger. Each transaction has a list of 'inputs' and 'outputs'. The output states the new owner's bitcoin address and the quantity to be transferred. The input will contain a signature to authorise the payment and a reference to a previous transaction whereby the user received the bitcoins. Figure 1 highlights how transactions are linked, which allows peers to perform the verification, by comparing the received transaction with their local copy of the ledger.

A special 'miner' will collect the most recent set of transactions from the network to form a 'block'. This block is appended to the longest chain of blocks (Blockchain) approximately every ten minutes by solving a computationally difficult problem (proof of work) in return for a subsidy of bitcoins. This append-only ledger has become a relatively secure time stamp server [7], since reversing transactions that are committed on the Blockchain is considered infeasible. Figure 1 demonstrates how transactions are stored aperiodically on the Blockchain.

2.2 Transaction Signature

Figure 1 presented earlier demonstrates that the signature is stored in the input of a transaction. This signature must be from the Bitcoin address mentioned in the previous transaction's output. Briefly, it is important to highlight that the user will create the transaction, specify the inputs and outputs, hash this transaction and then sign it using their private key. This prevents an adversary from modifying the contents of a transaction or claiming ownership of the bitcoins before it is accepted into the Blockchain.

Bitcoin incorporates the OpenSSL suite to execute the ECDSA algorithm. The NIST-P256 curve is used and all domain parameters over the finite field including group order n, generator P and modulus q can be found in [6]. An outline of the signature generation algorithm is presented in Algorithm 1 to highlight the usage of k as this will be required for the authenticated key exchange protocols. The verification algorithm follows what is defined in [13]. The notations and symbols used in our paper are summarised in Table 1.

3   Key Exchange Protocols

Key exchange protocols allow two or more participants to derive a shared cryptographic key, often used for authenticated encryption. In this section we will present two authenticated key exchange protocols: Diffie-Hellman-over-Bitcoin and YAK-over-Bitcoin. These protocols will take advantage of a random nonce k from an ECDSA signature. Our aim is to achieve transaction-level authentication by taking advantage of a secret that only exists due to the creation of a transaction that is stored on the Blockchain.

Both of these protocols will use k as a transaction-specific private key and $Q = kP$ as a transaction-specific public key. Diffie-Hellman-over-Bitcoin will be a non-interactive protocol without forward secrecy and YAK-over-Bitcoin will be an interactive protocol with forward secrecy. All domain parameters D for both protocols are the same as the ECDSA algorithm.

Table 1. Summary of notations and symbols

$ZKP\{w\}$	Zero knowledge proof of knowledge of w
(V, z)	Schnorr zero knowledge proof values
$KDF(.)$	Key derivation function
$Uncompress(x, sign)$	Uncompresses public key using x co-ordinate and $sign \in \{+, -\}$
(x, y)	Represents a point on the elliptic curve
P	Generator for the elliptic curve
(r, s)	Signature pair that is stored in a transaction
A, B	Alice and Bob's bitcoin addresses: $H(dP)$
d_A, d_B	Alice and Bob's private key for their Bitcoin address
k_A, k_B	Alice and Bob's transaction-specific private key
$\widehat{k}_A, \widehat{k}_B$	Alice and Bob's estimated transaction-specific private key
Q_A, Q_B	Alice and Bob's transaction-specific public key
$\widehat{Q}_A, \widehat{Q}_B$	Alice and Bob's estimated transaction-specific public key
w_A, w_B	Alice and Bob's ephemeral private keys used for YAK
κ_{AB}	Shared key for Alice and Bob

3.1   Setting the Stage

We will have two actors, Alice and Bob. A single transaction T_A is used by Alice to send her payment (anonymously or not) to Bob. For our protocols, we will assume that Bob has created a second transaction T_b using his ECDSA private key, so the Blockchain contains both Alice and Bob's ECDSA signature. This is a realistic assumption as Bob naturally needs to spend the money or re-organise his bitcoins to protect against theft. In one possible implementation, upon receiving Alice's payment, Bob can send back to Alice a tiny portion of the received amount as acknowledgement, so his ECDSA signature is published on the blockchain (the signature serves to prove that Bob knows the ECDSA private key). This is just one way to ensure that the Blockchain contains both actors' signatures, and there may be many other methods to achieve the same.

Blockchain contains (r_A, s_A) and (r_B, s_B) from T_A and T_B	
Alice (A, d_A)	**Bob (B, d_B)**
1. $k_A = (H(T_A) + d_A r_A) s_A^{-1}$	$k_B = (H(T_B) + d_B r_B) s_B^{-1}$
2. $\widehat{Q}_B = Uncompress(r_B, +)$	$\widehat{Q}_A = Uncompress(r_A, +)$
3. $k_A \widehat{Q}_B = (x_{AB}, \pm y_{AB})$	$k_B \widehat{Q}_A = (x_{AB}, \pm y_{AB})$
$\kappa = \mathrm{KDF}(x_{AB})$	$\kappa = \mathrm{KDF}(x_{AB})$

Fig. 2. The Diffie-Hellman-over-Bitcoin Protocol

The owner of a transaction will be required to derive the transaction-specific private key (random nonce) k from their signature before taking part in the key exchange protocols. For both protocols, we assume the transactions T_A, T_B between Alice and Bob have been sent to the network and accepted to the Blockchain with a depth of at least six blocks, which is considered the standard depth to rule-out the possibility of a double-spend attack [14].

In both protocols, each user will need to extract their partner's signature (r, s) and attempt to derive their partner's transaction-specific public key $Q = (x, y)$. Algorithm 1 demonstrates that the r value from the signature is equal to the x co-ordinate modulo n (note that there is a subtle difference in the data range, since $r \in Z_n$ and $x \in Z_q$, but this has an almost negligible effect on the working of the protocols as we will explain in detail in Sect. 5.2). However, the y co-ordinate of Q is not stored in the transaction, and it can be either of the two values (above/below the x axis).

We define the uncompression function as $Uncompress(x, sign)$ by using the x co-ordinate from their partner's signature and the y co-ordinate's $sign \in \{+, -\}$. Using point uncompression and assuming one of the two possible signs for the y co-ordinate, Alice or Bob will be able to derive a value $\widehat{Q}$ which we call the estimated transaction-specific public key for their partner. This $\widehat{Q}$ could be either $Q = (x, y)$ or its additive inverse $-Q = (x, -y)$. This $\widehat{Q}$ will correspond to the estimated transaction-specific private key $\widehat{k}$, which could be either k or $-k$.

3.2   Authentication

Our definition of authentication will refer to data origin authentication and we will use the Blockchain as a trusted platform for storing digital signatures. Knowledge of the private key d for a bitcoin address or the random nonce k in a signature will prove the identities of pseudonymous parties. We will define two concepts for authentication using Bitcoin:

1. **Bitcoin address authentication.** Knowledge of the discrete log d for a Bitcoin address.
2. **Transaction authentication.** Knowledge of the discrete log k from a single digital signature in a transaction.

Bitcoin address authentication is well-known in the community and has been used for other protocols. However, transaction authentication is a special case

Blockchain contains (r_A, s_A) and (r_B, s_B) from T_A and T_B	
Alice (A, d_A)	**Bob** (B, d_B)
1. $k_A = (H(T_A) + d_A r_A) s_A^{-1}$	$k_B = (H(T_B) + d_B r_B) s_B^{-1}$
2. $\quad Q_A = (r_A, y_A) = k_A P$	$Q_B = (r_B, y_B) = k_B P$
3. $\widehat{Q}_A = Uncompress(r_A, +)$	$\widehat{Q}_B = Uncompress(r_B, +)$
$\quad$ If $Q_A = \widehat{Q}_A$ then $\widehat{k}_A = k_A$	If $Q_B = \widehat{Q}_B$ then $\widehat{k}_B = k_B$
$\quad$ else $\widehat{k}_A = -k_A$	else $\widehat{k}_B = -k_B$
4. $\widehat{Q}_B = Uncompress(r_B, +)$	$\widehat{Q}_A = Uncompress(r_A, +)$
5. $\quad w_A \in_R [1, n-1],$	$w_B \in_R [1, n-1],$
$\quad W_A = w_A P$	$W_B = w_B P$
6. $\quad$ Verify ZKP$\{w_B\}$	Verify ZKP$\{w_A\}$
7. $\quad (x_{AB}, y_{AB}) =$	$(x_{AB}, y_{AB}) =$
$\quad (\widehat{k}_A + w_A)(\widehat{Q}_B + W_B)$	$(\widehat{k}_B + w_B)(\widehat{Q}_A + W_A)$
$\quad \kappa = \mathrm{KDF}(x_{AB})$	$\kappa = \mathrm{KDF}(x_{AB})$

Between steps 5 and 6: $W_A, \text{ZKP}\{w_A\} \longrightarrow$ and $\longleftarrow W_B, \text{ZKP}\{w_B\}$

Fig. 3. YAK-over-Bitcoin Protocol

that our protocols will exploit. Although k and d are equivalent in proving ownership of a Bitcoin address or transaction, k is randomly generated for every ECDSA signature and is unique for each new transaction.

We will show that Alice and Bob can authenticate each other based on the knowledge of the k. This relies on participants trusting the integrity of the Blockchain as the cornerstone for authentication. For an adversary to mount a man-in-the-middle attack in this scene, he would need to perform a history-revision attack to modify the ECDSA signatures stored in the Blockchain.

3.3 Diffie-Hellman-over-Bitcoin Protocol

Based on the concept of transaction authentication, the first protocol that we present is 'Diffie-Hellman-over-Bitcoin'. The protocol is non-interactive; the shared secret is generated using the signatures from two transactions and no additional information from the participants is required. However, forward secrecy is not provided, as we will illustrate in the security analysis.

Figure 2 presents an outline of the protocol. Initially, each user will derive the random nonce k from their own signatures and fetch their partner's transaction from the Blockchain. Each user will gain an estimation of their partner's public key $\widehat{Q}$ before using their own transaction-specific private key k to derive the shared secret $(x_{AB}, \pm y_{AB})$. Regardless of whether $\widehat{Q}_A = \pm Q_A$ (or $\widehat{Q}_B = \pm Q_B$), the x co-ordinate of $k_B \widehat{Q}_A$ will be the same as that of $k_A \widehat{Q}_B$. Following the Elliptic Curve Diffie Hellman (ECDH) [20] approach, the x_{AB} co-ordinate will be used to derive the key $KDF(x_{AB}) = \kappa$.

Algorithm 2. Schnorr Zero Knowledge Proof Generation Algorithm

Input: Domain parameters $D = (q, P, n, \text{Curve})$, signer identity ID, secret value w and public value W.
Output: (V, z)

1: Select $v \in_R [1, n - 1]$, Compute $V = vP$
2: Compute $h = H(D, W, V, ID)$
3: Compute $z = v - wh \mod n$
4: Return (V, z)

3.4 YAK-over-Bitcoin Protocol

The second protocol we present is 'YAK-over-Bitcoin'. This is based on adapting a PKI-based YAK key exchange protocol [12] to the Bitcoin application by removing the dependence on a PKI and instead relying on the integrity of the Blockchain. We chose YAK instead of others (e.g., station-to-station, MQV, HMQV, etc.), as YAK is the only PKI-based AKE protocol that requires each sender to demonstrate the proof of knowledge of both the static and ephemeral private keys. This requirement is important for the security proofs of our system as we will detail in Sect. 4. As well, we will show in the security analysis that the protocol allows the participants to have full forward secrecy.

An outline of our protocol is presented in Fig. 3. Initially, each user will follow the same steps as seen in the previous 'Diffie-Hellman-over-Bitcoin' protocol to derive their secret k and their partner's estimated public key $\widehat{Q}$. However, a subtle difference requires each user to compare their real public key Q with the estimation of their own key $\widehat{Q}$ to determine if they are equal. If these public keys are different, then the user will use the additive inverse of k as their estimated transaction-specific private key and we will denote this choice between the two keys as $\widehat{k}$. This subtle change will allow both parties to derive the same shared secret (x_{AB}, y_{AB}) which would be expected in an interactive protocol without exchanging their real y co-ordinates.

Each user generates an ephemeral private key w and computes the corresponding public key $W = wP$. As required in the original YAK paper [12], each user must also construct a zero knowledge proof to prove possession of the ephemeral private key w. These zero knowledge proofs can be sent over an insecure communication channel to their partners. Here, we will use the same Schnorr signature as in [12] to realise the ZKP. Details of the Schnorr ZKP are summarised in Algorithms 2 and 3. The definition of the Schnorr ZKP includes a unique signer identity ID, which prevents an attacker replaying the ZKP back to the signer herself [12]. In our case, we can simply use the unique r value from the user's ECDSA signature (r, s) in the associated Bitcoin transaction T as the user's identity.

Once the ZKPs have been verified, each user will derive (x_{AB}, y_{AB}) using their secret $w, \widehat{k}$, public value W and their partners' estimated transaction-specific public key $\widehat{Q}$. It should be easy to verify that although the shared secret has four different combinations $(\pm \widehat{k}_A + w_A)(\pm \widehat{k}_B + w_B)P$, the secret key derived between

Alice and Bob will always be identical (due to each participant predicting the estimated public key $\widehat{Q}$ that their partner will choose).

4   Security Analysis

Our protocols are based on reusing the signature-specific random value k in ECDSA as the transaction-specific secret on which the authenticated key exchange protocol is based. Hence, the security of both the ECDSA signature and the key exchange protocols needs to be analysed to make sure the reusing of k is sound in terms of security.

Algorithm 3. Schnorr Zero Knowledge Proof Verification Algorithm

---

Input: Domain parameters $D = (q, P, n, \text{Curve})$, signer identity ID, public value W, Schnorr zero knowledge proof values (V, z)
Output: Valid or invalid

1: Perform public key validation for W [13]
2: Compute partners $h = H(D, W, V, ID)$
3: Return $V \overset{?}{=} zP + hW \mod n$

For the AKE protocols, following the security analysis of YAK [12], we consider three security requirements, informally defined in the following:

– **Private key security:** The adversary is unable to gain any *extra*[2] information about the private key of an honest party by eavesdropping her communication with other parties, changing messages sent to her, or even participating in an AKE protocol with her.
– **Full forward secrecy:** The adversary is unable to determine the shared secret of an eavesdropped AKE session in the past between a pair of honest parties, even if their private keys are leaked subsequently.
– **Session key security:** The adversary is unable to determine the shared secret between two honest parties by eavesdropping their communication or changing their messages.

Note that in our security arguments we consider the security of *shared secrets* (x_{AB} in Figs. 2 and 3), as opposed to that of the subsequently calculated shared *session keys* (κ in the same figures). We henceforth denote the shared secret by K, i.e., $\kappa = \text{KDF}(K)$. We require the shared secret to be hard to determine for the adversary in the full forward secrecy and session key security requirements. A good key derivation function (KDF) derives from such a shared secret a session key which is indistinguishable from random. Our security proofs can be easily adapted to prove indistinguishability based on the decisional rather than computational Diffie-Hellman assumption.

---

[2] By "extra" information, we mean information other than what is derivable from the honest party's already available public key.

For ECDSA signature, we require that it remains unforgeable against chosen-message attacks despite the randomness k being reused in subsequent AKE protocols. Although ECDSA has withstood major cryptanalysis, the security of ECDSA has only been proven under non-standard assumptions or assuming modifications (see [28] for a survey of these results). In our analysis, we consider extra information available to an attacker as a result of k being reused in AKE protocols, and show that it does not degrade the security of ECDSA.

We assume ECDSA to be a (non-interactive honest-verifier) zero-knowledge proof of knowledge of the private key d. This is a reasonable assumption in the random oracle model which follows the work of Malone-Lee and Smart [16][3]. In practice, people accept bitcoin transactions only when the ECDSA signatures are verified successfully[4]. Verifying the ECDSA signature is tantamount to verifying the knowledge of the ECDSA private key d that should only be held by the legitimate bitcoin user.

We also note that given an ECDSA message-signature pair, $m, (r, s)$, knowledge of the private key d is equivalent to knowledge of the randomness k since given either the other can be calculated from $sk = H(m) + dr \bmod n$.

4.1   Security of Diffie-Hellman-over-Bitcoin

This protocol is an Elliptic Curve Diffie-Hellman key exchange and the public values are bound to two transactions in the Blockchain. Private key security considers a malicious active adversary "Mallory", and session key security considers an eavesdropper adversary "Eve". The protocol does not provide full forward secrecy. We will provide a sketch of the proof of security for each property in the following. In each proof, we follow the same approach as in [12] to assume an extreme adversary, who has all the powers except those that would allow the attacker to trivially break any key exchange protocol.

Theorem 1 (Private Key Security). *Diffie-Hellman-over-Bitcoin provides private key security under the assumption that ECDSA signature is a zero knowledge proof of knowledge of the ECDSA secret key.*

---

[3] Note that the results apply to a slightly modified version of ECDSA in which $e = H(r|m)$ where | denotes concatenation. Although the Bitcoin Core implementation is based on the original ECDSA standard, the above modification is included in more recent standards of ECDSA such as ISO/IEC 14888 [1]. Furthermore, as another option for signing, the Bitcoin community is considering including Schnorr signature [2], which is proven to be a zero-knowledge proof of knowledge of the private key.

[4] A bug in the Bitcoin implementation for the SIGHASH_SINGLE flag allows the message that is signed to authorise the transaction to be 1 instead of the hash of the transaction [8]. This bug is not likely to be fixed in the near-future as it is consensus-critical code. To address this bug, we assume that an implementation of our protocol properly checks that the message signed is a hash of a valid transaction as published on the Blockchain rather than 1.

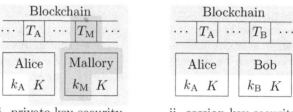

i. private key security   ii. session key security

Fig. 4. Security of Diffie-Hellman-over-Bitcoin. Light grey denotes what the adversary (Mallory in (i), Eve in (ii)) *knows*. Dark grey denotes what the adversary (Mallory) *chooses*.

Proof (sketch). The goal of an adversary Mallory is to be able to gain some *extra* information on Alice's transaction-specific private key k_A through the following attack. Mallory is given the public parameters of the system and access to the Blockchain which includes Alice's transaction T_A, then she provides a transaction T_M which is included in the Blockchain, then she carries out a Diffie-Hellman-over-Bitcoin protocol with Alice (which is non-interactive), and eventually is able to calculate the shared secret K. The attack is depicted in Fig. 4(i). Alice's ECDSA signature in T_A is assumed to be zero knowledge and hence does not reveal any information about her private key. Furthermore, since Mallory's transaction T_M includes an ECDSA signature by her, and ECDSA signature is a proof of knowledge of Mallory's ECDSA secret key d_M, Mallory must know d_M, and hence k_M. Hence, Mallory does not gain any extra knowledge from calculating K, since knowledge of k_M and Alice's public key enables her to simulate K on her own.   □

Theorem 2 (Session Key Security). *Diffie-Hellman-over-Bitcoin provides session key security based on the computational Diffie-Hellman assumption under the assumption that ECDSA signature is a zero knowledge proof of knowledge of the ECDSA secret key.*

Proof (sketch). Assume there is a successful adversary Eve that is able to calculate the shared secret K for a key exchange between two honest parties Alice and Bob, without knowing either Alice or Bob's transaction-specific secret keys, k_A or k_B. The attack is depicted in Fig. 4(ii). Note that since the protocol is non-interactive, the adversary is reduced to a passive adversary. A successful attack would contradict the computational Diffie-Hellman (CDH) assumption since given an instance of the CDH problem $(P, \alpha P, \beta P)$, one is able to leverage Eve and solve the CDH problem by setting up Alice and Bob's transaction-specific secrets as $k_A = \alpha$ and $k_B = \beta$, which results in $K = \alpha\beta P$. A successful Eve implies that CDH can be solved efficiently.   □

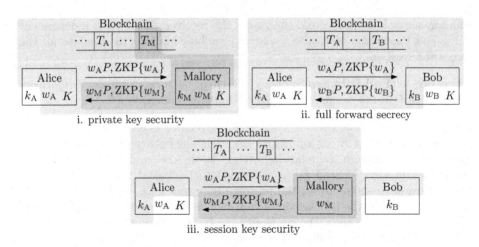

Fig. 5. Security of YAK-over-Bitcoin. Light grey denotes what the adversary (Mallory in (i) and (iii), Eve in (ii)) *knows*. Dark grey denotes what the adversary (Mallory) *chooses*.

4.2 Security of YAK-over-Bitcoin

This protocol is an Elliptic Curve YAK key exchange and the public values are bound to two transactions in the Blockchain. Private key security and session key security consider a malicious active adversary "Mallory", and full forward secrecy considers an eavesdropper adversary "Eve". Similar as before, we assume an extreme adversary who has all the powers except those that would trivially allow the attacker to break any key exchange protocol. Under this assumption, we provide a sketch of the proof of security for each property in the following.

Theorem 3 (Private Key Security). *YAK-over-Bitcoin provides private key security under the assumption that ECDSA signature is a zero knowledge proof of knowledge of the ECDSA secret key.*

Proof (sketch). The goal of an adversary Mallory is to be able to gain some *extra* information on Alice's transaction-specific private key k_A through the following attack. Mallory is given the public parameters of the system and access to the Blockchain which includes Alice's transaction T_A, then she provides a transaction T_M which is included in the Blockchain, then she carries out a YAK-over-Bitcoin protocol with Alice, in which Alice sends the message $(w_A P, \text{ZKP}\{w_A\})$ and Mallory sends the message $(w_M P, \text{ZKP}\{w_M\})$. Alice's ephemeral secret w_A is also assumed to be leaked to Mallory. The attack is depicted in Fig. 5(i). Alice's ECDSA signature in T_A is assumed to be zero knowledge and hence does not reveal any information about her private key. Furthermore, since the ECDSA signature in Mallory's transaction and her message in the protocol are proofs of knowledge of d_M (equivalently k_M) and w_M, respectively, Mallory must know both k_M and w_M. Note that she receives $(w_A P, \text{ZKP}\{w_A\})$ and w_A and hence will be able to calculate $K = (k_M + w_M)(k_A P + w_A P)$. Hence, Mallory does not gain any extra knowledge from the values she receives, since w_A is independent of k_A and knowledge of w_A, k_M, and w_M enables Mallory to simulate all the values she receives.   □

Theorem 4 (Full Forward Secrecy). *YAK-over-Bitcoin provides full forward secrecy based on the computational Diffie-Hellman assumption.*

Proof (sketch). Assume there is a successful adversary Eve that is able to calculate the shared secret K for a previous key exchange between two honest parties Alice and Bob through the following attack. Both Alice and Bob's transaction-specific secret keys k_A and k_B are assumed to be leaked to Eve. Eve is also assumed to have access to all the protocol messages exchanged between Alice and Bob, as well as the Blockchain of course. The attack is depicted in Fig. 5(ii). Given an instance of the CDH problem $(P, \alpha P, \beta P)$ one is able to leverage Eve and solve the problem as follows. The protocol is set up with the ephemeral secret values $w_A = \alpha$ and $w_B = \beta$ and all other parameters as per the protocol description. When Eve calculates K, the value $S = K - k_A k_B P - k_A(\beta P) - k_B(\alpha P)$ is calculated and returned as the solution to the CDH problem. Note that since $K = (k_A + w_A)(k_B + w_B)P$, we have $S = \alpha \beta P$. A successful Eve implies that CDH can be solved efficiently.   □

Theorem 5 (Session Key Security). *YAK-over-Bitcoin provides session key security based on the computational Diffie-Hellman assumption under the assumption that ECDSA signature is a zero knowledge proof of knowledge of the ECDSA secret key.*

Proof (sketch). Assume there is a successful adversary Mallory that is able to calculate the shared secret K for a key exchange between two honest parties Alice and Bob through the following attack by impersonating Bob to Alice. Alice believes she is interacting with Bob, whereas in reality she is interacting with an impersonator Mallory who replaces Bob's message in the protocol with her own $(w_M P, \mathrm{ZKP}\{w_M\})$. Alice's transaction-specific secret key k_A is assumed to be leaked to Mallory as well. However, Mallory does not know Bob's transaction-specific secret key k_B. The attack is depicted in Fig. 5(iii). Given an instance of the CDH problem $(P, \alpha P, \beta P)$ one is able to leverage Mallory and solve the problem as follows. The protocol is set up with Alice's ephemeral secret $w_A = \alpha$ and Bob's transaction-specific secret $k_B = \beta$ and all other parameters as per the protocol description. When Mallory calculates K, the value $S = K - k_A w_B P - w_A(\beta P) - w_B(\alpha P)$ is calculated and returned as the solution to the CDH problem. Note that since $K = (k_A + w_A)(k_B + w_B)P$, we have $S = \alpha \beta P$. A successful Mallory implies that CDH can be solved efficiently.   □

4.3   Security of ECDSA Signatures

Diffie-Hellman-over-Bitcoin is a non-interactive protocol and the protocol participants do not send any messages to each other that would potentially have an impact on the security of ECDSA signatures.

In 'YAK-over-Bitcoin', the messages that the protocol participants send each other include information about their ephemeral keys w_A and w_B only, which are chosen independently of all the secret values related to the ECDSA signatures in T_A and T_B. As shown in Theorem 1 in Sect. 4.2, the protocol does not reveal any information about the static private key (i.e., k), and hence not any information about the ECDSA private key (i.e., d) since the two values are linearly related. One can compute d from k, or vice versa. The key element in the proof of Theorem 3 is that each party is required to demonstrate knowledge of both the static and ephemeral keys. This also explains our choice of the YAK protocol, as YAK is the only PKI-based AKE protocol that has the requirement that each party must demonstrate the proof of knowledge for both the static and ephemeral keys (the former is realized by the Proof of Possession at the Certificate Authority registration while the later is achieved by Schnorr Non-interactive ZKP).

Table 2. Time performance for Alice executing Diffie-Hellman-over-Bitcoin and YAK-over-Bitcoin

Step	Description	Time
	Diffie-Hellman-over-Bitcoin	
1-2	Compute k_A and $\widehat{Q}_B$	0.08 ms
3	Compute shared secret K_{AB} and key κ_{AB}	0.51 ms
	Total:	0.59 ms
	YAK-over-Bitcoin	
1-4	Compute $k_A, Q_A, \widehat{Q}_A$ and $\widehat{Q}_B$	0.53 ms
5	Compute w_A, W_A and ZKP$\{w_A\}$	0.90 ms
6	Verify Bob's ZKP$\{w_B\}$	0.69 ms
7	Compute shared secret K_{AB} and key κ_{AB}	0.43 ms
	Total:	2.55 ms

5   Implementation

Our implementation is a modification of the Bitcoin Core client and is considered a proof of concept. We have included three new remote procedure commands (RPC) that will allow the client to perform a non-interactive Diffie-Hellman key exchange, generate a zero knowledge proof to be shared with their partner and verify a partner's zero knowledge proof before revealing the shared secret. Our modified implementation was executed using the -txindex parameter which allows us to query the Blockchain and retrieve the raw transaction data.

Two transactions were created using a non-modified implementation on the 10th December, 2013 to allow us to test our key exchange on the real network. All tests were carried out a MacBook Pro mid-2012 running OS X 10.9.1 with 2.3GHz Intel Core i7 and 4 cores and 16 GB DDR3 RAM. Each protocol is executed 100 times from Alice's perspective and the average times are reported.

5.1   Time Analysis

Preliminary steps for both protocols involve fetching the transactions from the Blockchain 0.04 ms and retrieving the signatures (r, s) stored in the transaction 0.08 ms. Overall, these steps on average require 0.12 ms.

This 'Diffie-Hellman-over-Bitcoin' protocol is non-interactive as participants are not required to exchange information before deriving the shared secret. Table 2 demonstrates an average time of 0.08 ms to derive Alice's transaction-specific private key k_A and Bob's estimated public key $\widehat{Q}_B$ and 0.051 ms to compute the shared key κ_{AB}. Overall, on average the protocol executes in 0.59 ms which is reasonable for real-life use.

The 'YAK-over-Bitcoin' protocol is interactive as it requires each party to send an ephemeral public key together with a non-interactive Schnorr ZKP to prove the knowledge of the ephemeral private key. Table 2 shows that computing and verifying zero knowledge proofs is the most time-consuming operation. However, a total execution time of 2.55 ms is still reasonable for practical applications.

5.2   Note About Domain Parameters

Our investigation highlighted that $q > n$ as seen in [6] which could obscure the relationship between k and r as the x co-ordinate can wrap around n. However, the probability that this may occur can be calculated as $(q - n)/q \approx 4 \times 10^{-39}$ and is unlikely to occur in practice. However, in the rare chance that this does happen then it is easily resolved by $r' = r + n$. This does not require any modification to the underlying signature code as it is simply an addition of the publicly available r with the modulus n. Once resolved, both parties can continue with the protocol. For reference, q and n are defined below:

```
q=FFFFFFFF FFFFFFFF FFFFFFFF FFFFFFFF FFFFFFFF FFFFFFFF FFFFFFFE FFFFFC2F
n=FFFFFFFF FFFFFFFF FFFFFFFF FFFFFFFE BAAEDCE6 AF48A03B BFD25E8C D0364141
```

6   Conclusion

In this paper, we have demonstrated transaction authentication by using the digital signatures stored in Bitcoin transactions to bootstrap key exchange. We proposed two protocols to allow for interactive and non-interactive key exchange, the latter offering an additional property of forward-secrecy. We encourage the community to try our proof-of-concept implementation and to take advantage of this new form of authentication to enable end-to-end secure communication between Bitcoin users.

Acknowledgements. The second, third and fourth authors are supported by the European Research Council (ERC) Starting Grant (No. 306994). We also thank Greg Maxwell for bringing the SIGHASH_SINGLE implementation bug to our attention.

References

1. ISO/IEC 14888: Information technology - Security techniques - Digital signatures with appendix (2008)
2. Andersen, G.: Conversation about OP_SCHNORRVERIFY. Freenode IRC bitcoin-wizards, October 2014. https://botbot.me/freenode/bitcoin-wizards/
3. Androulaki, E., Karame, G.O., Roeschlin, M., Scherer, T., Capkun, S.: Evaluating user privacy in bitcoin. In: Sadeghi, A.-R. (ed.) FC 2013. LNCS, vol. 7859, pp. 34–51. Springer, Heidelberg (2013)
4. Barber, S., Boyen, X., Shi, E., Uzun, E.: Bitter to better — how to make bitcoin a better currency. In: Keromytis, A.D. (ed.) FC 2012. LNCS, vol. 7397, pp. 399–414. Springer, Heidelberg (2012)
5. BBC: New Paypal partnership enables limited Bitcoin payments (2015). http://www.bbc.co.uk/news/technology-29341886. Accessed 06 January 2015
6. Research, Certicom: SEC 2: Recommended Elliptic Curve Domain Parameters. Standards for Efficient Cryptography Group, September 2000
7. Clark, J., Essex, A.: CommitCoin: carbon dating commitments with bitcoin. In: Keromytis, A.D. (ed.) FC 2012. LNCS, vol. 7397, pp. 390–398. Springer, Heidelberg (2012)
8. Corallo, M.: [Bitcoin-development] Warning to rawtx creators: bug in SIGHASH_SINGLE (2012). http://sourceforge.net/p/bitcoin/mailman/message/29699385/. Accessed 16 September 2015
9. Dell: Were Now Accepting Bitcoin on Dell.com (2015). http://en.community.dell.com/dell-blogs/direct2dell/b/direct2dell/archive/2014/07/18/we-re-now-accepting-bitcoin-on-dell-com. Accessed January 06 2015
10. Eyal, I., Sirer, E.G.: Majority is not enough: bitcoin mining is vulnerable (2013). arXiv preprint arXiv:1311.0243
11. Hankerson, D., Vanstone, S., Menezes, A.: Guide to Elliptic Curve Cryptography. Springer Professional Computing. Springer, New York (2004)
12. Hao, F.: On robust key agreement based on public key authentication. In: Sion, R. (ed.) FC 2010. LNCS, vol. 6052, pp. 383–390. Springer, Heidelberg (2010)
13. Johnson, D., Menezes, A., Vanstone, S.: The elliptic curve digital signature algorithm (ECDSA). Int. J. Inf. Secur. 1(1), 36–63 (2001)
14. Karame, G.O., Androulaki, E., Capkun, S.: Double-spending fast payments in bitcoin. In: Proceedings of the 2012 ACM Conference on Computer and Communications Security, pp. 906–917. ACM (2012)
15. Lo, S., Wang, J.: Bitcoin as money? current policy and perspectives, September 2014
16. Malone-Lee, J., Smart, N.P.: Modifications of ECDSA. In: Nyberg, K., Heys, H.M. (eds.) SAC 2002. LNCS, vol. 2595, pp. 1–12. Springer, Heidelberg (2003)
17. Maurer, B., Nelms, T., Swartz, L.: When perhaps the real problem is money itself!: the practical materiality of Bitcoin. Soc. Semiot. 23(2), 261–277 (2013)
18. Merrill, N.: The Calyx institute: privacy by design for everyone (2015). https://www.calyxinstitute.org/support-us/donate-via-bitcoin. Accessed January 06 2015
19. Miers, I., Garman, C., Green, M., Rubin, A.: Zerocoin: anonymous distributed E-cash from Bitcoin. In: 2013 IEEE Symposium on Security and Privacy (SP), pp. 397–411. IEEE (2013)
20. Miller, V.S.: Use of elliptic curves in cryptography. In: Williams, H.C. (ed.) CRYPTO 1985. LNCS, vol. 218, pp. 417–426. Springer, Heidelberg (1986)

21. Mozilla: Help protect the open Web (2015). https://sendto.mozilla.org/page/content/give-bitcoin/. Accessed January 06 2015
22. Nakamoto, S.: Bitcoin: a peer-to-peer electronic cash system (2008)
23. Reid, F., Harrigan, M.: An analysis of anonymity in the bitcoin system. In: 2011 IEEE Third International Conference on Social Computing (socialcom) Privacy, Security, Risk and Trust (Passat), pp. 1318–1326, October 2011
24. Rizzo, P.: Expedia exec says bitcoin spending has exceeded estimates (2015). http://www.coindesk.com/expedia-exec-bitcoin-payments-have-exceeded-estimates/. Accessed January 06 2015
25. Robleh, A., Barrdear, J., Clews, R., Southgate, J.: The economics of digital currencies. Q. Bull. **54**, Q3 (2014)
26. Ron, D., Shamir, A.: Quantitative analysis of the full bitcoin transaction graph. In: Sadeghi, A.-R. (ed.) FC 2013. LNCS, vol. 7859, pp. 6–24. Springer, Heidelberg (2013)
27. Tor: Make A Donation. 2015. https://www.torproject.org/donate/donate.html.en. Accessed January 06 2015
28. Vaudenay, S.: The security of DSA and ECDSA. In: Desmedt, Y.G. (ed.) PKC 2003. LNCS, vol. 2567, pp. 309–323. Springer, Heidelberg (2002)
29. Woo, D., Gordon, I., Iaralov, V.: Bitcoin: a first assessment. Bank of America Merrill Lynch, December 2013

Tap-Tap and Pay (TTP): Preventing the Mafia Attack in NFC Payment

Maryam Mehrnezhad[✉], Feng Hao, and Siamak F. Shahandashti

School of Computing Science, Newcastle University, Newcastle upon Tyne, UK
{m.mehrnezhad,feng.hao,siamak.shahandashti}@newcastle.ac.uk

Abstract. Mobile NFC payment is an emerging industry, estimated to reach \$670 billion by 2015. The Mafia attack presents a realistic threat to payment systems including mobile NFC payment. In this attack, a user consciously initiates an NFC payment against a legitimate-looking NFC reader (controlled by the Mafia), not knowing that the reader actually relays the data to a remote legitimate NFC reader to pay for something more expensive. In this paper, we present "Tap-Tap and Pay" (TTP), to effectively prevent the Mafia attack in mobile NFC payment. In TTP, a user initiates an NFC payment by physically tapping her mobile phone against the reader twice in succession. The physical tapping causes transient vibrations at both devices, which can be measured by the embedded accelerometers. Our experiments indicate that the two measurements are closely correlated if they are from the same tapping, and are different if obtained from different tapping events. By comparing the similarity between the two measurements, we can effectively tell apart the Mafia fraud from a legitimate NFC transaction. To evaluate the practical feasibility of this solution, we present a prototype of the TTP system based on a pair of NFC-enabled mobile phones and also conduct a user study. The results suggest that our solution is reliable, fast, easy-to-use and has good potential for practical deployment.

Keywords: Near Field Communication · Mobile NFC payment · Mafia attack · MITM attack · Mobile sensor · Accelerometer · Security · Usability

1 Introduction

NFC Payment: Near Field Communication (NFC) payment is an upcoming technology that uses Radio Frequency Identification (RFID) to perform contactless payments. An RFID system has two parts: the RFID tag (card) that can be attached to any physical object to be identified; and the RFID reader that can interrogate a tag within physical proximity, via radio frequency communication. An NFC-enabled payment card has an embedded RFID tag. To make an NFC payment, the user just needs to hold the card in front of an NFC reader for a short while and wait for confirmation. NFC payments are usually limited to rather small-value purchases[1].

---

[1] For instance, the contactless limit increased from £20 to £30 in 2015 in the UK.

© Springer International Publishing Switzerland 2015
L. Chen and S. Matsuo (Eds.): SSR 2015, LNCS 9497, pp. 21–39, 2015.
DOI: 10.1007/978-3-319-27152-1_2

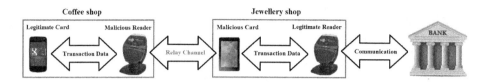

Fig. 1. The Mafia attack: a malicious reader colludes with a malicious card and fools the honest card to pay for something more expensive to a legitimate reader

A mobile phone can also be used as an NFC payment card. HSBC Hong Kong Mobile Payment[2], Google Wallet[3], Apple Pay[4], and Android Pay[5] are examples of NFC payment mobile apps. Using a mobile phone for NFC payment is considered convenient since people can save all of their cards in their phones. It is estimated that mobile payments using NFC will total 670 billion US dollars by 2015 [9]. To support this trend, new generations of smart phones have commonly been equipped with NFC sensors. In this paper, we focus on mobile payment using NFC. Hence unless stated otherwise, by "NFC card", we refer to an NFC-enabled mobile phone functioning as a payment card. By "NFC reader", we refer to a payment terminal that communicates with the card via NFC. A *legitimate* NFC reader is one that is authorised by the banking network and is connected to the back-end banking network for payment processing.

It is known that NFC payment is vulnerable to different types of Man-In-The-Middle (MITM) attacks [21], also known in the literature as relay, or wormhole attacks [19]. In a simple form of a relay attack known as ghost-and-leech attack [22], the attacker places an NFC reader so as to secretly interrogate the user's NFC card without the user's awareness, and relays the card response to a remote NFC reader to obtain a payment from the victim's account. Such an attack is demonstrated in [20,21].

Relay attacks can be countered in a number of ways. A simple solution is to put the NFC card within an NFC protective shield such as Id Stronghold[6]. Equivalently, one can add an activation button so that the NFC function on the phone is only turned on with an explicit user action. More advanced countermeasures are proposed in the literature, including *Secret Handshakes* [18], *UWave* [32], *Still and Silent* [37], and *Tap-Wave-Rub* [30]. However, none of these solutions can prevent a more severe type of attack as we explain below.

Mafia Attack: Another type of the MITM attack is called the Mafia attack, which is also known as Mafia fraud [19] or the reader-and-ghost attack [22,38]. In this more severe attack, the user consciously initiates an NFC payment with a legitimate-looking reader controlled by the Mafia; but the reader actually relays the card response to a remote legitimate NFC reader – via a malicious card – to

---

[2] www.hsbc.com.hk.

[3] http://wallet.google.com.

[4] www.apple.com/iphone-6/apple-pay.

[5] www.android.com/intl/en_us/pay.

[6] www.idstronghold.com.

pay for something more expensive. Figure 1 shows an example of such an attack. This attack has been shown to be feasible in [19].

Unlike simple relay attacks, the Mafia attack cannot be prevented by using a protective shield or an activation button since the user consciously initiates the payment. For the same reason, various user-movement-based unlocking mechanisms [18,30,32,37] cannot stop the attack either. We will explain the current countermeasures to this attack by first reviewing the NFC payment standards and specifications.

NFC Payment Standards and Specifications: EMV is the primary protocol standard for smart card payments in Europe. The EMV standards are managed by EMVCo[7], a consortium of multinational companies such as Visa, Mastercard, and American Express. These standards use smart-cards including contact and contactless cards and are based on ISO/IEC 7816 [4] and ISO/IEC 14443. Mobile NFC payment technologies, such as Android Host-based Card Emulation (HCE)[8], are also based on ISO/IEC 14443, which is an international standard in four parts, defining the technology-specific requirements for proximity cards used for identification [2,3,7,8].

The extensive EMV specifications—presented in 10 books: A [10], B [11], C1–C7 (e.g. [12,13]), and D [14]—provide the details of EMV-compliant payment system design. Furthermore, EMVCo provides a book on security and key management [1] as a part of EMV 4.3 specifications as well as additional security guidelines for acquirers [5] and issuers [6] of EMV payment cards.

The risk of MITM attacks in payment systems has been generally neglected in the above standards and specifications (except in a recent 2015 EMV Contactless payment specifications Book C-2 [12], as we will explain). As explained by Drimer et al. in [19], such attacks are commonly perceived to be too expensive to work. However, in the same paper, Drimer et al. show this is a misperception by demonstrating practical MITM attacks in a set of live experiments against the UK's EMV system. Given the practicality of deploying such attacks [19] and the projected rapid growth in the size of the contactless payment industry [9], we believe that it is important for the payment industry to seriously consider the security concerns posed by such attacks and the countermeasures that are needed.

Distance Bounding Protocols: Distance bounding protocols have been considered a potential solution to this problem. In the latest MasterCard EMV specifications (Book C-2 [12] released in March 2015), a distance bounding protocol (called the Relay Resistance Protocol in the specifications) is defined. This protocol starts with the reader sending the card a random challenge and the card replying with a digitally signed response. The reader verifies the digital signature and also checks the response time is within a specified range. This protocol requires an additional private key and a public key certificate installed on the card. Furthermore, the card needs to perform expensive public key operations,

---

[7] www.emvco.com.

[8] http://developer.android.com/guide/topics/connectivity/nfc/hce.html.

which may incur a notable processing delay. To minimize the processing delay on the card, most distance bounding protocols defined in the literature [16,19] resort to using only symmetric key operations, such as hash and symmetric-cipher encryptions. However, applying those solutions to NFC payment would require the card and the reader to have a pre-shared symmetric key. In the current practice, the card only has a pre-shared key with the issuing bank. By contrast, our solution does not require any additional cryptographic keys. In fact, it is orthogonal to distance bounding protocols and can be used in conjuction with any one of them.

Other Countermeasures: Other countermeasures to the MITM attack have been actively explored by a number of researchers. One straightforward solution is to require user vigilance at the time of making the NFC payment. However, it has been generally agreed that user vigilance alone is not sufficient [22,33,38]. It is desirable to design a countermeasure that can effectively prevent Mafia attacks without having to rely on user vigilance. Current solutions generally involve using ambient sensors to measure the characteristics of the surrounding environment, such as light [22], sound [22], location via GPS [33] and a combination of temperature, humidity, precision gas, and altitude [38]. The underlying assumption is that the malicious and legitimate readers will be in two different locations with distinct ambient environments. However, the validity of this assumption may be challenged in some situations where the two readers are in similar environments (e.g., nearby stalls in the same mall).

Overview of our Idea: Our idea is based on the following observation: as a result of the physical tapping between a pair of devices (a card and a reader quipped with accelerometers), the tapping creates transient vibrations, which can be measured using embedded accelerometer sensors. By comparing the similarity of the two measurements, we are able to determine if the two devices were involved in the same tapping event. This effectively distinguishes the Mafia attack from a normal NFC transaction.

In contrast to the mentioned solutions, we do not assume that the attacker's reader is in an environment different from that of the legitimate reader. Thus our threat model considers a more severe attack.

Contributions: Our main contributions are summarised below:

1. We propose "Tap-Tap and Pay" (TTP) as a new countermeasure to prevent Mafia attacks. Our solution is the first that works even if the malicious and legitimate readers are in similar environments.
2. We present a proof-of-concept implementation of TTP by using a pair of NFC-enabled smartphones. Experiments confirm that vibrations induced from the same tapping event are closely correlated between the card and the reader, while they are not if originating from different tapping events.
3. We conduct user studies to evaluate the usability of our TTP prototype. Based on the feedback, users generally find the suggested solution fast and easy to use.

A) Physically tapping the
mobile to the reader

B) Recording accelerometer
measurements on both sides

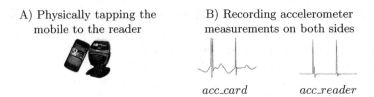

acc_card acc_reader

C) Sending the accelerometer measurements to the bank via the reader

1. $challenge\_card$

2. $response\_card = (challenge, acc\_card, \ldots)_{key}$

3. $response\_card,$
acc_reader

4. $result$

Fig. 2. Overview of the proposed solution: Tap-Tap and Pay

2 Our Solution: Tap-Tap and Pay (TTP)

2.1 Threat Model

We assume a user consciously initiates an NFC payment against a legitimate-looking NFC reader without realizing that it is a malicious one controlled by the Mafia. The difference between the malicious reader and the legitimate reader is that the former is not connected to the back-end banking network while the latter is. We assume the Mafia does not want to directly connect to the banking network, as that will run the risk of being caught by the bank. The malicious reader relays the victim's card to a remote legitimate reader to pay for something more expensive, through the help of an accomplice who holds a legitimate-looking NFC card (see Fig. 1). From the perspective of the legitimate merchant, there is nothing suspicious – a customer uses a mobile phone to make an NFC payment. The amount of the payment may be near the upper end of the limit, but that is perfectly acceptable (see [19] for a demonstration of successful Mafia attacks on the UK's EMV payment system using contact chip-and-PIN cards; the attacks on the contactless payment work in the same way).

Furthermore, we assume the attacker is able to put the NFC reader in an ambient environment that is very similar to the legitimate reader. In one scenario, the attacker may set up a mobile temporary stall near a shopping mall. He may pretend to sell cheap items such as coffee, tea or confectionery, and show the buyer a small amount on the reader's screen. While accepting the buyer's NFC payment, the attacker relays it to one of his accomplices in nearby shops to buy something more expensive. The attacker and his accomplices can avoid detection by constantly changing the location. Once they make enough profit in a day, they will disappear and repeat the same attack at a different place. Under the above threat model, previous ambient-sensor-based solutions may fail completely. However, despite the assumption of a stronger attacker, we will present a solution that can effectively prevent Mafia attacks under the same condition.

The practical feasibility of such Mafia attacks [19], compounded by the fact that they are undetectable by banks in the backend, can prove problematic. This can have serious implications on the liability if the security of the system only depends on user vigilance. In practice, if any dispute arises regarding the discrepancy of the amount charged for an NFC payment, users will be to blame by default since they are required to be "vigilant". We believe this is not fair to users. Our solution addresses this problem by providing banks more evidence so they can tell apart a legitimate NFC payment from a Mafia fraud. This is done at the minimum inconvenience to users, as we explain in the next section.

2.2   Overview of the Solution

An overview of our solution is shown in Fig. 2. First, the user physically taps the mobile phone against the reader twice to make an NFC payment. The tapping causes transient vibrations at both devices, which are measured by the embedded accelerometer sensors. The user then holds the card close to the reader. At this point, the reader detects the presence of an NFC card within physical proximity and starts a standard challenge-and-response process for the NFC payment. At a high level, this involves the reader sending a challenge to the card, and the card replying with a response authenticated by a pre-shared key via MAC with the issuing bank. Our solution does not alter this existing data flow; but within the card response, we propose to add an additional item acc_card to the items being sent by the card. This new item represents the measurement of the vibration by the card accelerometer. When the reader forwards the card's response to the issuing bank through a secure back-end network, it appends acc_reader, which is the measurement of the vibration by the reader accelerometer. The bank compares the two measurements and approves the transaction only if it finds the two sufficiently similar. Recall that in Fig. 1, the user's NFC card and the legitimate NFC reader are honest devices and can perform trustworthy measurements.

TTP suggests two taps because we found it to be the minimum number of taps needed to obtain both sufficiently correlated measurements of the same tapping, and at the same time sufficiently uncorrelated measurements of different tappings. Of course with more than two taps, more features can be extracted, but at the expense of user convenience. Hence, we chose double-tap as the default setting for our solution.

2.3   Sensor Data Preprocessing

To enable data collection, we developed two Android apps: *Card app* and *Reader app* and installed them on two NFC-enabled smartphones, two Nexus 5 devices[9], which are equipped with a range of different sensors.

---

[9] Prototyping of our TTP protocol requires the facility of bidirectional NFC using Host-based Card Emulation (HCE). At the time of experiments, Nexus 5 was the only device allowing that facility.

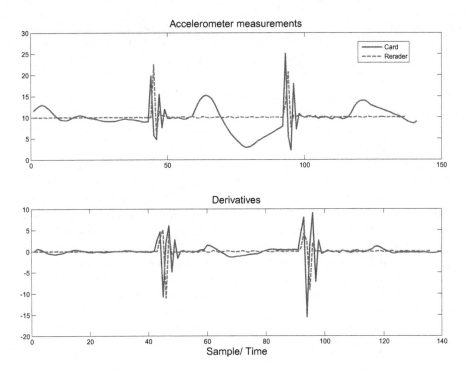

Fig. 3. Final sequences obtained from Eq. 1 (top), and their derivatives from Eq. 2 (bottom) of a sample of double tapping

Accelerometer Data: We use the embedded accelerometer sensor on the mobile phone to capture vibration changes during physical tapping. The accelerometer sensor returns acceleration data in three dimensions, obtained by measuring forces (including the force of gravity) applied to the sensor along the local x, y and z axes. The coordinate system is defined with reference to the phone screen in its portrait orientation; x is horizontal in the plane of the screen from left of the screen towards right, y vertical from the bottom of the screen towards up, and z perpendicular to the plane of the screen from inside the screen towards outside. We consider the sequence representing the length of the three-dimensional vector obtained through accelerometer measurements calculated from Eq. 1 where the components represent the i-th measurement in the three dimensions $(acc_{xi}, acc_{yi}, acc_{zi})$:

$$acc_i = \sqrt{acc_{xi}^2 + acc_{yi}^2 + acc_{zi}^2} \qquad (1)$$

Figure 3 (top) shows the above vector length sequences acc_i for a typical double-tapping as measured on a card and a reader. From now on, we refer to this vector length sequence acc_i simply as accelerometer measurement.

Derivatives: As shown in Fig. 3 (top), the accelerometer measurement made by the card is more vibrant than that by the reader, since the card is moving in the hand of the user. They are also different in scale, depending on the start status of accelerometers. In order to smooth out irrelevant movements specially on the card side, we apply the following equation (based on [26]) to approximate the first derivatives of the sequences. The results are displayed in Fig. 3 (bottom).

$$D_i = \frac{(acc_i - acc_{i-1}) + ((acc_{i+1} - acc_{i-1})/2)}{2} \tag{2}$$

Sequence Alignment: After obtaining the derivatives, we align the two sequences by identifying the peaks. This can be simply achieved by searching for the extreme values (max or min) with a minimum gap between them. The two sequences are then aligned based on the first peak (with a few linear shifts to get the best matching by trial-and-error). Based on our evaluation of the collected data, we found that this simple alignment algorithm is accurate and fast.

After the alignment of the two sequences, we cut a segment of each sequence, starting from 0.2 s before the first peak until 0.2 s after the second peak. This covers the whole significant variation of the accelerometer data. Our analysis shows that with this setting, the whole recording time is in the range of 0.6 and 1.5 s.

2.4 Similarity Comparison

Suggested sensor data comparison methods include correlation coefficients, covariance, cross covariance (e.g. [15]) and cross correlation (e.g. [18,22]) in the time domain, and coherence (e.g. [36]) in the frequency domain. We found the correlation coefficients in the time domain and the coherence in the frequency domain to be the two most effective ones on our collected data. Here we use them along with the energy of the series as well as the distance between the two peaks as the inputs of our suggest TTP decision maker.

Correlation Coefficient (Time Domain): The correlation coefficient is commonly used to compare the similarity of the shapes of two signals. The intuition is that if the two measurements originate from the same double-tap, their signal shapes, especially their tap shapes, would be highly correlated, and otherwise they would not be correlated. Given two sequences X and Y and $\text{Cov}(X, Y)$ denoting covariance between X and Y, the correlation coefficient is computed as below, where $\text{Cov}(X, X) = \sigma_X^2$ and $\text{Cov}(Y, Y) = \sigma_Y^2$:

$$R_{XY} = \frac{\text{Cov}(X, Y)}{\sqrt{\text{Cov}(X, X) \cdot \text{Cov}(Y, Y)}} \tag{3}$$

Coherence (Frequency Domain): To obtain a similarity measure in the frequency domain, we apply the coherence method which indicates the level of matching of features in the frequency domain between two time series. Given two sequences X and Y, we compute the magnitude squared coherence based on the following equation, where $P_{XX}(f)$ and $P_{YY}(f)$ are power spectral densities of X and Y, and $P_{XY}(f)$ the cross power spectral density between X and Y:

$$C_{XY}(f) = \frac{|P_{XY}(f)|^2}{P_{XX}(f) \cdot P_{YY}(f)} \tag{4}$$

We define the similarity rate between the two signals based on magnitude squared coherence as the sum of the squares of the magnitudes of coherence values at all frequencies as follows:

$$F_{XY} = \sum_f C_{XY}(f) \tag{5}$$

Energy Difference: Our analysis shows that different users tap devices with different strengths; some taps are very gentle, some are of medium strength, and some are very strong. We found that the total energy levels of the card and reader sequences of the same tap are strongly correlated, while they are distinctive if obtained from different taps. Hence, we use the following measure to capture the distance of two signals X and Y in term of the total signal energy levels:

$$D_{XY} = \left| \sum_t X(t)^2 - \sum_t Y(t)^2 \right| \tag{6}$$

Peak Gap Difference: Last but not least, the distance between the two peaks in each measured sequence is an important factor in deciding if two measurements come from the same double tapping or not. We define G_{XY} in Eq. 7 where Gap_X is the distance between the two extremums of sequence X and Gap_Y is the distance defined similarly for sequence Y:

$$G_{XY} = |Gap_X - Gap_Y| \tag{7}$$

TTP Decision Engine: Our TTP decision engine has two steps. First, we have an initial check according to the peak gap defined in Eq. 7 and then we use a combined method to include the other three similarity measures. We suggest a simple linear fusion method by using the weighted sum of the three measures: correlation coefficient, coherence, and the energy similarity. Therefore, the ultimate decision is made based on comparing the peak gap against a threshold and if successful comparing the weighted sum of the combined method against another threshold. Hence according to the output of the decision engine, the bank decides to authorize or decline the transaction.

We use a simple linear normalisation that maps the three values to the interval $[0, 1]$. Let us denote these normalised versions by $\bar{R}_{XY}$, $\bar{F}_{XY}$, and $\bar{D}_{XY}$, respectively. Since unlike the other two measures, $\bar{D}_{XY}$ decreases with similarity, we define $\bar{E}_{XY}$ as below. Note that $\bar{E}_{XY}$ is also a normalised value belonging to the interval $[0, 1]$.

$$\bar{E}_{XY} = 1 - \bar{D}_{XY} \tag{8}$$

Fig. 4. Data collection environment (left), Card app (centre), and Reader app (right)

Given $\bar{R}_{XY}$, $\bar{F}_{XY}$ and $\bar{E}_{XY}$, T_{XY} calculates the total similarity rate of two signals X and Y as below, where a, b and c are the weights of each method:

$$T_{XY} = a \cdot \bar{R}_{XY} + b \cdot \bar{F}_{XY} + c \cdot \bar{E}_{XY} \tag{9}$$

The weight parameters are determined through experiments based on the collected user data by testing all possible weights up to two decimal places for a, b, and c – under the condition that the sum of them is equal to 1 – and observing the equal error rate. The values which gave us the best error rate have been fixed as $a = 0.45$, $b = 0.21$, and $c = 0.33$.

3   System Evaluation

3.1   Experiment Setup and Data Collection

We implemented a proof-of-concept prototype for the TTP system by developing two Android apps (card and reader). When the user taps the reader, the two apps independently record the accelerometer data. Once the NFC card is detected by the reader in close proximity, the two devices start a two way NFC communication and simulate an NFC payment.

In order to evaluate the system performance based on real user data, we recruited 23 volunteers (university students and staff, 10 males and 13 females) to participate in the data collection, each performing five double tapping actions. We made a short self-explanatory training video to demonstrate how to do the double-tap and showed it to the users before the experiment. Users generally found the video guide useful in helping them quickly grasp the instruction of "Tap-Tap and Pay".

We fixed the reader phone to the table using double-sided tape, as shown in Fig. 4(left). The front of the phone faced downwards and the back was labelled

Table 1. Equal error rates for different suggested methods

Method	Equal error rate
Correlation coefficients	19.15 %
Coherence	27.91 %
Total energy	23.48 %
Peak gap	14.09 %
TTP decision algorithm	*9.99 %*

"Reader". We used MyMobiler[10] to operate the reader through a USB connection. The GUIs of the reader and card apps are shown in Fig. 4, right and centre, respectively. After launching the card app, the user just double tapped the phone to the reader and kept it close to complete an NFC payment. Once she was notified of a successful completion, she could repeat the experiment. The recorded sensor data were saved into a file for further analysis in Matlab.

3.2   Results

We use the False Negative Rate (FNR) and the False Positive Rate (FPR) to evaluate the performance. The FNR is the rate that two measurements from the same tap event are determined as not matching. The FPR is the rate that two measurements from two different tap events are determined as matching. FNR and FPR vary according to a threshold. The Equal Error Rate (EER) is the rate where the FNR and the FPR curves intersect. The EER is commonly used as a measure to evaluate the overall performance of a system. We computed the EERs based on the similarity comparison methods as described in Sect. 2.4. The results for EER are presented in Table 1. Overall, the Equal Error Rate of our prototype system is 9.99 % using the combined method (Table 1). Therefore with this setting, we have FNR= FPR= 9.99 %. Hence, a legitimate NFC transaction may be falsely rejected with a probability of 9.99 %. Then the user would need to try again. On average, it takes $1/(1 - 0.099) = 1.1$ attempts for a legitimate user to complete an NFC payment transaction. On the other hand, if the Mafia attack takes place during the NFC payment, the transaction is more likely to be denied by the bank due to inconsistent data measurements. The Mafia may trick the user to try again, but it would require on average $1/0.099 = 10$ attempts to get a fraudulent transaction to come through. However, consecutively failed verifications for a single NFC transaction will likely trigger an alert at the back-end banking network, prompting an investigation. Furthermore, when the user gets repeated denials from the NFC payment (say three times), she might not try further and may choose to query her bank instead. All this can significantly increase the chance of having the Mafia attack exposed.

---

[10] www.mymobiler.com.

3.3   Online and Offline Modes

So far, the description of our TTP solution assumes that the NFC transaction is *online* i.e., the reader is connected to the banking network, so that the backend system is able to evaluate the received measurements and authorize the payment in real-time. The same assumption is made in other researchers' solutions [22, 33,38] (which we will detail in Sect. 5).

However in practice, an NFC transaction may be performed *offline*. According-ing to the EMV specifications, an EMV transaction flow includes several steps including *offline* data authentication and *online* transaction authorisation. Depending on the result of the negotiation between the card and the reader, the card may decide to go with offline authorisation. This decision is based on different factors including the transaction value, the type, and the card's record of recent offline transactions. Our solution will be less effective in the offline mode, however, we believe it still provides important added value in preserving critical evidence when a dispute regarding Mafia attacks occurs and a retrospective fraud investigation is needed.

4   Usability Study

4.1   Experiment Setup and Data Collection

We performed a second experiment to evaluate usability aspects of the system. We asked 22 different users (partially overlapped the previous user set, university students and staff, 15 males and 7 females) to perform two NFC payments; first by using the contactless method, and second by using TTP. We developed two Android apps (card and reader) to simulate the two tasks. Before the experiment, we presented users a study description, including a short introduction of mobile contactless payment using NFC, followed by a general description of mobile payment using TTP. In the first task, the user was asked to hold the phone near the reader and wait for the confirmation message. In the second task, the user was asked to double-tap the reader, keep the phone near the reader and wait for the confirmation. Figure 5 shows the GUIs of the two tasks in this experiment.

4.2   Findings

After completing the two tasks, the users were asked to fill in a questionnaire and rate the level of convenience, speed, and feeling of the security of each payment method in a Likert scale from level 5 to 1 (corresponding to "strongly agree", "agree", "neutral", "disagree", and "strongly disagree"). They were also asked to write free comments about their experience in this experiment. Figure 6 shows the average user rating of using the contactless payment and the TTP method.

As shown in Fig. 6, users generally found contactless payment more convenient than TTP. Including a physical action makes it less convenient for some users. As one user commented: "... the fact that I need to keep the device close to the reader after tapping made the experience less convenient".

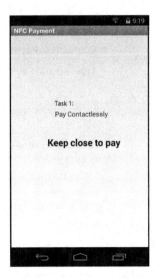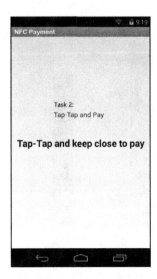

Fig. 5. User study Card app; Task 1: Contactless payment (left), Task 2: TTP (right)

However, in contrast to convenience, many users considered TTP faster than the contactless method, since they were able to precisely sense the start of the action by tapping, while it took them some time to find the proper distance in contactless payment. The uncertainty about when contactless payment would start made some people feel that the process took longer than how long it actually took. As one user commented: "Even [though] I had to tap twice, but the process felt faster comparing to the first one. I feel after tapping I automatically bring the phone close enough to the reader, but in first task, my phone was not close for a while and it took longer".

Moreover, users felt TTP is more secure than contactless payment. By performing a physical tapping action, users felt in control of the transaction and would worry less about accidental payments. As one of the users commented: "As before [i.e. task 1] payment is very easy. I like the action of tapping the reader as this made me feel more in control of when the transaction took place. I felt this method [TTP] was more secure due to the action of tapping to start the transaction. This meant I know when the transaction took place". A similar view was expressed by another user in the comment: "The payment [in task 1] is very easy, but I don't know when the connection between wallet and reader is made; range or time, so I would keep my payment device away from the reader to be sure until I want to pay."

5   Comparison with Previous Works

Table 2 briefly compares TTP with previous ambience sensing based solutions. In terms of security, TTP is the first solution able to prevent the Mafia attack even when both readers share the same ambient environment. Ambience sensing solutions are inherently incapable of detecting the attack in this condition.

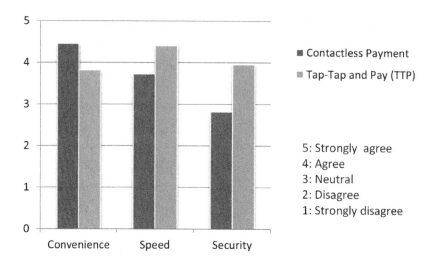

Fig. 6. User study: average user rating of contactless payment and TTP

We now review the error rates reported in the previous works based on measuring the ambient environment. Halevi et al. [22] (sensors: audio and light) report false positive and false negative rates of 0 % for audio sensor, and around 5 % for light sensor for distinguishing different business types (such as library, concert hall, restaurant, etc.). Ma et al. [33] (sensor: GPS) report a 0 % false negative rate under the assumption that the attacker is located 20 meters or farther, 67.5 % when the distance is more that 5 meters, and 100 % when the distance is less than one meter. False positive rates are not reported in their work. Shrestha et al. [38] (sensors: multiple sensors) report false negative rates approximately in the range of 10 %–25 % and false positive rates approximately in the range of 15 %–30 % for individual sensors. By combining the sensor readings, they achieve a false negative rate of about 3 % and a false positive rate of about 6 %.

The equal error rate of 9.99 % in our result is comparable to those reported in the previous works. However, when the two readers are in nearby locations and share the same or similar ambient environments, the reported error rates in [22,33,38] are no longer meaningful and all previous ambient-sensor based solutions may fail completely. By contrast, our TTP solution works regardless of whether or not the two readers share similar ambient environments.

In terms of usability, our protocol needs a sensor recording of only 0.6 to 1.5 s which is sufficiently fast for contactless payment. Schemes based on audio and light sensors [22] achieve similar timings. However, the GPS-based protocol [33] requires 10 s of sensor recording which makes the system not suitable for contactless payment. Our scheme is based on accelerometer which is readily available on most mobile devices, as are microphones (audio), light sensors, and GPS. However, meteorological sensors [38] are only available on specialised devices which is a barrier in adopting such protocols in practice.

Table 2. Comparing TTP with ambient sensors based solutions

Sensor/ Solution	Prevents attacker at same environment	Recording duration (sec)	Embedded mobile sensors	Based on ambience or device
Audio [22]	✗	1	✓	Ambience
Light [22]	✗	2	✓	Ambience
GPS [33]	✗	10	✓	Ambience
Temperature (T) [38]	✗	instant	✗	Ambience
Precision Gas (G) [38]	✗	instant	✗	Ambience
Humidity (H) [38]	✗	instant	✗	Ambience
Altitude (A) [38]	✗	instant	✗	Ambience
THGA [38]	✗	instant	✗	Ambience
Accelerometer (TTP)	✓	0.6–1.5	✓	Device

In summary, our solution presents a new approach in tackling the Mafia attack with promising initial results in terms of security, efficiency and usability. Being orthogonal ways to solve the same problem, TTP and ambient-sensor-based solutions could potentially be combined to achieve even better results. We leave this as a subject for further investigation in future.

6   Further Related Works

In this section, we present some other related works that either use *Tap* gesture, or accelerometer sensor data for other security purposes, and explain how TTP differs from them.

Bump. Using the tap gesture to establish device to device communication has been suggested before. Bump[11] is probably the most well-known example in this category. Two users bumps their mobile phones together to exchange contacts, photos and files. Each phone sends a set of data to a remote server, including the device's location (via GPS), the IP address, the timestamp of bumping and the accelerometer measurement. The server matches the devices based on the received data and transfers the data between the two matched devices. Bump and TTP are clearly distinct as they solve different problems and they assume different threat models. Our threat model assumes a malicious reader, whereas in the Bump model, the two devices bumped to each other are assumed to be both legitimate. Consequently, our main goal is to protect against MITM adversaries whereas Bump's main goal is to identify devices being bumped together. In fact, it has been shown that Bump is vulnerable to MITM attacks [39] due to timing issues. It is worth mentioning that privacy concerns that arise from environment sensing also apply to Bump since at least the locations and IP addresses of all

---

[11] www.bu.mp.

users in the system are communicated with the Bump server each time the app is used. Since January 2014, Bump has been discontinued with all apps removed from App Store and Google Play [31].

Tap Identification Proposals. Performing a tap gesture in order to synchronise multiple devices has been proposed in *Synchronous Gestures* [24]. Tap identification using mobile accelerometer is another problem which could also be applied for security purposes. For example *Tap-Wave-Rub* [30] suggests a system for malware prevention for smarphones. Although similar sensors are used in these proposals, they are in general orthogonal to our solution since they are designed to solve an identification problem for legitimate devices, whereas our solution is designed to resist Mafia attacks in an environment where one of the devices behaves maliciously. Consequently, these solutions can be used alongside our proposal to provide a system in which tapping is used to both unlock the device and secure the payment.

Shake to Pair. The idea of shaking two devices for device pairing has been suggested by multiple works [15,27,28,34–36]. While both TTP and the mentioned works use accelerometer, the amount of entropy produced by shaking, the eventual application, the threat model, and the problem solved by these works are all different from ours. In these works, the user needs to shake the two devices together for a while until both devices generate and agree on a shared key, whereas in our work we do not aim to generate shared keys and we only need the user to tap her device to the reader twice. Device pairing, and more generally key exchange cannot prevent Mafia attacks due to the involvement of the malicious reader. Device pairing and securing NFC payments are distinct security problems. While the former has been explored by researchers for a long time [17,25,29], the latter is less explored. However, with the impending global deployment of NFC payments, we believe the security of NFC payments deserves more attention by the security community.

7   Conclusion

In this paper, we have proposed a simple and effective solution, called "Tap-Tap and Pay" (TTP), to prevent the Mafia attack in NFC payment by sing mobile sensors. Our solution leverages the characteristics of vibration when an NFC card is physically tapped on an NFC reader. We observed that the accelerometer measurements produced by both devices were closely correlated within the same tapping, while they were different if obtained from different tapping events. The experimental results and the user feedback suggest the practical feasibility of the proposed solution. As compared with previous ambient-sensor based solutions, ours has the advantage that it works even when the attacker's reader and the legitimate reader are in nearby locations or share similar ambient environments.

The TTP solution can be easily integrated into existing EMV standards and requires minimal infrastructural change to the EMV system. The structure of the payment protocol remains the same; only an extra string of accelerometer measurement is added in the transmitted message. In terms of hardware, deploying TTP requires the integration of accelerometer sensors in contactless readers. This can be done progressively by equipping the next generation of the readers with accelerometer sensors which are quite inexpensive (e.g., iPhone 4 accelerometers are estimated to cost 65 cents each [23]). Furthermore, TTP can be rolled out gradually since the protocols remain backward compatible.

In future work, we plan to investigate how to further improve system performance by e.g., combining different sensor measurements and using more precise sensors on newer mobile phones. Moreover, it will also be interesting to explore if it is feasible to apply TTP to other NFC-based payment solutions such as NFC-enabled credit/debit cards, and Barclays bPay band[12] to defend against the Mafia attack by retrofitting accelerometers to such devices.

Acknowledgements. We thank all the participants who contributed to our experiments. We also thank the anonymous reviewers of this paper. The second and the third authors are supported by ERC Starting Grant No. 306994.

References

1. Book 2 - Security and Key Management (2011). http://www.emvco.com/specifications.aspx?id=223
2. International Organization for Standardization, BS ISO/IEC 14443–1:2008+A1:2012 Identification cards. Contactless integrated circuit cards. Proximity cards. Physical characteristics (2012). http://www.bsol.bsigroup.com
3. International Organization for Standardization, BS ISO/IEC 14443–2:2010+A2:2012 Identification cards. Contactless integrated circuit cards. Proximity cards. Radio frequency power and signal interface (2012). http://www.bsol.bsigroup.com
4. International Organization for Standardization, BS ISO/IEC 7816–4:2013, Identification cards. Integrated circuit cards. Organization, security and commands for interchange (2013). http://www.bsol.bsigroup.com
5. EMV Acquirer and Terminal Security Guidelines (2014). http://www.emvco.com/specifications.aspx?id=71
6. EMV Issuer and Application Security Guidelines (2014). http://www.emvco.com/specifications.aspx?id=71
7. International Organization for Standardization, BS ISO/IEC 14443–3:2011+A6:2014 Identification cards. Contactless integrated circuit cards. Proximity cards. Initialization and anticollision (2014). http://www.bsol.bsigroup.com
8. International Organization for Standardization, BS ISO/IEC 14443–4:2008+A4:2014 Identification cards. Contactless integrated circuit cards. Proximity cards. Transmission protocol (2014). http://www.bsol.bsigroup.com
9. Mobile payment strategies: Remote, contactless & money transfer 2014–2018. Market leading report by Juniper Research, July 2014. http://www.juniperresearch.com/reports.php?id=726
10. EMV Contactless Specifications for Payment Systems, Book A: Architecture and General Requirements (2015). http://www.emvco.com/specifications.aspx?id=21

---

[12] www.bpayband.co.uk.

11. EMV Contactless Specifications for Payment Systems, Book B: Entry Point (2015). http://www.emvco.com/specifications.aspx?id=21
12. EMV Contactless Specifications for Payment Systems, Book C2: Kernel 2 Specification (2015). http://www.emvco.com/specifications.aspx?id=21
13. EMV Contactless Specifications for Payment Systems, Book C3: Kernel 3 Specification (2015). http://www.emvco.com/specifications.aspx?id=21
14. EMV Contactless Specifications for Payment Systems, Book D: Contactless Communication Protocol (2015). http://www.emvco.com/specifications.aspx?id=21
15. Bichler, D., Stromberg, G., Huemer, M., Löw, M.: Key generation based on acceleration data of shaking processes. In: Krumm, J., Abowd, G.D., Seneviratne, A., Strang, T. (eds.) UbiComp 2007. LNCS, vol. 4717, pp. 304–317. Springer, Heidelberg (2007)
16. Brands, S., Chaum, D.: Distance bounding protocols. In: Helleseth, T. (ed.) EUROCRYPT 1993. LNCS, vol. 765, pp. 344–359. Springer, Heidelberg (1994)
17. Chong, M.K., Gellersen, H.: How users associate wireless devices. In: Proceedingsof the SIGCHI Conference on Human Factors in Computing Systems, CHI 2011, pp. 1909–1918. ACM, New York, (2011)
18. Czeskis, A., Koscher, K., Smith, J.R., Kohno, T.: RFIDs and secret handshakes: defending against ghost-and-leech attacks and unauthorized reads with contextawarecommunications. In: Proceedings of the 15th ACM conference on Computerand communications security, pp. 479–490. ACM (2008)
19. Drimer, S., Murdoch, S.J.: Keep your enemies close: distance bounding against smartcard relay attacks. In: Proceedings of 16th USENIX Security Symposium on USENIX Security Symposium, SS 2007, pp. 7:1–7:16. USENIX Association, Berkeley (2007)
20. Emms, M., van Moorsel, A.: Practical attack on contactless payment cards. In: HCI2011 Workshop-Heath, Wealth and Identity Theft (2011)
21. Francis, L., Hancke, G.P., Mayes, K., Markantonakis, K.: Practical relay attack on contactless transactions by using nfc mobile phones. In: IACR Cryptology ePrint Archive, p. 618 (2011)
22. Halevi, T., Ma, D., Saxena, N., Xiang, T.: Secure proximity detection for nfc devices based on ambient sensor data. In: Foresti, S., Yung, M., Martinelli, F. (eds.) ESORICS 2012. LNCS, vol. 7459, pp. 379–396. Springer, Heidelberg (2012)
23. Hesseldahl, A.: Apple iPhone 4 parts cost about $188. Bloomberg Business, June 2010. http://www.bloomberg.com/bw/technology/content/jun2010/tc20100627_763714.htm
24. Hinckley, K.: Synchronous gestures for multiple persons and computers. In: Proceedings of the 16th Annual ACM Symposium on User Interface Software and Technology, UIST 2003, pp. 149–158. ACM, New York (2003)
25. Ion, I., Langheinrich, M., Kumaraguru, P., Čapkun, S.: Influence of user perception, security needs, and social factors on device pairing method choices. In: Proceedings of the Sixth Symposium on Usable Privacy and Security, SOUPS 2010, pp. 6:1–6:13. ACM, New York (2010)
26. Keogh, E.J., Pazzani, M.J.: Derivative dynamic time warping. In: The 1st SIAM International Conference on Data Mining (SDM-2001). SIAM, Chicago (2001)
27. Kirovski, D., Sinclair, M., Wilson, D.: The martini synch. Technical report MSR-TR-2007-123, Microsoft Research, September 2007
28. Kirovski, D., Sinclair, M., Wilson, D.: The martini synch: device pairing via joint quantization. In: IEEE International Symposium on Information Theory, 2007. ISIT 2007, pp. 466–470, June 2007

29. Kobsa, A., Sonawalla, R., Tsudik, G., Uzun, E., Wang, Y.: Serial hook-ups: a comparative usability study of secure device pairing methods. In: Proceedings of the 5th Symposium on Usable Privacy and Security, SOUPS 2009, pp. 10:1–10:12. ACM, New York (2009)
30. Li, H., Ma, D., Saxena, N., Shrestha, B., Zhu, Y.: Tap-Wave-Rub: lightweight malware prevention for smartphones using intuitive human gestures. In: Proceedings of the Sixth ACM Conference on Security and Privacy in Wireless and Mobile Networks, WiSec 2013, pp. 25–30. ACM, New York (2013)
31. Lieb, D.: All good things (2014). http://blog.bu.mp/post/71781606704/all-good-things
32. Liu, J., Zhong, L., Wickramasuriya, J., Vasudevan, V.: uWave: accelerometer-based personalized gesture recognition and its applications. Pervasive Mob. Comput. **5**(6), 657–675 (2009)
33. Ma, D., Saxena, N., Xiang, T., Zhu, Y.: Location-aware and safer cards: enhancing RFID security and privacy via location sensing. IEEE Trans. Dependable Secure Comput. **10**(2), 57–69 (2013)
34. Mayrhofer, R.: The candidate key protocol for generating secret shared keys from similar sensor data streams. In: Stajano, F., Meadows, C., Capkun, S., Moore, T. (eds.) ESAS 2007. LNCS, vol. 4572, pp. 1–15. Springer, Heidelberg (2007)
35. Mayrhofer, R., Gellersen, H.-W.: Shake well before use: authentication based on accelerometer data. In: LaMarca, A., Langheinrich, M., Truong, K.N. (eds.) Pervasive 2007. LNCS, vol. 4480, pp. 144–161. Springer, Heidelberg (2007)
36. Mayrhofer, R., Gellersen, H.: Shake well before use: intuitive and secure pairing of mobile devices. IEEE Trans. Mob. Comput. **8**(6), 792–806 (2009)
37. Saxena, N., Voris, J.: Still and silent: motion detection for enhanced RFID security and privacy without changing the usage model. In: Ors Yalcin, S.B. (ed.) RFIDSec 2010. LNCS, vol. 6370, pp. 2–21. Springer, Heidelberg (2010)
38. Shrestha, B., Saxena, N., Truong, H.T.T., Asokan, N.: Drone to the rescue: relay-resilient authentication using ambient multi-sensing. In: Christin, N., Safavi-Naini, R. (eds.) FC 2014. LNCS, vol. 8437, pp. 344–359. Springer, Heidelberg (2014)
39. Studer, A., Passaro, T., Bauer, L.: Don't bump, shake on it: the exploitation of a popular accelerometer-based smart phone exchange and its secure replacement. In: Proceedings of the 27th Annual Computer Security Applications Conference, ACSAC 2011, pp. 333–342. ACM, New York (2011)

Protocol and API

Robust Authenticated Key Exchange Using Passwords and Identity-Based Signatures

Jung Yeon Hwang[1], Seung-Hyun Kim[1], Daeseon Choi[2], Seung-Hun Jin[1], and Boyeon Song[3(✉)]

[1] Electronics and Telecommunications Research Institute (ETRI), Daejeon, Korea
{videmot,ayo,jinsh}@etri.re.kr
[2] Department of Medical Information, Kongju University, Nonsan-si, Korea
sunchoi@kongju.ac.kr
[3] Korea Institute of Science and Technology Information (KISTI), Daejeon, Korea
bysong@kisti.re.kr

Abstract. In the paper we propose new authenticated key exchange (AKE) protocols from a combination of identity-based signature (IBS) and a password-based authentication. The proposed protocols allows for a client to execute a convenient authentication by using only a human-memorable password and a server's identity. The use of an IBS gives security enhancements against threats from password leakage. A server authentication method is based on an IBS, which is independent of a password shared with a client. Even if a password is revealed on the side of a client protected poorly, server impersonation can be prevented effectively. In addition, our protocols have resilience to server compromise by using 'password verification data', not a true password at the server. An adversary cannot use the data revealed from server compromise directly to impersonate a client without additional off-line dictionary attacks. We emphasize that most of existing password-based AKE protocols are vulnerable to subsequent attacks after password leakage.

Our first hybrid AKE protocol is constructed using concrete parameters from discrete logarithm based groups. It is designed to give resilience to server compromise. Our second protocol is a simplified version of the first protocol where the computation cost of a client is cheap. Generalizing the basic protocols, we present a modular method to convert Diffie-Hellman key exchange into an AKE protocol based on a password and an IBS. Finally, we give performance analysis for our protocols and comparison among known hybrid AKE protocols and ours. As shown later in the paper, our protocols provide better performance. Our experimental results show that the proposed protocols run in at most 20 ms. They can be widely applied for information security applications.

Keywords: Authentication · Password · Identity-based signature · Key exchange

This work was supported by the ICT R&D program of MSIP/IITP [B1206-15-1007, Development of Universal Authentication Platform Technology with Context-Aware Multi-Factor Authentication and Digital Signature].

© Springer International Publishing Switzerland 2015
L. Chen and S. Matsuo (Eds.): SSR 2015, LNCS 9497, pp. 43–69, 2015.
DOI: 10.1007/978-3-319-27152-1_3

1   Introduction

Explosive growth of computing environment is opening the era of big data. Application domains of the pervasive computing are connected densely, and entities with smart devices exchange information frequently in the advanced network. Massive amounts of data can be collected from various sources and processed for user-centric services. Furthermore, automated analysis and deep learning technologies are actively developed to extract highly valuable knowledge from a huge amount of data. In the upcoming future, a user will be able to enjoy unprecedented convenience timely from the advanced information service.

These services will be available when not only feasibility but also security properties are well provided. For secure data transmission between entities, key exchange is one of the most crucial and fundamental protocols. That is, to access valuable data resources on a server (or a service provider), a client should establish a temporal digital key with the server securely. The shared key builds a *secure channel* between a client and a server, which provides security properties including data integrity and confidentiality. Diffie-Hellman (DH) key exchange (KE) [19] is a well-known popular protocol for key establishment between two entities.

In general, a network used to establish a shared key is public and insecure, where an adversary may control exchanged messages in an adaptive way. An adversary is able to perform impersonation attacks to a user by intercepting and/or modifying messages. Thus, authentication of legal participants is necessary for secure key exchange. An authenticated key exchange (AKE) protocol has been widely studied and developed for a long time, to achieve key establishment and authentication of participants over a public network. AKE is mainly constructed by combining the DH KE and an authentication method, for which entities' computing capability and users' convenience should be considered. In a client-server model, a client is a typically human who has a device(s) with limited computing resources, while a server is a powerful machine which can store high-entropy secret numbers.

In a client-server model, a password is a commonly used authentication factor, because a client can generate and memorize it easily. In practice, most of IT services use ID/password as a log-in method. To construct a secure password-based AKE (PAKE) lots of research have been performed [7,32,40]. As a password used for authentication is low-entropy, PAKE is vulnerable to *dictionary* attacks which systematically check all possible passwords from a password space of small size until the correct one is found. Dictionary attacks can be mounted in two types, i.e., on-line and off-line. An on-line dictionary attack is mounted by using guessed passwords. It is easy to prevent it just by limiting the number of on-line password trials. An off-line dictionary attack is performed without interaction with a server, when some information to confirm a true password is obtained. Thus, to resist an off-line dictionary attack, a PAKE protocol makes sure not to reveal any information related to a password. when sending a message including the password.

For this reason, Encrypted Key Exchange (EKE) was introduced in [7], where at least one party encrypts a key value using a password and sends it to a second party who decrypts it to negotiate a shared key with the first party. Since then, EKE has been modified and extended in various ways. Some PAKE protocols have been developed as international standards by IEEE [31], IETF, ISO/IEC JTC 1/SC 27 [32], and ITU-T [33]. Most PAKE protocols including EKE are constructed in a shared-password authentication model where a client and a server identifies each other by using a shared password (or variant of a password). Intrinsically, the model is vulnerable to threats from password exposure, in the sense that a password stolen from one party can be used to impersonate the other party. There are various possibilities to leak a password, for example, by malware, hacking, shoulder surfing attacks or from lost/stolen portable devices.[1] Since a client's device may be insufficiently protected, password exposure could be more realistic. When a password of a client is revealed, it is inevitable that an adversary impersonate the client. But, if a server can be impersonated to the client, it will bring more dangerous and serious risks. For example, malicious modification of critical information such as clients' financial services or healthcare will be possible.

As a solution to the above issue, a *hybrid* AKE has been introduced which is constructed in conjunction of an asymmetric encryption scheme and password-based authentication. The protocols by [23,25] make use of a public key encryption, and the protocol of [53] is based on an identity-based encryption (IBE). The intuition to prevent server impersonation attacks is to use an independent decryption key for a server. For password-based authentication, the protocols take a simple approach to encrypt a password with a server's public key. To guarantee confidentiality on a password, they apply a highly secure encryption to meet so-called CCA-security.[2]

Hybrid protocols based on IBE may be preferable in a client-server model because a client is assumed to be a human who can merely memorize limited information such as a password or server's ID. However, CCA-secure IBE encryption entails relatively complex computation of parameters. It will impose expensive computation or communication costs on a client side using a device with a limited resource. In addition, the above-mentioned hybrid protocols do not consider server compromise. A server manages a password file for a large number of clients. The file contains secret values to be used to authenticate clients. If the password file is revealed, it will cause disastrous results, because any client can be immediately subject to impersonation.

1.1 Our Contributions

In the paper, we propose new efficient yet robust AKE protocols from a combination of a password-based authentication and an identity-based signature (IBS).

---

[1] These are different from dictionary attacks to reveal a password.

[2] The notion of CCA security means that a PKE scheme should reveal no meaningful information about the original message from public ciphertexts to attackers who can probe the decryption oracle with chosen ciphertexts.

For distinction, they are called *IBS-PAKE* protocols. An IBS can be used by not only a powerful server but also a client, while a client is enough to memorize a password to invoke the protocol. The adoption of an IBS gives desired solutions to party compromise issues as follows.

- Basically, a server executes an independent authentication based on an IBS. Even if a password is revealed from a client, server impersonation is impossible without access to the server's IBS key. It will be reasonable for a server to manage a sinlge key secretly, rather than the whole password file of large size.
- In an IBS scheme, a public key may be defined by an arbitrary public string such as an e-mail address or a company/brand name. A client can authenticate a server by verifying a server's IBS with a server's publicly known identity. For himself or herself, the client executes password-based authentication. Thus he or she can do a convenient authentication based on only a human-memorable password and a server's identity without holding a high entropy secret key.
- Finally, our protocol allows for a client to use an IBS by accessing an IBS key stored at a server. More concretely, the client receives an encryption of the IBS key from the server and decrypt it with his or her password. The client should know the original password to obtain the correct signing key. It involves a kind of a knowledge proof of a password. This idea can be applied to achieve resilience to server compromise [24]. Assume that the knowledge proof is required to a client for each login. An adversary cannot use the password file stolen the server directly for impersonation attacks but additionally making an off-line dictionary attack to extract clients' real passwords.

In order to construct an IBS-PAKE protocol, we take a modular approach first, that is, present two modular methods to yield IBS-PAKE protocols generically when a symmetric PAKE and an IBS are given. The first method is designed to achieve resilience to server compromise. The underlying idea to achieve resilience to server compromise is similar to that of [24] based on a normal signature scheme. However, there is a difference between the ideas because a public verification key, i.e., a client's identity is known to an adversary in our protocol while hidden in [24]. Our second one is a simplified version of the first method, to handle the situation that server compromise is not mainly considered due to strong security at the side of a server. Compared to the first method, it can run in a single round which consists of two passes independently sent from a client and a server. In addition, the computation cost at the side of a client is quite cheap.

In the modular methods, IBS schemes can be selected flexibly and independently according to a design strategy. For example, we can pick an IBS scheme with low signing (or verifying) cost for a client. Also, an IBS-PAKE protocol can be constructed from more realistic hardness assumptions. As instances of IBS-PAKE resulting from the modular methods, two protocols are presented by using concrete discrete logarithm parameters. They are built on the IBS scheme constructed from the Schnorr signature.

Finally, we give performance analysis for our protocols and comparison among known AKE protocols using an identity-based cryptosystem and a password [53], and ours. As shown in the comparison table, our AKE protocols provide better performance with a robust security property. We also present experimental results to show that the proposed protocol runs in at most 20 ms.

1.2   Related Work

Since the introduction of Diffie-Hellman KE protocol [19], KE protocols have been widely studied to achieve various authentication goals [10,11,37]. AKE protocols have been developed largely according to two authentication types, i.e., symmetric and asymmetric. Symmetric authentication type assumes that participants share a secret key before running a protocol [10,11]. A password-based KE (PAKE) protocol is a primary example of symmetric authentication where a client and a server share a password as an authentication factor. The formal treatment for PAKE was given in [8,10]. Refer to [40] for a survey of PAKE. Some research proposes PAKE protocols with security under standard assumptions [3,35]. Recently, research on PAKE protocols [3] focuses on meeting highly theoretical security requirement such as UC model [16]. Asymmetric authentication type assumes that a participant uses a secret key and its corresponding public key. The secret key is kept secret by the participant while the public key is set to be public and so anyone can access it. Since the secret key is a random long bit string, a client needs a mean to store it. For example, we can consider KE based on a standard public key digital signature [37] and identity-based KE [17]. A hybrid authentication type combines symmetric and asymmetric types to gain merits of the two types [25]. In contrast to the symmetric and asymmetric types, hybrid authentication and KE have not been studied intensively.

1.3   Organization

The remainder of this paper is organized as follows. In Sect. 2, we briefly review some preliminaries. In Sect. 3 we give a security model for an IBS-PAKE. In Sect. 4 we present an IBS scheme based on Schnorr signature. In Sect. 5 we present modular methods to yield IBS-PAKE protocols, and also concrete IBS-PAKE protocols using discrete logarithm parameters and prove the security. In Sect. 6 we give a performance analysis and comparison among known AKE protocols. Finally, we conclude in Sect. 7.

2   Preliminaries

In this section, we review some background knowledge for our construction.

Let $\mathsf{poly}(\lambda)$ denote a polynomial in variable λ. We define that $\nu(\lambda)$ is a negligible function if $\nu(\lambda) < 1/\mathsf{poly}(\lambda)$ for any $\mathsf{poly}(\lambda)$ and sufficiently large λ. We denote by $A \stackrel{?}{=} B$ the equality test between two group elements, A and B. We denote by $s \stackrel{R}{\leftarrow} S$ the operation that picks an element s of set S uniformly at random. We denote by '$||$' the concatenation operation on strings.

Computational Assumptions. For the security of our construction, computational assumptions such as discrete logarithm, decisional Diffie-Hellman, and computational Diffie-Hellman assumptions, are needed. For more details, refer to Appendix.

Symmetric Encryption. A symmetric encryption (SE) scheme consists of two functions, $\mathcal{E}$ and $\mathcal{D}$ associated to key space $\mathcal{K}_{SE}=\{0,1\}^{\lambda}$.

- $\mathcal{E}_k(m)$. It takes as input a key $k \in \mathcal{K}_{SE}$ and a message $m \in \{0,1\}^n$ and outputs a ciphertext $\chi \in \{0,1\}^n$.
- $\mathcal{D}_k(\chi)$. It takes as input a key k and a ciphertext $\chi \in \{0,1\}^n$, and then outputs a message $m \in \{0,1\}^n$.

To define *one-time indistinguishability* for SE, we consider the following game:

$$
\begin{aligned}
Challenge &: (m_0, m_1, \eta) \leftarrow \mathcal{A}(1^{\lambda}) \\
Response &: k \xleftarrow{R} \mathcal{K}_{SE},\, b \xleftarrow{R} \{0,1\},\, \chi^* \leftarrow \mathcal{E}_k(m_b) \\
Guess &: b' \leftarrow \mathcal{A}(\eta, \chi^*)
\end{aligned}
$$

Assume that m_0 and m_1 has a same length. We define $\mathsf{Adv}_{\mathcal{A},\mathsf{SE}}^{IND\text{-}OTK}(t)=\mid \Pr[b = b'] - 1/2\mid$, where $\mathcal{A}$ runs in time t, and $\mathsf{Adv}_{\mathsf{SE}}^{IND\text{-}OTK}(t) = \max_{\mathcal{A}}[\mathsf{Adv}_{\mathcal{A},\mathsf{SE}}^{IND\text{-}OTK}(t)]$ where the maximum is taken over all $\mathcal{A}$. We say that SE is *one-time secure* if $\mathsf{Adv}_{\mathsf{SE}}^{IND\text{-}OTK}(t)$ is negligible.

Identity-Based Signature. An IBS scheme consists of four algorithms for setup, private key extraction, signing, and verifying [9,45]. These are denoted by Setup, KeyExt, Sign, and Vrfy, respectively.

- Setup takes as input a security parameter λ, and outputs a master secret key, msk and a set of public parameters, pp.
- KeyExt takes as input (pp,msk) and an identity $ID \in \{0,1\}^{\ell_{id}}$ for $\ell_{id} \in \mathbb{N}$, and then outputs a private key, sk_{ID}.
- Sign takes as input pp, an identity ID, a key sk_{ID}, a message $m \in \{0,1\}^*$, and then outputs a signature, σ.
- Vrfy takes as input pp, ID, a signature σ, and a message m, then outputs 0 (meaning 'invalid') or 1 (meaning 'valid').

We say that an IBS scheme is *correct* if the following condition holds: $1 \leftarrow \mathsf{Vrfy}(pp, ID, \sigma, m)$ for any pair of (m, ID) where $(pp, \mathsf{msk}) \leftarrow \mathsf{Setup}(1^{\kappa})$, $sk_{ID} \leftarrow \mathsf{KeyExt}(pp, \mathsf{msk}, ID)$, and $\sigma \leftarrow \mathsf{Sign}(pp, ID, sk_{ID}, m)$.

Next we consider a game to define the existential unforgeability under *chosen message and identity attacks (CMIDA)* for IBS = (Setup, KeyExt, Sign, Vrfy). The game consists of SETUP, QUERY, and FORGE phases. Let EQ and SQ denote an extraction query and a signing query, respectively. In the QUERY phase, $\mathcal{F}$ is allowed to make extraction and signing queries to the key extraction and signing oracles adaptively.

$Setup$: $(pp, \mathsf{msk}) \leftarrow \mathsf{Setup}(1^\kappa)$

$Query$: $EQ(ID) \leftarrow \mathcal{F}, \; sk_{ID} \leftarrow \mathsf{KeyExt}(pp, \mathsf{msk}, ID)$

$SQ(ID', m) \leftarrow \mathcal{F}, \; \sigma \leftarrow \mathsf{Sign}(pp, ID', sk_{ID'}, m)$

$Forge$: $(ID^*, m^*, \sigma^*) \leftarrow \mathcal{F}$

Let L_{Ex} and L_{Sign} denote a list of all extraction and signing queries that $\mathcal{F}$ have made. We say that the forgery, (ID^*, m^*, σ^*) is *valid* if $ID^* \notin L_{Ex}$ and $(ID^*, m^*) \notin L_{Sign}$, and $\mathsf{Vrfy}(pp, ID^*, \sigma^*, m^*) = 1$. The EUF-CMIDA-advantage of adversary $\mathcal{F}$, denoted by $\mathsf{Adv}_{\mathsf{IBS}}^{EUF\text{-}CMIDA}(\mathcal{F})$, is defined as the probability that $\mathcal{F}$ outputs a valid forgery in the above experiment. That is, $\mathsf{Adv}_{\mathcal{F},\mathsf{IBS}}^{EUF\text{-}CMIDA}(t) = \Pr[ID^* \notin L_{Ex} \wedge \mathsf{Vrfy}(pp, m^*, ID^*, \sigma^*) = 1]$, where $\mathcal{F}$ runs in time t. We also define that $\mathsf{Adv}_{\mathsf{IBS}}^{EUF\text{-}CMIDA}(t) = \max_{\mathcal{F}}[\mathsf{Adv}_{\mathcal{F},\mathsf{IBS}}^{EUF\text{-}CMIDA}(t)]$ where the maximum is taken over all $\mathcal{F}$. We say that IBS is *existentially unforgeable* if $\mathsf{Adv}_{\mathsf{IBS}}^{EUF\text{-}CMIDA}(t)$ is negligible.

Various IBS schemes have been proposed from various mathematical parameters such as DL, composite numbers, and bilinear maps [9,45].

3 Security Model

We present a security model for a hybrid AKE protocol based on a password and an IBS by modifying known security models for AKE such as [10,24,37] and extending the model of [15,53]. Our model captures security by considering resilience to server compromise.

PARTICIPANTS. Let **Clients** and **Servers** denote sets of clients and servers, respectively. Let $\mathcal{U}$ be a set of all principals and defined by **Clients** ∪ **Servers**. The number of principals is bounded by a polynomial in a security parameter. We assume that each principal is labeled by a unique identity of a ℓ_{id}-bit string for a positive integer ℓ_{id}. For example, ID_C and ID_S are used to denote client C and server S, respectively.

In the model, we assume that there is a trusted third party, called KGA (Key Generation Authority) who manages the key extraction algorithm, KeyExt of an IBS scheme and keeps the master secret key, msk.[3] Whenever a principal requests, KGA issues a long-term secret signing key corresponding to the identity of the principal. A principal with ID obtains a long-term secret signing key sk_{ID} from KGA. We note that a client is enough to use only a human-memorable password to execute the protocol. The password or its verifier is shared between a client and a server. However, though the client holds no long-term secret key of high-entropy, he or she can use the IBS key transmitted from the server during a run of a protocol.

---

[3] To prevent misuse of the master secret key, the authority of KGA can be distributed into multiple authorities by using known threshold techniques [4,12].

INITIALIZATION. In the initialization phase, Setup, Extract, and Registration processes are executed. Setup generates global system parameters and keys including the master secret key, msk of IBS. The global system parameters, denoted by pp and identities, ID_U are publicly known to a client and a server, and also an adversary. Through Extract, a principal obtains a signing key corresponding to an identity. An IBS generated by the signing key can be verified by the identity. We assume that a principal is identified in a pre-defined way before issuing a key.

Whenever a client of ID_C wants to join as a valid user (of a service) to a server of ID_S, Registration is executed between them. Let $\mathcal{D}_{PW}$ denote a password space, i.e., a dictionary of passwords. Assume that a password, $pw_C \in \mathcal{D}_{PW}$ is generated by the client according to a pre-defined password creation policy. After completing the registration, the server stores a password verifier, pv_C which is derived from pw_C. For example, pv_C can be computed from a hash function or some deterministic function with input pw_C. Let $\pi_S[C] = (ID_C, pv_C)$ and $\mathcal{PF}_S = \{\pi_S[C]\}_{C \in \mathbf{Clients}}$, which is called a password file, including all authentication information registered for clients.

PROTOCOL EXECUTION. A principal is allowed to invoke the protocol (multiple times) with a partner principal to establish a session key. It can be run in a concurrent way. Multiple executions of a principal is modeled via instance. Let s be a positive integer. The sth instance of a principal with ID_U is represented by Π_U^s where $U \in \mathcal{U} = \mathbf{Clients} \cup \mathbf{Servers}$. The index, s sequentially increases according to the number of executions of the principal [11,37].

PARTNERING. A session id of instance Π_U^s is defined as the concatenation of all transcripts sent and received between an instance of a client and an instance of a server during the execution of the protocol. For $U \in \mathcal{U}$, let sid_U^s denote a session id for instance sid_U^s.

Partner identifier pid_U^s for instance Π_U^s is defined by a set of the identities of protocol participants who intend to establish a session key. We say that instance Π_U^s *accepts* when it computes a session key, sk_U^s. Let acc_U^s denote an boolean variable to show whether a given instance has accepted or not. Assume that, if an instance computes a session key, sk_U^s, it outputs $(\mathsf{sid}_U^s, \mathsf{sk}_U^s)$. For $C \in \mathbf{Clients}$ and $S \in \mathbf{Servers}$, we say that Π_C^i and Π_S^j are *partnered* if and only if (1) $\mathsf{pid}_C^i = \mathsf{pid}_S^j$; (2) $\mathsf{sid}_C^i = \mathsf{sid}_S^j$ and (3) they have both accepted.

ADVERSARIAL MODEL. An adversary $\mathcal{A}$ is a PPT algorithm that has complete control over all the communications. Attacks that $\mathcal{A}$ can make are modeled via the following queries.

- Extract(ID_U): By this query, $\mathcal{A}$ is given the long-term secret key of ID_U where $U \in \mathcal{U} = \mathbf{Clients} \cup \mathbf{Servers}$.
- Execute(ID_C, ID_S): $\mathcal{A}$ is given the complete transcripts of an honest execution between C and S. It models passive attacks eavesdropping an execution of the protocol.

- Send(Π_U^s, m): $\mathcal{A}$ is given the response generated by Π_U^s according to the protocol, when m is given to Π_U^s. It models an active attack where the adversary controls messages elaborately. By this query, a message m can be sent to instance Π_U^s.
- Reveal(Π_U^s): $\mathcal{A}$ is given the session key that instance Π_U^s has generated. It models a *known key* attack.
- Corrupt(ID_U): It models exposure of the long-term secret key held by ID_U where $U \in \mathcal{U}$. $\mathcal{A}$ is given $\mathcal{PF}_S = \{\pi_S[C]\}_{C \in}$ **Clients** if $U \in$ **Servers**, and otherwise pw_U.
- Test(Π_U^s): It is used to define the advantage of $\mathcal{A}$. When $\mathcal{A}$ asks this query to an instance Π_U^s, a random bit b is chosen; if $b = 1$ then the session key is returned. Otherwise, a random string is drawn from the space of session keys, and returned. $\mathcal{A}$ is allowed to make a Test query once, at any time.

In the model we consider two types of adversaries according to their attack types. The attack types are simulated by the queries issued by an adversary. A *passive adversary* is allowed to issue Execute, Reveal, Corrupt, and Test queries, while an *active adversary* is additionally allowed to issue Send and Extract queries. Even though Execute query can be using Send queries repeatedly, we use Execute query for more concrete analysis.

FRESHNESS. An instance Π_U^s is said *fresh* (or holds a *fresh* key ssk) if the following conditions hold:

1. Reveal(Π_U^s) has not been asked for all $U \in \mathrm{pid}_U^s$,
2. Corrupt($ID_{U'}$) has not been asked for $U' \in \mathrm{pid}_U^s$.

An instance Π_U^s is said *semi-fresh* (or holds a *semi-fresh* key ssk) if the following conditions hold:

1. Reveal(Π_U^s) has not been asked for all $U \in \mathrm{pid}_U^s$,
2. Corrupt(ID_U) has not been asked, and
3. Corrupt(ID_S) has not been asked for $U \in$ **Clients**.

AKE SECURITY. Let $\mathcal{P}$ be an IBS-PAKE protocol and $\mathcal{A}$ an adversary to attack it. $\mathcal{A}$ is allowed to make oracle queries in an adaptive way and receives the corresponding responses from oracles. At some point during the game a Test query is asked to a fresh or semi-fresh oracle, and $\mathcal{A}$ may continue to make other queries. Finally $\mathcal{A}$ outputs its guess b' for the bit b used by the Test oracle, and terminates. Let Succ denote an event that $\mathcal{A}$ correctly guesses the bit b. The advantages of $\mathcal{A}$ must be measured in terms of the security parameter λ.

The IBS-PAKE-advantage, $\mathrm{Adv}_{\mathcal{P}}^{IBS-PAKE}(\mathcal{A})$ of $\mathcal{A}$ is defined as the probability that $\mathcal{A}$ correctly guesses the bit in the above experiment. That is, $\mathrm{Adv}_{\mathcal{A},\mathcal{P}}^{IBS-PAKE}(\lambda) = 2 \cdot \Pr[\mathrm{Succ}] - 1$, where $\mathcal{A}$ runs in time t. We also define that $\mathrm{Adv}_{\mathcal{P}}^{IBS-PAKE}(\lambda) = \max_{\mathcal{A}}[\mathrm{Adv}_{\mathcal{A},\mathcal{P}}^{IBS-PAKE}(\lambda)]$ where the maximum is taken over all $\mathcal{A}$.

Let $|\mathcal{D}_{\mathcal{PW}}|$ be the size of the password space. Let q_s be the maximum number of Send queries. We say a protocol $\mathcal{P}$ is a *secure IBS-PAKE* protocol if the following condition hold: For a negligible function $\epsilon(\lambda)$, $\mathsf{Adv}_{\mathcal{P}}^{IBS\text{-}PAKE}(\lambda)$ is bounded by $\frac{q_s}{|\mathcal{D}_{\mathcal{PW}}|} + \epsilon(\lambda)$.

4 Our Identity-Based Signature Scheme

In this section, we present an IBS scheme that works with discrete logarithm parameters. It will be used for our AKE as a building block in the next section.

- Setup. For a given security λ, it generates a cyclic group, $\mathbb{G}$ of prime order q and a random generator, g of $\mathbb{G}$. It picks $\theta \in \mathbb{Z}_q^*$ uniformly at random and computes $u = g^\theta$. It also generates independent cryptographic hash functions, $H : \{0,1\}^* \to \{0,1\}^\ell$. It outputs the system public parameters $pp = (\mathbb{G}, q, g, u, H)$ and the corresponding master secret key $\mathsf{msk} = \theta$.
- KeyExt(pp, msk, ID). It picks $r \in \mathbb{Z}_q^*$ uniformly at random. It computes $R = g^r$, $w = H(ID, R)$, and $v = r + \theta w \pmod{q}$. It then returns $sk_{ID} = (R, v)$.
- Sign(pp, sk_{ID}, ID, m). It takes as input pp, a message $m \in \{0,1\}^*$, an identity ID and a key $sk_{ID} = (R, v)$. It picks $e \in \mathbb{Z}_q^*$ and compute $E = g^e$, $z = H(m, ID, E)$ and $d = e - vz \pmod{q}$. The signature on m is $\sigma = (d, z, R)$.
- Vrfy(pp, ID, σ, m). It takes as inputs pp, a message m, ID and a signature $\sigma = (d, z, R)$. It computes $E' = g^d \cdot (R \cdot u^w)^z$ and $w = H(ID, R)$. Finally, it checks if $z = H(m, ID, E')$ holds. If the equality holds, it outputs 1, and otherwise, 0.

The above IBS scheme is correct because we have $g^d \cdot (R^z \cdot u^{wz}) = g^{e-(r+\theta w)z}$. $(g^{rz} \cdot g^{\theta wz}) = g^e$ and so $z = H(m, ID, E) = H(m, ID, g^d \cdot R^z \cdot u^{wz})$ where $w = H(ID, R)$ and $(R = g^r, v = r + \theta w)$ is generated from KeyExt for ID.

The idea behind the above construction is to combine two Schnorr signatures [46] sequentially. Similar constructions are known in the literature [22]. Using the DL assumption in a group $\mathbb{G}$, we can formally prove the security of the above IBS scheme, that is, existential unforgeability against adaptive chosen message and identity attacks in the random oracle model. A proof idea is actually similar to that of [22] with a slight modification on the so-called Forking Lemma [42]. For more details, refer to the full version of this paper. Our IBS consists of one group element and two hash outputs, while the IBS of [22] consists of two group elements and one hash output. Since the size of a hash output is smaller than that of an group element, our scheme gives a shorter IBS.

5 Our IBS-PAKE Protocols

In this section, we present IBS-PAKE protocols, i.e., AKE protocols using a password and an IBS as authentication means. First we present two generic methods to construct IBS-PAKE protocols. We then present two concrete IBS-PAKE protocols using DL parameters.

5.1   Generic Construction

Our generic methods are presented by using two-party PAKE and an IBS in a modular way. The first generic method gives resilience to server compromise. That is, even if a password file is revealed from a server compromised, an adversary can only obtain password verification information, not a real password. Thus, to impersonate a client, he or she must mount an offline dictionary attack additionally. In the protocol, an IBS scheme is used for both of a client and a server. A server uses an IBS to authenticate itself to a client. A client uses an IBS to prove the possession of his or her own password.

Let PAKE denote a two-party PAKE protocol. For example, we can consider EKE [7], PAK [8], SRP, SPEKE, and AMP [32].

Assume that an IBS scheme, IBS = (Setup, KeyExt, Sign, Vrfy) and a symmetric encryption scheme, $SE = (\mathcal{E}, \mathcal{D})$ are given. A client or a server obtains a signing key from KGA running KeyExt in the initialization phase. For distinction, a set of public parameters of IBS is denoted by pp_{IBS}.

The protocol consists of two phases, initialization and key establishment as follows.

Initialization Phase. Three processes Setup, Extract, and Registration are executed as follows. Let $pw_U \in \mathcal{D}_{PW}$ denote a password chosen by a user, U.

– Setup: For a given security parameter λ, it generates pp_{PAKE} for the given PAKE protocol. It also generates independent cryptographic hash functions, $H_i : \{0,1\}^* \rightarrow \{0,1\}^{\ell_i}$ for $i = 1, 2, 3$. It runs Setup of the IBS scheme to generate (msk, pp_{IBS}). It outputs the system public parameters, $pp = (pp_{\mathsf{PAKE}}, pp_{\mathsf{IBS}}, H_{i=1,2,3}, SE)$. The master secret key, msk is kept secret.
– Extract. For a given identity ID, KeyExt is run (by KGA) to output a private key, sk_{ID}. Assume that the private key is transmitted to the user ID via a secure channel.
– Registration. A client, C generates a password, pw_C according to a pre-defined password creation policy. Let sk_{ID_C} be a signing key of C. Assume that a secure channel is established between the client and a server, S. To register a service, C sends (Register-Request, $ID_C, \pi_1 = H_1(pw_C), ESK = \mathcal{E}_{H_2(pw_C)}(sk_{ID_C}))$) to the server via a secure channel.[4]

Key Establishment Phase. A client, C and a server, S execute a run of the protocol to agree on a temporal key to be used for a session as follows. See Fig. 1.

1. PAKE-Client. The client C computes $\pi_1 = H_1(pw_C)$. Using π_1 instead of pw_C, the client performs its part in PAKE with the following modification: Whenever the client receives a pair of a message and a signature, $(\overline{m}_S, \sigma_S)$ from the server S, the client verifies the signature, σ_S on $\overline{m}_S$, that is, checks if $1 = \mathsf{Vrfy}(pp_{\mathsf{IBS}}, ID_S, \sigma_S, \overline{m}_S)$. If the signature is valid then C performs its part of PAKE for $\overline{m}_S$. Finally, C obtains a common key K.

---

[4] Note that a secure channel is needed because $\pi_1 = H_1(pw_C)$ or $ESK = \mathcal{E}_{H_2(pw_C)}(sk_{ID_C})$ can be used to mount off-line dictionary attacks by an adversary.

Client C		Server S
$pp = \{pp_{\mathsf{PAKE}}, pp_{\mathsf{IBS}}, H_{i=1,2,3}, SE\}$		$pp = \{pp_{\mathsf{PAKE}}, pp_{\mathsf{IBS}}, H_{i=1,2,3}, SE\}$
$[ID_C, pw_C]$		$[ID_S, sk_{ID_S}]$
		$\pi_S[C] = (ID_C, \pi_1 = H_1(pw_C),$
		$ESK = \mathcal{E}_{H_2(pw_C)}(sk_{ID_C}))$
Using $\pi_1 = H_1(pw_C)$ instead of pw_C,	Modified	perform its part in PAKE
perform its part in PAKE	Execution	with the following modification:
with the following modification:	of PAKE	For each $\overline{m}_S$ to be sent to C in PAKE,
Whenever $(\overline{m}_S, \overline{\sigma}_S)$ is received,	with $H_1(pw_C)$	$\overline{\sigma}_S \leftarrow \mathsf{Sign}(pp_{\mathsf{IBS}}, ID_S, sk_{ID_S}, \overline{m}_S),$
if $0 = \mathsf{Vrfy}(pp_{\mathsf{IBS}}, ID_S, \overline{\sigma}_S, \overline{m}_S)$, abort. ⟵		and then send $(\overline{m}_S, \overline{\sigma}_S)$.
Otherwise, perform the client's part		
for given $\overline{m}_S$ in PAKE.		
Output K		Output K
$ek = H_3(K)$, $ESK\|\sigma_S = \mathcal{D}_{ek}(CT_S)$ ⟵ ID_S, CT_S		$ek = H_3(K)$, $CT_S = \mathcal{E}_{ek}(ESK\|\sigma_S)$
$M_S = ID_S\|\mathcal{T}_{\mathsf{PAKE}}\|ESK$		$\sigma_S \leftarrow \mathsf{Sign}(pp_{\mathsf{IBS}}, ID_S, sk_{ID_S}, M_S),$
If $0 = \mathsf{Vrfy}(pp_{\mathsf{IBS}}, ID_S, \sigma_S, M_S)$, abort.		$M_S = ID_S\|\mathcal{T}_{\mathsf{PAKE}}\|ESK$
Otherwise, $sk_{ID_C} = \mathcal{D}_{H_2(pw_C)}(ESK)$		
$\sigma_C \leftarrow \mathsf{Sign}(pp_{\mathsf{IBS}}, ID_C, sk_{ID_C}, M_C)$		
where $M_C = ID_C\|\mathcal{T}_{\mathsf{PAKE}}\|CT_S$		
$CT_C = \mathcal{E}_{ek}(\sigma_C)$ ⟶ ID_C, CT_C		$\sigma_C = \mathcal{D}_{ek}(CT_C)$
		$M_C = ID_C\|\mathcal{T}_{\mathsf{PAKE}}\|CT_S$
		If $0 = \mathsf{Vrfy}(pp_{\mathsf{IBS}}, ID_C, \sigma_C, M_C)$, abort.
		Otherwise,
$\mathsf{pid}_C = ID_C\|ID_S$		$\mathsf{pid}_S = ID_C\|ID_S$
$\mathsf{sid}_C = \mathcal{T}_{\mathsf{PAKE}}\|ID_S\|CT_S\|ID_C\|CT_C$		$\mathsf{sid}_S = \mathcal{T}_{\mathsf{PAKE}}\|ID_S\|CT_S\|ID_C\|CT_C$
$ssk = H_3(\mathsf{pid}_C\|\mathsf{sid}_C\|K)$		$ssk = H_3(\mathsf{pid}_S\|\mathsf{sid}_S\|K)$

Fig. 1. Generic construction of an IBS-PAKE protocol.

2. **PAKE-Server.** The server S performs its part in PAKE with the following modification: For each $\overline{m}_S$ to be sent to C in PAKE, S generates $\sigma_S \leftarrow \mathsf{Sign}(pp_{\mathsf{IBS}}, ID_S, sk_{ID_S}, \overline{m}_S)$, and then sends $(\overline{m}_S, \sigma_S)$. Finally, S obtains a common key K.

3. **Server.** Let $\mathcal{T}_{\mathsf{PAKE}}$ denote a concatenation of all transcripts generated from a run of PAKE. S generates $\sigma_S \leftarrow \mathsf{Sign}(pp_{\mathsf{IBS}}, ID_S, sk_{ID_S}, M_S)$ for $M_S = ID_S\|\mathcal{T}_{\mathsf{PAKE}}\|ESK$. It computes $ek = H_3(K)$ and $CT_S = \mathcal{E}_{ek}(ESK\|\sigma_S)$, and then sends (ID_S, CT_S).

4. **Client.** Upon receiving $[ID_S, CT_S]$, the client C computes $ek = H_3(K)$ and obtains $ESK\|\sigma_S = \mathcal{D}_{ek}(CT_S)$. Then C checks if σ_S is valid, i.e., $1 = \mathsf{Vrfy}(pp_{\mathsf{IBS}}, ID_S, \sigma_S, M_S)$ for $M_S = ID_S\|\mathcal{T}_{\mathsf{PAKE}}\|ESK$. If the validity does not hold then the session is aborted. Otherwise, C computes $\pi_2 = H_2(pw_C)$ and $sk_{ID_C} = \mathcal{D}_{\pi_2}(ESK)$. Using sk_{ID_C}, the client generates a signature, $\sigma_C \leftarrow \mathsf{Sign}(pp_{\mathsf{IBS}}, ID_C, sk_{ID_C}, M_C)$ on $M_C = ID_C\|\mathcal{T}_{\mathsf{PAKE}}\|CT_S$. Then the client generates $CT_C = \mathcal{E}_{ek}(\sigma_C)$ and sends $[ID_C, CT_C]$ to S.

Finally, the client computes a secret session key, $ssk = H_3(\mathsf{pid}_C||\mathsf{sid}_C||K)$ where $\mathsf{pid}_C = ID_C||ID_S$ and $\mathsf{sid}_C = \mathcal{T}_{\mathsf{PAKE}}||ID_S||CT_S||ID_C||CT_C$.

5. **Server.** Upon receiving $[ID_C, CT_C]$, the server computes $\sigma_C = \mathcal{D}_{ek}(CT_C)$ and checks if σ_C is valid, i.e., $1 = \mathsf{Vrfy}(pp_{\mathsf{IBS}}, ID_C, \sigma_C, M_C)$ for $M_C = ID_C$ $||\mathcal{T}_{\mathsf{PAKE}}||CT_S$. If it is not valid then the session is aborted. Otherwise, C computes a secret session key $ssk = H_3(\mathsf{pid}_S||\mathsf{sid}_S||K)$ where $\mathsf{pid}_S = ID_C||ID_S$ and $\mathsf{sid}_S = \mathcal{T}_{\mathsf{PAKE}}||ID_S||CT_S||ID_C||CT_C$.

In the above construction, different IBS schemes can be used for participants, to gain advantages. Assume that an IBS scheme has an efficient verifying algorithm and another IBS scheme has an efficient signing algorithm. A client's performance can be significantly improved if a server and a client use the first and the second IBS schemes, respectively.

The second generic method is a simplified version of the first method to omit executing a knowledge proof that a client is aware of the original password. In certain applications, a server can be managed systematically and sufficiently protected from a well-organized security architecture. For the situation, we can relax the security requirement on server compromise. An IBS-PAKE protocol is constructed by eliminating the third flow of the first method. The resulting protocol reduces the client's computation significantly. See Fig. 3.

In the above generic methods, a server authenticates himself to a client using two factors, i.e., a password of low entropy and an IBS of high entropy.

5.2   Instances

As instances resulting from the generic methods, we present two IBS-PAKE protocols using PAK [8] and the IBS scheme in Sect. 4. The instances are constructed not exactly following the generic methods but with a modification where a server use a single authentication factor, i.e., an IBS.[5] The first protocol is, for short called PWIBS-AKE. Each phase of PWIBS-AKE is given as follows.

Initialization Phase. Three processes Setup, Extract, and Registration are executed as follows. Let $pw_C \in \mathcal{D}_{PW}$ denote a password chosen by a client, C.

- Setup(λ): For a given security parameter λ, it generates a cyclic group, $\mathbb{G}$ of prime order q and two random generators, g and g_1 of $\mathbb{G}$. It generates $\theta \in \mathbb{Z}_q^*$ uniformly at random and computes $u = g^\theta$. It also generates independent cryptographic hash functions, $H : \{0,1\}^* \rightarrow \{0,1\}^\ell$, $H_1 : \{0,1\}^* \rightarrow \mathbb{Z}_q^*$, $H_i : \{0,1\}^* \rightarrow \{0,1\}^{\ell_i}$ for $i = 2,3$. It outputs public parameters $pp = (\mathbb{G}, g, g_1, u, H, H_{i=1,2,3})$ and the corresponding master secret key $\mathsf{msk} = \theta$.
- Extract($\mathsf{msk}=\theta$, ID). For a given identity ID, it picks $r \in \mathbb{Z}_q^*$ uniformly at random. It computes $R = g^r$, $w = H(ID, R)$, and $v = r + \theta w \pmod{q}$. It then returns $sk_{ID} = (R, v)$. Assume that the private key is transmitted via a secure channel.

---

[5] It is not difficult to fix the instances to follow the generic methods.

- Registration(C, S). First, a client, C generates its password, pw_C according to a pre-defined password creation policy. Also, C obtains a signing key, $sk_{ID_C} = (v_C, R_C)$ from Extract. Assume that a secure channel is established in advance between C and S. To register a service, C sends (Register-Req, ID_C, $g_1^{-H_1(pw_C)}$, $\mathcal{E}_{H_2(pw_C)}(v_C), R_C$) to the server, S over the secure channel. The server appends $\pi_S[C] = (ID_C, g_1^{-H_1(pw_C)}, \mathcal{E}_{H_2(pw_C)}(v_C), R_C)$ to $\mathcal{PF}$.

Key Establishment Phase. A client, C and a server, S execute a run of PWIBS-AKE to agree on a temporal session key. The concrete protocol is described as follows (See Fig. 2).

1. C picks $x \in \mathbb{Z}_q^*$ uniformly at random and computes $W = g^x g_1^{H_1(pw_C)} \in \mathbb{G}$ using the password, pw_C. Then, C sends $[ID_C, W]$ to S.

2. Upon receiving $[ID_C, W]$, S picks $y \in \mathbb{Z}_q^*$ uniformly at random and computes $Y = g^y \in \mathbb{G}$, and also $X' = W g_1^{-H_1(pw_C)}$ and $K' = (X')^y$. It finds $\pi_S[C]$ corresponding to ID_C, i.e., $[ID_C, g_1^{-H_1(pw_C)}, ESK = \mathcal{E}_{H_2(pw_C)}(v_C), R_C]$ from a database. Using its signing key, $sk_{ID_S} = (R_S, v_S)$, the server generates a signature, $\sigma_S = (d_S, z_S, R_S)$ on $M_S = ID_S \| W \| Y \| ESK$, where $E_S = g^{e_S}$, $z_S = H(M_S, ID_S, E_S)$ and $d_S = e_S - v_S z_S \pmod{q}$ for random $r_S, e_S \in \mathbb{Z}_q^*$. Also, using $ek = H_1(K')$ as an encryption key, S generates a ciphertext, $CT_S = \mathcal{E}_{ek}(ESK \| \sigma_S)$. Then S sends $[ID_S, Y, CT_S]$ to C.

3. Upon receiving $[ID_S, Y, CT_S]$, the client C computes $K = Y^x$ and $ek = H_1(K)$, and $ESK \| \sigma_S = \mathcal{D}_{ek}(CT_S)$. It checks if the signature, σ_S is valid, i.e., the equality of $z_S = H(M_S, ID_S, g^{d_S} \cdot (R_S^{z_S} \cdot u^{w_S \cdot z_S}))$ holds for $M_S = ID_S \| W \| Y \| ESK$ and $w_S = H(ID_S, R_S)$. If the validity does not hold then the session is aborted. Otherwise, the client computes $\pi_1 = H_2(pw_C)$ and decrypts ESK to obtain $v_C = \mathcal{D}_{\pi_1}(ESK)$. Using v_C, the client generates a signature share, $\sigma_C = (d_C, z_C)$ where $z_C = H(M_C, ID_C, E_C)$ and $d_C = e_C - v_C z_C \pmod{q}$ for random $r_C, e_C \in \mathbb{Z}_q^*$, $M_C = ID_C \| W \| Y \| CT_S$. Let $\sigma'_C = (z_C, d_C)$. The client computes $CT_C = \mathcal{E}_{ek}(\sigma'_C)$. Finally, the client sends $[ID_C, CT_C]$ to S.

   Then the client computes a secret session key, $ssk = H_3(\mathsf{pid} \| \mathsf{sid}_C \| K)$ where $\mathsf{pid}_C = ID_C \| ID_S$ and $\mathsf{sid}_C = ID_C \| W \| Y \| CT_S \| CT_C$.

4. Upon receiving $[ID_C, CT_C]$, the server decrypts CT_C to obtain $(z_C, d_C) = \sigma'_C = \mathcal{D}_{ek}(CT_C)$ and checks if the signature is valid, i.e., the equality of $z_C = H(M_C, ID_C, g^{d_C}(R_C u^{w_C})^{z_C})$ holds. Here R_C is the value stored at the database, and $M_C = ID_C \| W \| Y \| CT_S$ and $w_C = H(ID_C, R_C)$. If the validity does not hold then the session is aborted. Otherwise, the server computes a secret session key, $ssk = H_3(\mathsf{pid} \| \mathsf{sid}_S \| K')$ where $\mathsf{pid}_S = ID_C \| ID_S$ and $\mathsf{sid}_S = ID_C \| W \| Y \| CT_S \| CT_C$.

At Step 3, $\sigma'_C = (z_C, d_C)$ is generated by the client. It does not consist of a full IBS because R_C is not given, and thus nobody can check its validity. Instead of an encryption of σ'_C, we can send σ'_C to the server. However, since R_C is stored at the server, the server is able to verify it. Note that R_C is a global value included in all signatures generated by a client. This modification will alleviate the computation and communication cost on the client side.

A simplified IBS-PAKE protocol can be constructed from PWIBS-AKE by omitting a knowledge proof for an original password. The resulting protocol can run in two independent passes. For more details, refer to the appendix.

5.3   Security Proofs

In this section we prove the security of the proposed protocols in the model of Sect. 3. We prove that our first protocol provides AKE security and resilience to server compromise. That is, an adversary attacking the protocol cannot obtain useful information about session keys of fresh and semi-fresh instances with greater advantages than that of an on-line dictionary attack.

Theorem 1. *Assume that the IBS scheme, DL-IBS is used for PWIBS-AKE. Also, assume that the CDH assumption holds in* $\mathbb{G}$. *The proposed AKE protocol, PWIBS-AKE is AKE-secure in the random oracle model under the security model of Sect. 3.*

Proof. In the proof we consider a series of protocols, P_i ($i = 0, 1, .., 7$) which are modified sequentially from the original protocol, $P_0 = $ PWIBS-AKE. For each modification, we shall show that the advantage of an adversary increases with a negligible fraction. In the final protocol, P_7, the adversary will be able to get only an advantage from an on-line-guessing attack.

Assume that an adversary $\mathcal{A}$ can make at most q_{ex} and q_s queries to the Execute and the Send oracles. In the original protocol, P_0, we consider the random oracle model where hash functions are considered random functions. For each new hash query, a fresh random output is returned. To make consistent responses to hash queries, lists L_H and $L_{H_{i=1,2,3}}$ are maintained.

– Hash query. On a $\mathcal{H}(m)$-query for $\mathcal{H} = H, H_i$, returns h as follows. Let $h = \rho$ if (m, ρ) exists in $L_{\mathcal{H}}$, and otherwise, let $h = \rho' \overset{R}{\leftarrow} D$ where D is the domain of $\mathcal{H}$. $L_{\mathcal{H}}$ is updated with (m, ρ').

Next we describe P_i for $i = 0, 1, ..., 7$ concretely.

P_0: It is the original protocol, PWIBS-AKE defined in Subsect. 5.2 under the random oracle model.

P_1: It is modified from P_0 as follows. Let Rep denote the event that honest parties do not generate $W = g^x g_1^{H_1(pw_C)}$ or $Y = g^y$ twice. In P_1, we assume that Rep does not occur. Let q be the order of the group $\mathbb{G}$. We have $\Pr[\text{Rep}] \leq \frac{(q_s + q_{ex})^2}{q}$ by using a similar analysis of [37]. It is negligible because q is assumed to be sufficiently large. P_0 and P_1 are indistinguishable except the negligible probability.

P_2: It is modified from P_1 as follows. In P_2, we assume that the adversary cannot generate a valid server's signature without making a Corrupt query to a server. As shown in Sect. 4, the given IBS scheme is existentially unforgeable. It is obvious that by this assumption, P_1 and P_2 are indistinguishable except a negligible probability from existential unforgeability, i.e., $\mathsf{Adv}_{\mathcal{A},\mathsf{DL-IBS}}^{EUF\text{-}CMIDA}$. An adversary is able to use only (ID_S, Y, CT_S) which has been generated by a server, in order to make a Send query to a client instance.

In the protocol, a password-based authentication and an IBS works independently. Even if a password is revealed on a client side, a server's IBS key cannot be compromised.

P_3: It is modified from P_2 as follows. When an H_3 query is issued, P_3 does not check consistency against Execute queries, but returns a random output. The response to an Execute query is a collection of transcripts generated from an honest execution of the protocol. That is, it has the form, $[(ID_C, W), (ID, Y, CT_S), (ID_C, CT_C)]$, where $CT_S = \mathcal{E}_{ek}(ESK\|\sigma_S)$, $CT_C = \mathcal{E}_{ek}(\sigma_C')$ and $ek = H_3(g^{ab})$.

The way that an adversary know the inconsistency can be used to solve a CDH problem as follows. For a given CDH problem $(A = g^a, B = g^b)$, we plug in A and B for X and Y, respectively. We have $ek = H_3(g^{ab})$. The adversary would have made a H_3 query with g^{ab} to distinguish the distribution of ek. We can get g^{ab} for the solution to the CDH problem.

In other words, P_2 and P_3 are indistinguishable except the negligible probability to solve the CDH problem. Since a random output is used as an encryption key, ek, the ciphertext $\mathcal{E}_{ek}(\cdot)$ looks random from a viewpoint of an adversary. Also, in the above case, a session key is defined by a random value because it is an output of H_3. Hence, if Test query is asked to an instance which was initialized via an Execute query, $\mathsf{Adv}_{\mathcal{A},\mathsf{PWIBS\text{-}AKE}}^{IBS\text{-}PAKE}$ is upper bounded by a negligible probability.

P_4: It is modified from P_3 as follows. We assume that P_4 halts if a correct guess for a password is made against a client instance or a server instance before a Corrupt query. We can determine whether a password is correctly guessed or not, by an H_3 query using a correct input to compute ek and ssk. In this case, P_3 and P_4 are identical.

In the case of semi-freshness, the following assumption is added. If a correct password guess is made against a server instance before a Corrupt query to a client instance, P_4 halts. We can determine whether a password is correctly guessed or not, by an $H_{i=1,2}$ query with the correct password and a Corrupt query to a server. P_4 is identical to P_3 except that off-line dictionary attacks occurs.

P_5: It is modified from P_4 as follows. We assume that the adversary cannot make a password guess against client and server instances which are partnered. To argue the impossibility, we show that the capability to make a password guess can be used to solve a CDH problem. Let (ID_C, W) be the first protocol transcript which is generated via a Send query to a client instance or directly given by the adversary. Let $W = g^\alpha$. Then, to a Send query with $(ID_C, W = g^\alpha)$ to a server instance, the response, which is also the second transcript, (ID, Y, CT_S) is generated as follows. Let $(g, A = g^a, B = g^b)$ be a given CDH problem. We plug in A for g_1 and B for Y and returns a random value for ek. Define $ek = H_3(g^{ab}g_1^{-\gamma b})$ where $\gamma = H_2(pw_C)$. Let $CT_S = \mathcal{E}_{ek}(ESK\|\sigma_S)$. We have $g_1^{-\gamma y} = (g^{ab})^{-\gamma}$. To guess a password, the adversary must know g^{ab} which is the CDH solution. Hence, a correct password guess from client and server instances which are partnered, is impossible under the CDH assumption.

P_6: It is modified from P_5 as follows. We assume that the adversary cannot generate a client's valid signature share, (z_C, d_C) without making H_2 query with the correct password and a Corrupt query to a server. It is easy to see that the adversary get no useful information about the secret key. Note that $R_C = g^{r_C}$ for $r_C \in \mathbb{Z}_q^*$ is used as a signature part but is stored by a server and so the adversary cannot be aware of it. Thus, the probability that the adversary generates (z_C, d_C) verified with R_C is upper bounded by $1/q$.

P_7: It is modified from P_6 as follows. We assume that a password guess can be checked by a *password* oracle, that is, whether it is correct or not. Let cguess denote an event that the adversary guesses a password correctly. P_7 accepts a Corrupt(U) query and returns $\mathcal{PF}_S = \{\pi_S[C]\}_{C\in\mathsf{Clients}}$ if $U \in \mathbf{Servers}$, and otherwise pw_U. For freshness, there are at most q_s queries before a Corrupt query. Thus, we have $\Pr[\mathsf{cguess}] \leq \frac{q_s}{|\mathcal{D}_{\mathcal{PW}}|}$. For semifreshness, $q_{H_1} + q_{H_2}$ more queries are considered before a Corrupt query to a client. Here, let q_{H_i} denote the maximum number of H_i queries. Since these occur if there has been a Corrupt query to a server, we have $\Pr[\mathsf{cguess}] \leq \frac{q_s + q_{H_1} + q_{H_2}}{|\mathcal{D}_{\mathcal{PW}}|}$.

Overall, the success probability of the adversary can be evaluated by expanding with the event of cguess. That is, we have $\Pr[\mathsf{Succ}_{P_7}] = \Pr[\mathsf{cguess}] + \Pr[\mathsf{Succ}_{P_7}|\overline{\mathsf{cguess}}]$ where $\overline{\mathsf{cguess}}$ denotes the negation of cguess. As we shown in a series of the protocol modification above, $\Pr[\mathsf{Succ}_{P_7}|\overline{\mathsf{cguess}}]$ can be upper bounded by the negligible probability under the CDH assumption and existential unforgeability of the given IBS scheme. Therefore we obtain the desired results.

Based on the security of a given PAKE and an IBS scheme, we can prove that the first modular method (defined in Fig. 1) to yield an IBS-PAKE protocol is AKE secure and resilient to server compromise. That is, an adversary attacking the protocol cannot obtain useful information about session keys of fresh and semi-fresh instances with greater advantages than that of an on-line dictionary attack. The security proof for the modular method can be completed by following the security proof of PAKE with consideration for unforgeability of the underlying IBS scheme. As in the PWIBS-AKE, a client's signature, σ_C is encrypted and so not revealed to an adversary in the modular method. An adversary gains no meaningful advantage for a password guess from $CT_C = \mathcal{E}_{ek}(\sigma_C)$.

Also, a similar proof idea can be applied with a slight modification, to prove the AKE security of the simplified **PWIBS-AKE** and AKE protocols generated from our second modular method (defined in Fig. 3), equivalently that an adversary attacking the protocol cannot obtain useful information about session keys of fresh instances with greater advantages than that of an on-line dictionary attack. Actually, the proofs can be completed by simplifying the security proof of Theorem 1, i.e., P_4 because the protocols do not consider semi-fresh to capture 'server compromise'.

6   Performance Analysis

In this section, we compare performance between our protocols and other known AKE protocols using a combination of a password and an asymmetric techniques [53]. We also give experimental results for our protocols.

6.1   Performance Comparison

The performance is analyzed in terms of communication and computation overhead. Our protocols work with discrete logarithm parameters. Let $\mathbb{G}$ be a group of prime order q. Let ℓ_q and $\ell_\mathbb{G}$ denote the bit-length of the order of $\mathbb{G}$ and an element of $\mathbb{G}$, respectively. Let ℓ_H denote the bit-length of a hash output. Let Exp_t denote simultaneous multi-exponentiation (or scalar multiplication) using t group elements. In the communication analysis, we exclude identifiers commonly required for every protocol.

In **PWIBS-AKE**, a client transmits ID_C, W in the first round, and ID_C, $CT_C = \mathcal{E}_{ek}(\sigma_C')$ in the third round where $\sigma_C' = (z_C, d_C)$. Since W is an element of $\mathbb{G}$, and z_C and d_C are elements of $\mathbb{Z}_q$, a client transmits a $(\ell_\mathbb{G} + 2\ell_q)$-bit string. A client executes two Exp_1, one Exp_2, and one Exp_3 and two decryption, i.e., $\mathcal{D}$ of a symmetric encryption scheme. A server transmits $ID_S, Y, CT_S = \mathcal{E}_{ek}(ESK\|\sigma_S)$ where $\sigma_S = (z_S, d_S, R_S)$ and ESK is a ciphertext of v_C, i.e., $ESK = \mathcal{E}_{H_1(pw_C)}(v_C)$. Since v_C is an elements of $\mathbb{Z}_q$, we can assume that the bit length of CT_S is $2\ell_\mathbb{G} + 3\ell_q$. Note that W and R_S are elements of $\mathbb{G}$, and z_S and d_S are elements of $\mathbb{Z}_q$. Thus a server must transmit a $(2\ell_\mathbb{G} + 3\ell_q)$-bit string. Also, a server executes three Exp_1, one Exp_3 and one encryption, i.e., $\mathcal{E}$ of a symmetric encryption scheme. When an elliptic curve group with $\ell_p = 192$ and $\ell_\mathbb{G} = 192$ is considered, a client's transcript length is about 576 bits or 72 bytes. In the simplified **PWIBS-AKE**, reduced computation and smaller transcripts are required for a client and a server.

In Table 1, we summarizes comparison results among PAKE, IBE-PAKE, and our IBS-PAKE protocols, **PWIBS-AKE** and the simplified **PWIBS-AKE** (denoted by Sim-PWIBS in the table). For PAKE, we consider SPEKE [28,30], J-PAKE [1,27], SRP6 [51], AMP [34], and SK [49] which are presented in ISO/IEC 11770-4 [32] or IEEE P1363.2 [31]. For IBE-PAKE, we consider the protocols of [53] constructed from two different IBE schemes, i.e., the pairing-based Boneh-Franklin (BF) IBE [4] and TDL-based IBE [43] with the CCA-security. In the table,

Table 1. Comparison of AKE protocols

Protocol		RSI	RSC	Round (Pass)	Client Comm.	Client Comp.	Server Comm.	Server Comp.
PAKE (ISO/IEC) [32]	SPEKE	X	X	1(2)	ℓ_G	$2\mathsf{Exp}_1$	ℓ_G	$2\mathsf{Exp}_1$
	J-PAKE	X	X	2(4)	$6\ell_G + 3\ell_H$	$6\mathsf{Exp}_1 + 4\mathsf{Exp}_2$	$6\ell_G + 3\ell_H$	$6\mathsf{Exp}_1 + 4\mathsf{Exp}_2$
	SRP6	X	O	2(4)	$1\ell_{G''} + 1\ell_H$	$2\mathsf{Exp}_1$	$1\ell_{G''} + 1\ell_H$	$1\mathsf{Exp}_1 + 1\mathsf{Exp}_2$
	AMP	X	O	2(4)	$1\ell_G + 1\ell_H$	$2\mathsf{Exp}_1$	$1\ell_G + 1\ell_H$	$2\mathsf{Exp}_2$
	SK	X	O	2(4)	$1\ell_G + 1\ell_H$	$2\mathsf{Exp}_1$	$1\ell_G + 1\ell_H$	$1\mathsf{Exp}_1 + 1\mathsf{Exp}_2$
IBE-PAKE [53]	w/BF [4]	O	X	2(2)	$2\ell_{G_1} + 2\ell_H$	$1\mathsf{P} + 6\mathsf{Exp}$	$2\ell_{G_1}$	$1\mathsf{P} + 4\mathsf{Exp}$
	w/TDL [43]	O	X	2(2)	$2\ell_{G'} + 2\ell_H$	$6\mathsf{Exp}$	$2\ell_{G'}$	$5\mathsf{Exp}$
IBS-PAKE	PWIBS-AKE	O	O	3(3)	$1\ell_G + 2\ell_q$	$2\mathsf{Exp}_1 + \mathsf{Exp}_2 + \mathsf{Exp}_3 + 2\mathcal{D} + \mathcal{E}$	$2\ell_G + 3\ell_q$	$3\mathsf{Exp}_1 + \mathsf{Exp}_3 + \mathcal{D} + \mathcal{E}$
	Sim-PWIBS	O	X	1(2)	$1\ell_G$	$\mathsf{Exp}_1 + \mathsf{Exp}_2 + \mathsf{Exp}_3$	$2\ell_G + 2\ell_q$	$3\mathsf{Exp}_1$

let 'RSI' and 'RSC' denote 'Robustness to Server Impersonation (when a password is revealed)' and 'Resilience to Server Compromise', respectively. Also, let 'Comm.' and 'Comp.' denote the communication length and the computation cost, respectively. Let $\mathbb{G}'$ and $\mathbb{G}''$ be a TDL group defined with RSA parameters and a multiplicative group of a finite field, respectively. Thus $\ell_{G'}$ or $\ell_{G''}$ should be larger than at least 1024. 'P' denotes a pairing operation and let $\mathbb{G}_1$ be a bilinear group. It is known that a pairing operation is more expensive than (or comparable to) an exponentiation or a scalar multiplication when a similar security level is assumed [5,6,20]. Our protocols can be efficiently performed without requiring any pairing operation. Similarly, our generic construction can be efficiently performed.

6.2 Experimental Results

The test for our experimental results has been performed on an Intel Pentium model CPU clocked at 2.40GHz. The algorithms were written in Python 2.7 and based on Charm-Crypto [14] and PyCrypto [41] libraries.[6] Each result is the average of 1,000 tests.

For a mathematical group in the protocol, we use four elliptic curve groups, 'prime192v1', 'sect193r1', 'secp224r1', and 'sect163k1'. 'prime192v1' represents NIST/X9.62/SECG curve over a 192 bit prime field [48]. 'sect193r1' represents SECG curve over a 193 bit binary field [47]. 'secp224r1' represents NIST/SECG curve over a 224 bit prime field [48]. 'sect163k1' represents NIST/SECG/WTLS curve over a 163 bit binary field [48]. As a symmetric encryption scheme we use AES with the CBC mode. The bit sizes of a key used for AES are 128 and 256.

Table 2 shows the running time of PWIBS-AKE. According to the distinct tasks by a communication round, the protocol can be divided into four sub-modules, Client.s1, Server.s2, Client.s3, and Server.s4. Client.s1 represents the

---

[6] Even though Charm is not optimised, our results are enough to show feasible and efficient implementation of our protocols.

Table 2. Experimental results of PWIBS-AKE (time:msec)

EC Group	AES key(bit)	Client.s1	Server.s2	Client.s3	Server.s4	Total
prime192v1	256	1.28	2.21	3.80	2.24	9.53
	128	1.30	2.25	3.90	2.27	9.66
sect193r1	256	2.28	4.01	6.58	3.83	16.7
	128	2.51	4.21	6.86	4.01	17.59
secp224r1	256	1.64	3.01	5.03	2.92	12.6
	128	1.64	3.01	5.01	2.90	12.56
sect163k1	256	1.85	2.87	4.84	2.87	12.43
	128	1.74	2.71	4.54	2.71	11.7

generation of $W = g^x g_1^{H_1(pw_C)}$ by a client. Server.s2 represents the generation of $(Y = g^y, \sigma_S = (z_S, d_S, R_S))$ and $CT_S = \mathcal{E}_{ek}(ESK \| \sigma_S)$ by a server. Client.s3 represents the verification of σ_S, the computation of $K = Y^x$, the generation of $\sigma_C = (z_C, d_C)$, and the computation of a session key by a client. Server.s4 represents the verification of $\sigma_C = (z_C, d_C)$ and the computation of a session key by a server.

Table 3 shows the running time of the simplified PWIBS-AKE. In the experiment, the protocol is divided into four submodules, Client.s1, Server.s2, Client.s3, and Server.s4. As a client does not generate a signature for the possession proof of a password, Client.s3 of the simplified PWIBS-AKE is faster than that of PWIBS-AKE.

In both of the tables, 'Total' represents the sum of running time of all submodules. As shown in the tables, the protocols give different experimental results according to elliptic curve groups used. The most time-consuming task occurs in Client.s3. However, the total running time is only at most 0.02 s.

7   Conclusion

We have proposed efficient AKE protocols based on a password and an IBS. A client is able to do an easy authentication using a human-memorable password and an ID-based signature as authentication means. The use of an IBS gives two security enhancements against party compromise, i.e., resistance to sever impersonation attacks from client compromise and resilience to client impersonation attacks from server compromise. The proposed protocols also give good performance compared to known AKE protocols. They can be applied for various applications.

A   Bilinear Maps [21, 39]

Let $\mathbb{G}_1$ and $\mathbb{G}_2$ be additive groups and $\mathbb{G}_T$ a multiplicative group. Assume that the groups have the same prime order, q. We say that $e: \mathbb{G}_1 \times \mathbb{G}_2 \to \mathbb{G}_T$ is an

admissible bilinear map (or a pairing) if the following properties are satisfied:
(1) Bilinearity: $e(aP, bQ) = e(P, Q)^{ab}$ for all $P \in \mathbb{G}_1$ and $Q \in \mathbb{G}_2$, and $a, b \in \mathbb{Z}_q^*$.
(2) Non-degeneracy: There exist $P \in \mathbb{G}_1$ and $R \in \mathbb{G}_2$ such that $e(P, R) \neq 1$. (3)
Computability: There exists an efficient algorithm to compute $e(P', Q')$ for all
$P' \in \mathbb{G}_1$ and $Q' \in \mathbb{G}_2$.

Bilinear maps can be classified in three types, i.e., Type I, II, III accord-
ing to the existence of morphisms between $\mathbb{G}_1$ and $\mathbb{G}_2$. Type I pairings, called
'*symmetric*', have $\mathbb{G}_1 = \mathbb{G}_2$. Type II pairings have an efficiently computable isomorphism from $\mathbb{G}_1$ to $\mathbb{G}_2$ or from $\mathbb{G}_2$ to $\mathbb{G}_1$ but none in the reverse direction.
Type III pairings have no efficiently computable isomorphism between $\mathbb{G}_1$ and
$\mathbb{G}_2$. Type II and III pairings are called '*asymmetric*'. For more details, refer to
[21, 39].

B   Computational Assumptions

DISCRETE LOGARITHM (DL) ASSUMPTION. Assume that a group $\mathbb{G}$ of order q
and a generator g of $\mathbb{G}$ are given. To define *Discrete Logarithm (DL) problem*,
we consider the following game:

$$Initialization: x \xleftarrow{R} \mathbb{Z}_q,\ g_1 = g^x,$$
$$Output: r \leftarrow \mathcal{A}(\mathbb{G}, g, g_1 = g^x)$$

We define $\mathsf{Adv}_{\mathcal{A}, \mathbb{G}}^{DL}(t) = \Pr[r = x]$, where $\mathcal{A}$ runs in time t. We define that $\mathsf{Adv}_{\mathbb{G}}^{DL}(t)$
$= \max_{\mathcal{A}}[\mathsf{Adv}_{\mathbb{G}}^{DL}(t)]$ where the maximum is taken over all $\mathcal{A}$. We say that DL
assumption holds for $\mathbb{G}$ if $\mathsf{Adv}_{\mathbb{G}}^{DL}(t)$ is negligible.

DECISIONAL DIFFIE-HELLMAN (DDH) ASSUMPTION. Assume that a group $\mathbb{G}$
of order q and a generator g of $\mathbb{G}$ are given. To define *Decisional Diffie-Hellman
(DDH) problem*, we consider the following distinguishability game:

$$Initialization: x, y, r \xleftarrow{R} \mathbb{Z}_q,\ e_0 = xy,\ e_1 = r,$$
$$b \xleftarrow{R} \{0, 1\},\ h = g^{e_b}$$
$$Guess: b' \leftarrow \mathcal{A}(\mathbb{G}, g, g^x, g^y, h)$$

We define $\mathsf{Adv}_{\mathcal{A}, \mathbb{G}}^{DDH}(t) = |\Pr[b = b'] - 1/2|$, where $\mathcal{A}$ runs in time t. We define
that $\mathsf{Adv}_{\mathbb{G}}^{DDH}(t) = \max_{\mathcal{A}}[\mathsf{Adv}_{\mathbb{G}}^{DDH}(t)]$ where the maximum is taken over all $\mathcal{A}$.
We say that DDH assumption holds for $\mathbb{G}$ if $\mathsf{Adv}_{\mathbb{G}}^{DDH}(t)$ is negligible.

COMPUTATIONAL DIFFIE-HELLMAN (CDH) ASSUMPTION. Assume that a
group $\mathbb{G}$ of order q and a generator g of $\mathbb{G}$ are given. To define *Computational
Diffie-Hellman (CDH) problem*, we consider the following game:

$$Initialization: x, y \xleftarrow{R} \mathbb{Z}_q,\ g_1 = g^x,\ g_2 = g^y,$$
$$Output: g^z \leftarrow \mathcal{A}(\mathbb{G}, g, g^x, g^y)$$

We define $\mathsf{Adv}_{\mathcal{A}, \mathbb{G}}^{CDH}(t) = \Pr[z = xy]$, where $\mathcal{A}$ runs in time t. We define that
$\mathsf{Adv}_{\mathbb{G}}^{CDH}(t) = \max_{\mathcal{A}}[\mathsf{Adv}_{\mathbb{G}}^{CDH}(t)]$ where the maximum is taken over all $\mathcal{A}$. We
say that CDH assumption holds for $\mathbb{G}$ if $\mathsf{Adv}_{\mathbb{G}}^{CDH}(t)$ is negligible.

Table 3. Experimental results of our simplified PWIBS-AKE (time:msec)

EC Group	Client.s1	Server.s2	Client.s3	Server.s4	Total
prime192v1	1.33	1.52	3.01	0.84	6.7
sect193r1	2.36	2.58	5.11	1.36	11.41
secp224r1	1.66	2.01	3.82	1.09	8.58
sect163k1	1.75	1.76	3.53	0.92	7.96

C   Simplified IBS-PAKE Protocols

Initialization Phase. Three processes Setup, Extract, and Registration are executed as follows.

- Setup and Extract are the same to those of PWIBS-AKE.
- Registration(C, S). First, a client, C generates his or her password, pw_C according to a pre-defined password creation policy. To register a service, C sends (Register-Req, ID_C, $g_1^{-H_1(pw_C)}$) to the server, S over a secure channel. The server appends $\pi_S[C] = (ID_C, g_1^{-H_1(pw_C)})$ to $\mathcal{PF}$.

Key Establishment Phase. A client, C and a server, S execute the protocol to agree on a temporal key to be used for a session. The concrete protocol is described as follows (See Fig. 4).

1. C picks $x \in \mathbb{Z}_q^*$ uniformly at random and computes $W = g^x g_1^{H_1(pw_C)} \in \mathbb{G}$ using the password, pw_C. Then, C sends $[ID_C, W]$ to S.

2. S picks $y \in \mathbb{Z}_q^*$ uniformly at random and computes $Y = g^y \in \mathbb{G}$. Also, using its signing key, $sk_{ID_S} = (R_S, v_S)$, the server generates a signature, $\sigma_S = (d_S, z_S, R_S)$ on $M_S = ID_S\|Y$, where $E_S = g^{e_S}$, $z = H(m, ID_S, E_S)$ and $d_S = e_S - v_S z_S \pmod{q}$ for random $r_S, e_S \in \mathbb{Z}_q^*$. Then S sends $[ID_S, Y, \sigma_S]$ to C. From the receipt message $[ID_C, W]$, the server finds authentication information corresponding to ID_C, i.e., $[ID_C, g_1^{-H_1(pw_C)}]$ from a database. It then computes $X' = W g_1^{-H_1(pw_C)}$ and $K' = (X')^y$. Finally, S computes $ssk = H_3(\mathsf{pid}_S\|\mathsf{sid}_S\|K')$, where $\mathsf{sid}_S = ID_C\|W\|Y\|\sigma_S$.

3. Upon receiving $[ID_S, Y, \sigma_S]$, the client C checks if the signature, σ_S is valid, i.e., the equality of $z_S = H(M_S, ID_S, g^{d_S} \cdot (R_S \cdot u^{w_S})^{z_S})$ holds. Here $M_S = ID_S\|Y$ and $w_S = H(ID_S, R_S)$. If the validity does not hold then the session is aborted. Otherwise, the client computes $K = Y^x$. Finally, S computes $ssk = H_3(\mathsf{pid}_S\|\mathsf{sid}_S\|K')$, where $\mathsf{sid}_S = ID_C\|W\|Y\|\sigma_S$.

Client C		Server S
$pp = \{\mathbb{G}, g, g_1, u, H, H_1, H_2, H_3\}$		$pp = \{\mathbb{G}, g, g_1, u, H, H_1, H_2, H_3\}$
$[ID_C, pw_C]$		$[ID_S, sk_{ID_S} = (v_S, R_S)]$
		$\pi_S[C] = (ID_C, g_1^{-H_1(pw_C)},$
		$ESK = \mathcal{E}_{H_2(pw_C)}(v_C), R_C))$

$\pi_1 = H_1(pw_C)$		
$x \xleftarrow{R} \mathbb{Z}_q^*, \;\; W = g^x g_1^{\pi_1}$	$\xrightarrow{\;\;ID_C, W\;\;}$	$y \xleftarrow{R} \mathbb{Z}_q^*, \;\; Y = g^y$
		$X' = W g_1^{-H_1(pw_C)}, \;\; K' = (X')^y$
		$e_S \xleftarrow{R} \mathbb{Z}_q^*, \; E_S = g^{e_S}$
		$z_S = H(M_S, ID_S, E_S)$
		$d_S = e_S - v_S z_S \pmod{q}$
		where $M_S = ID_S\|W\|Y\|ESK$
		$\sigma_S = (z_S, d_S, R_S)$
$K = Y^x, \; ek = H_3(K)$	$\xleftarrow{\;\;ID_S, Y, CT_S\;\;}$	$ek = H_3(K'), \;\; CT_S = \mathcal{E}_{ek}(ESK\|\sigma_S)$
$ESK\|\sigma_S = \mathcal{D}_{ek}(CT_S)$		
$M_S = ID_S\|W\|Y\|ESK$		
$h_S = g^{d_S}(R_S u^{w_S})^{z_S}$		
If $z_S \neq H(M_S, ID_S, h_S)$, abort		
Otherwise, proceed as follows		
$v_C = \mathcal{D}_{H_2(pw_C)}(ESK)$		
$e_C \xleftarrow{R} \mathbb{Z}_q^*, \; E_C = g^{e_C}$		
$z_C = H(M_C, ID_C, E_C)$		
$d_C = e_C - v_C z_C \pmod{q}$		
where $M_C = ID_C\|W\|Y\|CT_S$		
$\sigma_C' = (z_C, d_C)$		
$CT_C = \mathcal{E}_{ek}(\sigma_C')$	$\xrightarrow{\;\;ID_C, CT_C\;\;}$	$\sigma_C' = \mathcal{D}_{ek}(CT_C)$
		$M_C = ID_C\|W\|Y\|CT_S$
		$h_C = g^{d_C}(R_C u^{w_C})^{z_C}$
		If $z_C \neq H(M_C, ID_C, h_C)$, abort
		Otherwise, proceed as follows
$\mathsf{pid}_C = ID_C\|ID_S$		$\mathsf{pid}_S = ID_C\|ID_S$
$\mathsf{sid}_C = ID_C\|W\|Y\|CT_S\|CT_C$		$\mathsf{sid}_S = ID_C\|W\|Y\|CT_S\|CT_C$
$ssk = H_3(\mathsf{pid}_C\|\mathsf{sid}_C\|K)$		$ssk = H_3(\mathsf{pid}_S\|\mathsf{sid}_S\|K')$

Fig. 2. PWIBS-AKE: AKE from a combination of PAK and a Schnorr-based IBS

Client C		Server S
$pp = \{\mathbb{G}, g, pp_{\mathsf{IBS}}, H_{i=1,2,3}, SE\}$		$pp = \{\mathbb{G}, g, pp_{\mathsf{IBS}}, H_{i=1,2,3}, SE\}$
$[ID_C, pw_C]$		$[ID_S, sk_{ID_S}]$
		$\pi_S[C] = (ID_C, \pi_1 = H_1(pw_C))$
Using $\pi_1 = H_1(pw_C)$ instead of pw_C,	Modified	perform its part in PAKE
perform its part in PAKE	Execution	with the following modification:
with the following modification:	of PAKE	For each $\overline{m}_S$ to be sent to C,
Whenever $(\overline{m}_S, \overline{\sigma}_S)$ is received,	with $H_1(pw_C)$	$\overline{\sigma}_S \leftarrow \mathsf{Sign}(pp_{\mathsf{IBS}}, ID_S, sk_{ID_S}, \overline{m}_S)$,
if $0 = \mathsf{Vrfy}(pp_{\mathsf{IBS}}, ID_S, \overline{\sigma}_S, \overline{m}_S)$, abort. $\longleftarrow$	$\xrightarrow{\hspace{2cm}}$	and then send $(\overline{m}_S, \overline{\sigma}_S)$.
Otherwise, perform the client's part		
for given $\overline{m}_S$ in PAKE.		
Output K		Output K
$\mathsf{pid}_C = ID_C \| ID_S$, $\mathsf{sid}_C = \mathcal{T}_{\mathsf{PAKE}}$		$\mathsf{pid}_S = ID_C \| ID_S$, $\mathsf{sid}_S = \mathcal{T}_{\mathsf{PAKE}}$
$ssk = H_3(\mathsf{pid}_C \| \mathsf{sid}_C \| K)$		$ssk = H_3(\mathsf{pid}_S \| \mathsf{sid}_S \| K)$

Fig. 3. Generic construction of a simplified IBS-PAKE protocol

Client C		Server S
$pp = \{\mathbb{G}, g, g_1, u, H, H_1, H_2\}$		$pp = \{\mathbb{G}, g, g_1, u, H, H_1, H_2\}$
$[ID_C, pw_C]$		$[ID_S, sk_{ID_S} = (d_S, z_S, R_S)]$
		$\pi_S[C] = (ID_C, g_1^{-H_1(pw_C)})$
$x \xleftarrow{R} \mathbb{Z}_q^*$, $W = g^x g_1^{H_1(pw_C)}$	$\xrightarrow{ID_C, W}$	$y \xleftarrow{R} \mathbb{Z}_q^*$, $Y = g^y$
If $0 = \mathsf{Vrfy}(pp, ID_S, \sigma_S, ID_S \| Y)$, abort	$\xleftarrow{ID_S, Y, \sigma_S}$	$\sigma_S \leftarrow \mathsf{Sign}(pp, ID_S, sk_{ID_S}, M_S)$
		where $M_S = ID_S \| Y$
Otherwise, $K = Y^x$		$X' = W g_1^{-H_1(pw_C)}$, $K' = (X')^y$
$\mathsf{pid}_C = ID_C \| ID_S$		$\mathsf{pid}_S = ID_C \| ID_S$
$\mathsf{sid}_C = ID_C \| W \| Y \| \sigma_S$		$\mathsf{sid}_S = ID_C \| W \| Y \| \sigma_S$
$ssk = H_2(\mathsf{pid}_C \| \mathsf{sid}_C \| K)$		$ssk = H_2(\mathsf{pid}_S \| \mathsf{sid}_S \| K')$

Fig. 4. Simplified PWIBS-AKE

References

1. Abdalla, M., Benhamouda, F., Mackenzie, P.: Security of the J-PAKE password-authenticated key exchange protocol. In: IEEE Symposium on Security and Privacy 2015, pp. 571–587. IEEE Computer Society (2015)

2. Boyarsky, M.K.: Public-key cryptography and password protocols: the multi-user case. In: ACMCCS 1999, pp. 63–72. ACM, New York (1999)

3. Benhamouda, F., Blazy, O., Chevalier, C., Pointcheval, D., Vergnaud, D.: New techniques for SPHFs and efficient one-round PAKE protocols. In: Canetti, R., Garay, J.A. (eds.) CRYPTO 2013, Part I. LNCS, vol. 8042, pp. 449–475. Springer, Heidelberg (2013)

4. Boneh, D., Franklin, M.: Identity-based encryption from the weil pairing. In: Kilian, J. (ed.) CRYPTO 2001. LNCS, vol. 2139, pp. 213–229. Springer, Heidelberg (2001)

5. Barreto, P.S.L.M., Galbraith, S.D., hÉigeartaigh, C.Ó., Scott, M.: Efficient pairing computation on supersingular abelian varieties. In: Yung, M. (ed.) CRYPTO 2002. LNCS, vol. 2442, pp. 354–368. Springer, Heidelberg (2002)
6. Barreto, P.S.L.M., Lynn, B., Scott, M.: Efficient implementation of pairing based cryptosystems. J. Cryptol. **17**, 321–334 (2004). Springer-Verlag
7. Bellovin, S.M., Merritt, M.: Encrypted key exchange: Password-based protocol secure against dictionary attack. In: IEEE Symposium on Research in Security and Privacy, pp. 72–84 (1992)
8. Boyko, V., MacKenzie, P.D., Patel, S.: Provably secure password-authenticated key exchange using Diffie-Hellman. In: Preneel, B. (ed.) EUROCRYPT 2000. LNCS, vol. 1807, pp. 156–171. Springer, Heidelberg (2000)
9. Bellare, M., Namprempre, C., Neven, G.: Security proofs for identity-based identification and signature schemes. In: Cachin, C., Camenisch, J.L. (eds.) EUROCRYPT 2004. LNCS, vol. 3027, pp. 268–286. Springer, Heidelberg (2004)
10. Bellare, M., Pointcheval, D., Rogaway, P.: Authenticated key exchange secure against dictionary attacks. In: Preneel, B. (ed.) EUROCRYPT 2000. LNCS, vol. 1807, pp. 139–155. Springer, Heidelberg (2000)
11. Bellare, M., Rogaway, P.: Entity authentication and key distribution. In: Stinson, D.R. (ed.) CRYPTO 1993. LNCS, vol. 773, pp. 232–249. Springer, Heidelberg (1994)
12. Chen, L., Harrison, K., Soldera, D., Smart, N.P.: Applications of multiple trust authorities in pairing based cryptosystems. In: Davida, G.I., Frankel, Y., Rees, O. (eds.) InfraSec 2002. LNCS, vol. 2437, pp. 260–275. Springer, Heidelberg (2002)
13. Clancy, T.: Eap password authenticated exchange, draft archive (2005). http://www.cs.umd.edu/clancy/eap-pax/
14. Akinyele, J.A., et al.: Charm: a framework for rapidly prototyping cryptosystems. J. Crypt. Eng. **3**(2), 111–128 (2013)
15. Choi, K.Y., Hwang, J.Y., Cho, J., Kwon, T.: Constructing efficient PAKE protocols from identity-based KEM/DEM, Cryptology ePrint Archive, Report 2015/606 (2015). http://eprint.iacr.org/2015/606. (To appear in WISA 2015)
16. Canetti, R., Halevi, S., Katz, J., Lindell, Y., MacKenzie, P.: Universally composable password-based key exchange. In: Cramer, R. (ed.) EUROCRYPT 2005. LNCS, vol. 3494, pp. 404–421. Springer, Heidelberg (2005)
17. Choi, K.Y., Hwang, J.Y., Lee, D.-H.: Efficient ID-based group key agreement with bilinear maps. In: Bao, F., Deng, R., Zhou, J. (eds.) PKC 2004. LNCS, vol. 2947, pp. 130–144. Springer, Heidelberg (2004)
18. Dent, A.W., Galbraith, S.D.: Hidden pairings and trapdoor DDH groups. In: Hess, F., Pauli, S., Pohst, M. (eds.) ANTS 2006. LNCS, vol. 4076, pp. 436–451. Springer, Heidelberg (2006)
19. Diffie, W., Hellman, M.: New directions in cryptography. IEEE Trans. Inf. Theory **22**(6), 644–654 (1976)
20. Elashry, I., Mu, Y., Susilo, W.: Jhanwar-Barua's identity-based encryption revisited. In: Au, M.H., Carminati, B., Kuo, C.-C.J. (eds.) NSS 2014. LNCS, vol. 8792, pp. 271–284. Springer, Heidelberg (2014)
21. Gallbraith, S.: Pairings, Advances in Elliptic Curve Cryptography, vol. 317, Chapter IX, pp. 183–213. Cambridge University Press (2005)
22. Galindo, D., Garcia, F.D.: A schnorr-like lightweight identity-based signature scheme. In: Preneel, B. (ed.) AFRICACRYPT 2009. LNCS, vol. 5580, pp. 135–148. Springer, Heidelberg (2009)
23. Gong, L.A., Lomas, T.M., Needham, R., Saltzwe, J.: Protecting poorly chosen secrets from guessing attacks. IEEE J. Sel. Areas Commun. **11**(5), 648–656 (1993)

24. Gentry, C., MacKenzie, P.D., Ramzan, Z.: A method for making password-based key exchange resilient to server compromise. In: Dwork, C. (ed.) CRYPTO 2006. LNCS, vol. 4117, pp. 142–159. Springer, Heidelberg (2006)
25. Halevi, S., Krawczyk, H.: Public-key cryptography and password protocols. ACM Trans. Inf. Syst. Secur. 2(3), 230–268 (1999)
26. Housley, R., Polk, T.: Planning for PKI: Best Practices Guide for Deploying Public Key Infrastructure. Wiley, Chichester (2001)
27. Hao, F., Ryan, P.Y.A.: Password authenticated key exchange by juggling. In: Christianson, B., Malcolm, J.A., Matyas, V., Roe, M. (eds.) Security Protocols 2008. LNCS, vol. 6615, pp. 159–171. Springer, Heidelberg (2011)
28. Hao, F., Shahandashti, S.F.: The SPEKE protocol revisited. In: Chen, L., Mitchell, C. (eds.) SSR 2014. LNCS, vol. 8893, pp. 26–38. Springer, Heidelberg (2014). Cryptology ePrint Archive, Report 2014/585. http://eprint.iacr.org/2014/585
29. Internet Engineering Task Forces, Eap password authenticated exchange (2005). http://www.ietf.org/internet-drafts/draft-clancy-eap-pax-03.txt
30. Jablon, D.: Strong password-only authenticated key exchange. ACM SIGCOMM Comput. Commun. Rev. 26(5), 5–26 (1996)
31. IEEE 1363.2:2008 Specification For Password-based Public-key Cryptographic Techniques
32. ISO/IEC 11770–4:2006 Information technology - Security techniques - Key management - Part 4: Mechanisms based on weak secrets
33. ITU-T Recommendation X. 1035: Password-Authenticated Key Exchange (PAK) Protocol. https://www.itu.int/rec/T-REC-X.1035/en
34. Kwon, T.: Addendum to Summary of AMP, In Submission to the IEEE P1363 study group for future PKC standards (2003)
35. Katz, J., Ostrovsky, R., Yung, M.: Efficient password-authenticated key exchange using human-memorable passwords. In: Pfitzmann, B. (ed.) EUROCRYPT 2001. LNCS, vol. 2045, pp. 475–494. Springer, Heidelberg (2001)
36. Kolesnikov, V., Rackoff, C.: Key exchange using passwords and long keys. In: Halevi, S., Rabin, T. (eds.) TCC 2006. LNCS, vol. 3876, pp. 100–119. Springer, Heidelberg (2006)
37. Katz, J., Yung, M.: Scalable protocols for authenticated group key exchange. In: Boneh, D. (ed.) CRYPTO 2003. LNCS, vol. 2729, pp. 110–125. Springer, Heidelberg (2003)
38. Lee, H.T., Cheon, J.H., Hong, J.: Accelerating ID-based Encryption Based on Trapdoor DL Using Pre-computation. Cryptology ePrint Archive, Report 2011/187 (2011). http://eprint.iacr.org/2011/187
39. Paterson, K.: Cryptography from pairings, Advances in Elliptic Curve Cryptography, vol. 317, Chap. X, pp. 215–251. Cambridge University Press, Cambridge (2005)
40. Pointcheval, D.: Password-based authenticated key exchange. In: Fischlin, M., Buchmann, J., Manulis, M. (eds.) PKC 2012. LNCS, vol. 7293, pp. 390–397. Springer, Heidelberg (2012)
41. Litzenberger, D.C.: Pycrypto-the python cryptography toolkit (2014). https://www.dlitz.net/software/pycrypto
42. Pointcheval, D., Stern, J.: Security arguments for digital signatures and blind signatures. J. Cryptol. 13(3), 361–396 (2000)
43. Paterson, K.G., Srinivasan, S.: On the relations between non-interactive key distribution, identity-based encryption and trapdoor discrete log groups. Des. Codes Crypt. 52(2), 219–241 (2009)

44. Rivest, R., Shamir, A., Adleman, L.: A method for obtaining digital signatures and public-key cryptosystems. Commun. ACM **21**(2), 120–126 (1976)
45. Shamir, A.: Identity-based cryptosystems and signature schemes. In: Blakely, G.R., Chaum, D. (eds.) CRYPTO 1984. LNCS, vol. 196, pp. 47–53. Springer, Heidelberg (1985)
46. Schnorr, C.-P.: Efficient signature generation by smart cards. J. Cryptol. **4**(3), 161–174 (1991)
47. Certicom, S.E.C.: SEC 2: Recommended elliptic curve domain parameters. In: Proceeding of Standards for Efficient Cryptography, Version 1 (2000)
48. Brown, D.: SEC 2: Recommended Elliptic Curve Domain Parameters, Version 2 (2010). http://www.secg.org/sec2-v2.pdf
49. Shin, S., Kobara, K.: Efficient Augumented Password-only Authentication and Key Exchange for IKEv2, RFC 6628, ISSN 2070–1721, IETF (2012)
50. Sakai, R., Kasahara, M.: ID Based Cryptosystems with Pairing over Elliptic Curve, Cryptology ePrint Archive, Report 2003/054. http://eprint.iacr.org/2003/054
51. Wu, T.: SRP-6: Improvements and Refinements to the Secure Remote Password Protocol, In Submission to the IEEE P1363 Working Group (2002)
52. Yi, X., Tso, R., Okamoto, E.: ID-based group password-authenticated key exchange. In: Takagi, T., Mambo, M. (eds.) IWSEC 2009. LNCS, vol. 5824, pp. 192–211. Springer, Heidelberg (2009)
53. Yi, X., Tso, R., Okamoto, E.: Identity-based password-authenticated key exchange for client/server model. In: SECRYPT 2012, pp. 45–54 (2012)
54. Yi, X., Hao, F., Bertino, E.: ID-based two-server password-authenticated key exchange. In: Kutyłowski, M., Vaidya, J. (eds.) ICAIS 2014, Part II. LNCS, vol. 8713, pp. 257–276. Springer, Heidelberg (2014)

Non-repudiation Services for the MMS Protocol of IEC 61850

Karl Christoph Ruland and Jochen Sassmannshausen$^{(\boxtimes)}$

Chair for Data Communications Systems, Faculty of Science
and Engineering, University of Siegen, 57076 Siegen, Germany
Christoph.Ruland@uni-siegen.de,
Jochen.Sassmannshausen@student.uni-siegen.de

Abstract. In Smart Grids various processes can be automated using communication between the components of the grid. The standard IEC 61850 defines, among other requirements and parts of the system, different communication protocols, that shall be used for different purposes. Although the scope of IEC 61850 is the automation of substations, there are also use cases beyond that can be addressed by IEC 61850. The standard IEC 62351 sets the focus on security in Smart Grids and lists various security requirements, that should be met, and further a series of measures to accomplish the required level of security. However, there are additional security requirements, such as non-repudiation and traceability of transactions, which cannot be sufficed using only the mechanisms provided by IEC 62351. In this paper a security solution will be presented that meets these additional requirements. Basically, it uses certificates for the proof of identity of the system participants and provides the two non-repudiation services Non-repudiation of Origin and Non-repudiation of Delivery using mechanisms described by the standard ISO 13888-3. The focus is set on the MMS protocol that is used for end-to-end communication between client and server. However, due to the flexibility of the mechanisms used, the security solution can also be transferred to different protocols. Finally, this paper describes a way to implement the solution using XML signatures and X.509 certificates.

Keywords: Smart grid security · IEC 61850 · MMS protocol · ISO 9506 · IEC 62351 · Non-repudiation · ISO 13888-3

1 Introduction

In a Smart Grid several processes can be automated using communication between the single components of the system. For this purpose, various communication protocols are used, each for different applications. With the decentralized and automated control of networks, there are new requirements for the safety and security of the system. In this paper the focus will be set on the MMS protocol, the use of which is intended by the standard IEC 61850 for the communication between client and server. Initially, this standard specified the automation

© Springer International Publishing Switzerland 2015
L. Chen and S. Matsuo (Eds.): SSR 2015, LNCS 9497, pp. 70–85, 2015.
DOI: 10.1007/978-3-319-27152-1_4

of substations. However, the protocols are very flexible and there are new use cases that can also be addressed by the standard IEC 61850. This leads to new security requirements for the data transfer within the .communication network. Among the Standard IEC 62351, which sets the focus on security aspects for certain standards including IEC 61850, there are several documents that deal with security in Smart grid security in general. One of these documents are the "NISTIR 7628 guidelines for Smart Grid Cyber Security" which define additional security requirements that have to be met. One of these requirements is non-repudiation of transactions, which cannot be achieved using the existing security measures. This paper presents a security solution for the MMS protocol using non-repudiation mechanisms defined by ISO 13888-3.

2   The State of the Art

This paper sets the focus on two key standards for Smart Grids: The standards IEC 61850 [7] and IEC 62351 [10]. While IEC 61850 deals with the control of the substations in electric power networks, the scope of IEC 62351 is security in Smart Grids, but the main focus is set on security for communication protocols specified in IEC 61850. In addition to the standard IEC 62351 there are several other documents that deal with security aspects in Smart Grids, pointing out additional security requirements like non-repudiation of transactions, that also have to be met.

2.1   The standard IEC 61850

The standard IEC 61850 focuses on the automation of substations and introduces rules for the modeling of the system, standardized data models, various communication protocols for different requirements and applications, communication interfaces and a configuration language for the description of the system components. The data transfer is performed via Ethernet using different protocols for different purposes. For example, a TCP/IP based protocol is used for the communication between clients and server.

Although the scope of IEC 61850 is the automation of substations, there are new use cases that can be addressed by IEC 61850, like the integration of Distributed Energy Resources (DER) into the electric power system or the communication of Intelligent Electronic Devices within buildings. Due to its flexibility, the standard IEC 61850 can also be used for wide area communication beyond the substation. This flexibility is enabled with standardized data models and services that can be mapped to various protocols. Currently, the standard defines a mapping to the MMS protocol, but there are other mappings planned, like a mapping to webservices [2]. With these new use cases beyond substations there are new requirements to the communication structure, like scalability or increased privacy and security. One of the well-known parts of 61850 is GOOSE for the fast transmission of status information and events. For GOOSE events already MAC mechanisms are specified by the standard IEC 62351-6 [13], which support data integrity and data authentication. However, in this paper we focus on the MMS protocol.

The standard introduces so called logical nodes, which represent certain functionalities. A logical node can represent, for example, switches in the grid, sensors, communication interfaces or it can simply contain descriptions of devices. Logical nodes can be summarized to logical devices which represent a real, physical device and its functionality.

The standard IEC 61850 also introduces the Abstract Communication Service Interface (ACSI), which is defined in IEC 61850-7-2, and describes services and functions, which can be used to interact with devices. For example, there are functions to read or write values, to obtain information about the data model, etc. The ACSI does not specify, how the functionality is implemented. There are various communication protocols to which a mapping of the ACSI can be implemented. For IEC 61850, a mapping to MMS (Manufacturing Message Specification) is defined in IEC 61850-8-1. A mapping of the ASCI to web services is also planned, but not yet published [2].

The MMS protocol was originally developed by General Motors for the purpose of communication in automated systems in the manufacturing scenario. MMS is defined in ISO 9506. The MMS protocol is based on the OSI protocol stack with ACSE and the Session Control Protocol ISO 8326 using TCP/IP [18]. IEC 61850 and IEC 62351 distinguish between two categories of the protocol stack: The T-Profile is related to the transport system, whose services are offered by the transport layer, and the A-Profile, which is related to the application layer on top of presentation layer and session layer. MMS provides a set of functions that allow the client to obtain the data model of the server, read or modify individual values or even delete entries. In addition, MMS also provides functionality to transfer files.

2.2   The Standard IEC 62351

The standard IEC 62351 is dedicated to the security aspects in smart grids. There are existing security requirements discussed, which have to be met. These are particularly data integrity, confidentiality, availability and non-repudiation. Furthermore, the standard presents a number of measures, by which these objectives are to be achieved. The recommended security measures vary according to the treated protocol. For example, for time-critical systems there shall be no time consuming encryption of the messages, instead the focus shall be put on data integrity and authenticity.

The standard IEC 62351 introduces security measures for both the T-Profile and the A-Profile: The measures for the T-Profile security cannot meet security requirements such as end-to-end authentication at the application level, non-repudiation, timeliness and traceability of transactions. These security mechanisms should be implemented within the A-Profile. Today, the security measures for the A-Profile support only the use of authenticated ACSE (Association Control Service Element), a mechanism that can be used to setup or close an association between two application entities. The following figure illustrates the communication between server and client and the messages that are sent during the communication (Fig. 1).

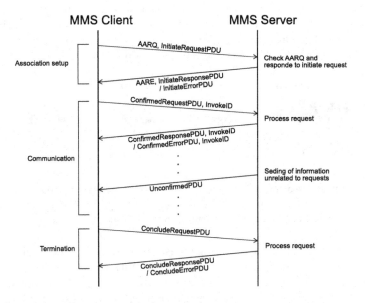

Fig. 1. The MMS protocol and typical data packets that are sent during a communication.

The association setup includes an authentication using ACSE and the association will be closed with a conclude request sent by the client. To perform an action such as changing or receiving a given value, the client sends a request to the server, which processes the request and sends a response back to the client, stating whether the request was processed successfully or not [9]. Further, the server can send data without a corresponding request, like periodic status information or reports that are sent when previously defined events occur [18].

2.3 The Weak Point of IEC 62351

Based on the specifications in IEC 62351, an authentication of the communication partners is only performed at the association establishment. However, this does not provide authentication of origin of the data exchanged via this association. During data transfer the security of the A Profile relies on the T Profile. Consider the following setup: Two Instances, the Control Center and the Substation Controller are permitted to issue commands on the same device. Both of them establish a MMS association and communicate with the device without gateways. In this scenario, the security is only based on TLS. Since an authentication of the communication partners only takes place at the setup of the association, services like nonrepudiation cannot be realized, because the data packets are not provided by digital signatures by the sender and no timing information.

In a different scenario shown in Fig. 2, where the communication takes place across several gateways, each using TLS connections, the mentioned problem becomes more severe and TLS does not provide a sufficient level of security since TLS only affects the data transmission between two transport layer entities, and the data can be manipulated by intermediate stations/gateways.

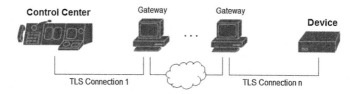

Fig. 2. The communication takes place across several gateways using different TLS connections. Note that the data could be manipulated between these connections. Both authenticity and integrity cannot be ensured in this scenario.

From the viewpoint of the receiving transport layer entity, the origin of incoming data packets cannot be determined (see [3]). With the expansion of the scope of IEC 61850 to use cases beyond substations, the use of these topologies will become necessarily. This is precisely the weak point, where the following security solution treats with.

2.4 Additional Security Requirements

As a summary of the guidelines mentioned in the introduction and IEC 62351, five important security requirements can be pointed out: Confidentiality, Integrity, Availability, and Nonrepudiation of actions and Traceability of actions In Smart Grids it is important that all transactions, that take place, and status messages which are sent can be traced and assigned at a later time. For example in case of a damage caused by a switching error or incorrect settings, it can be determined when and especially by whom certain actions that led to the error or damage were initialized (see [12, 20]). As it was described before, the existing security measures cannot provide authenticity for every single data packet, which is mandatory to provide non-repudiation of transactions.

The NISTIR 7628 guidelines for Smart Grid security also classify nonrepudiation as an important goal for smart grid security: *"Integrity, including non-repudiation, is critical since energy and pricing data will have financial impacts"*[16]. The main precondition for traceability of actions is also non-repudiation of actions that took place.

3 A Security Solution for the A-Profile

The security issues lead to the conclusion that the A-Profile has to be extended to suffice additional security requirements like non-repudiation and also authenticity

for every data packet as authenticity is a mandatory precondition for nonrepudiation. There are several ways to achieve authenticity for every data packet, resp.
command. For example, a digital signature can be added to every single APDU
(Application Protocol Data Unit). However, the ASN.1 definition of the MMS
APDUs does not intend the use of signatures for all data packets. In this paper
the focus will be on a different approach to achieve authenticity for each APDU.
Instead of modifying the APDUs the proof of origin will be provided by so called
tokens that are associated with a particular message.

The security solution is based on the approach that is described by the standard IEC 13888-3. Two non-repudiation services are be provided: non-repudiation
of origin and non-repudiation of delivery. These services can be realized with two
different tokens that are introduced by IEC 13888-3: The NROT (Non-repudiation
of Origin Token) and the NRDT (Non-repudiation of Delivery Token). The NROT
is added to every message sent by both server and client as proof of the authorship of the message. The NRDT stating that a particular message was received is
added to every response that is sent by the server to the client. Figure 3 illustrates
the exchange of messages between client and server using the tokens described in
IEC 13888-3.

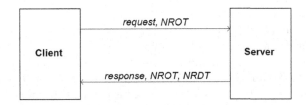

Fig. 3. MMS request and response with Non-repudiation Tokens

The standard ISO 13888-3 also describes various other communication scenarios, it is also possible that there is a trustworthy third party involved in the
communication. In this case, the tokens are generated by this instance. In the
communication scenarios MMS is used, this may sometimes be difficult, so this
security solution chooses the first approach in which the tokens are generated
by the participants of the communication. The tokens NRO and NRD are very
similar, the standard defines the following form for the tokens:

$$\text{Token} = \text{text}\|z\|S(K, z) \text{ with}$$

$$z = \text{Policy}\|\text{flag}\|A\|B\|C\|T_g\|T\|Q\|\text{Imp}(m)$$

($\|$ means concatenation).

The element "text" is optional, it can contain additional information about
the token. $S(K,z)$ is a signature, the element z is signed with the private key K
of the instance that generates the token. The element z contains several elements
that can also be optional, depending on the type of the token. The whole list of

elements can be viewed in the standard IEC 13888-3 [14]. In this section only the required and finally used elements shall be explained.

- **Policy:** An identifier of a non-repudiation policy under which the token was generated.
- **flag:** A flag that indicates the type of the token (so it can be distinguished between NROT and NRDT.)
- **A:** For a NRO token this field contains the name of the originator of the corresponding message and therefore also the creator of the token. For a NRDT, this field is optional and contains the name of the originator of the corresponding message for which the token was created.
- **B:** For a NRO token this field is optional and contains the name of the addressee of the corresponding message. For a NRDT, this field contains the name of the originator of the NRD token, and therefore the receiver of the associated message for which the proof of receipt is delivered.
- **C:** This field is optional and contains the identifier of an authority, eg. a trust party, that is involved. This security solution does not intend the participation of a trust party
- $\mathbf{T}_g$: The date and time the token was generated. This timestamp is provided by the token generator.
- **T:** NRO token: The date and time the token was created according to the originator of the token. NRD token: The date and time the associated message for which the token is generated, was received.
- **Q:** This field is optional and can contain any additional information, like the used hash algorithm or a distinguishing identifier of the related message m.
- $\mathbf{Imp}(m)$: The imprint of the message m. This can be the whole message itself or only a hash value of the Message. This security solution will store only a hash value of the message in the tokens. The standard IEC 13888-3 does not specify a certain hash algorithm that should be used.

3.1 Difference Between NROT and NRDT

Whereas the NROT provides proof that A is the originator of the Message m, the NRDT is generated by B and provides the proof that B received the message m. The main difference between a NROT and a NRDT are the fields containing the addressee and the originator. According to the standard IEC 13888-3 the element containing the name of the tokens addressee is optional. However, for NROT it is important to indicate the addressee, so an attacker cannot just copy and forward the message and the associated request to different participants that also communicate within the same system.

3.2 Generation of NROT and NRDT

Consider the following setup: Client **A** wants to issue a command on the server **B**. Further, **A** wants to receive a proof that the server actually received the command. First, **A** generates a MMS APDU containing the request and the

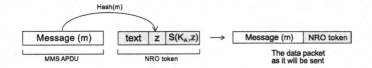

Fig. 4. The NROT is generated for and appended to a MMS APDU

corresponding NRO token. After the element z is initialized, it is signed with the private Key K_A (Fig. 4).

Note that the element z that is shown contains more elements than only the hash value. The list of elements can be viewed in the previous section. The NROT provides proof that the message m was sent by **A**. After the server **B** receives the message, it creates the NRDT and signs it with the private key K_B. The NRDT will later be transmitted together with the response message. The following figure shows the procedure on the server side (Fig. 5):

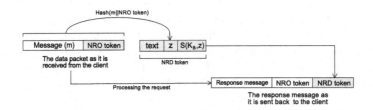

Fig. 5. The server generates and stores the NRDT till sending the Response

The NROT generated by the client is related to the request, the NRDT generated by the servers is related to the response. After receiving the response, A is able to verify the NRDT, as the hash value of the message and NRO token sent previously can be calculated and compared with the value contained by the NRDT. A stores NROT and NRDT received from B, B stores NROT received from A.

3.3   The Verification of the APDUs

Received data packets can be verified using the associated NROT. The receiver first calculates the hash value of the message, using the hash algorithm specified by the token. After that, the signature of the token will be verified using the corresponding public key, rsp. certificate. The used certificate is specified by the element of the token, which contains the distinguishing identifier of the instance that issued the token. As the number of authorized clients and their identities will be known in advance and limited, the certificate management should be easier than in open PKI systems.

3.4   Checking the NRDT

When receiving a response from the server, the client searches for the NRDT that was sent by the server together with the response. Since the client may have calculated and retained the hash value of the information that was sent before, the verification of the NRDT is straightforward. NRDT will be logged to be able to prove that the request was received by the application of the server.

3.5   NRD Tokens for the Server

Up to this point, the focus was set only on a single request to the server: The client sends a request and receives a response together with a NRD token indicating that the server received the previous request. From this arises the realization that not only the client, but also the server should get a NRD token signaling that the last message sent was received by the communication partner. Whereas this mechanism is easy to realize for a client (the NRD token can be sent along with the response), it is more difficult to achieve for the server: When the server has sent the response there is no reply from the client to which a NRD token indicating the receipt of the servers message can be added. There may be several ways to implement this functionality, the implementation of this security solution stores the NRD token on the client side and sends it to the server along with the next request to the server. This way, there are no additional messages required and the protocol sequence of the MMS protocol does not need to be changed.

3.6   The Application Security Sublayer

The extensions for the MMS protocol are integrated into the system in a way that the functionalities of the server and the client do not need to be changed. Both for client and server the security sublayer is transparent. In the MMS stack this requirement can be met inserting an intermediate layer (Sub layer principle) to the protocol stack which contains the functionality mentioned above. Figure 6 shows the principle of the sublayer.

3.7   Providing the APDUs with Tokens

The intermediate application security sublayer forwards outgoing packets to the lower layers after generating a NROT for the message. For this purpose, a signature algorithm based on elliptic curve cryptography, like ECDSA can be used for the signature used in the token. On the server side, the response will also be provided by the NRD token that was generated when the associated request was received. In general the field Q of the tokens will be used to store a distinguishing identifier of the associated message, so it can be determined more easily for which message the token was generated, after the messages and tokens were written to a log file.

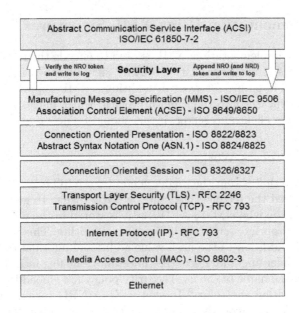

Fig. 6. MMS protocol stack with security sublayer and TLS

3.8   Access Control Lists

Since every participant of the communication has an identity provided by the corresponding certificate every single data packet is provided with a digital signature, the implementation of access control lists is straightforward. The *NISTIR 7628 Guidelines for Smart Grid Cyber Security* propose the use of role-based access control. A *role* is an accumulation of permissions and every instance can hold one or more roles [16]. In addition to verifying the NROT for incoming messages, server-side access control lists can be used to classify clients not only in trustworthy/non trustworthy, but to realize further authorization levels. For example, some clients can be granted full access, whereas other clients are only permitted to read certain values. This measure also increases the security level, because not all clients will need full access to the server to work properly.

3.9   Logging of Events

The system for logging of the messages is relatively simple: both the incoming as well as the outgoing APDUs need to be written to a log file. Due to the associated tokens that are also logged, the authorship of the data packet can be traced back without ambiguity and, the moment the packet was sent can be determined precisely since the NROT contains the timestamp (Note: the difficulties of provision of true timestamps are not addressed by this paper). The servers response should also be logged with the associated NROT and NRDT. The implementation of the system has to ensure that the log files are stored in a protected memory where the files are safe from unauthorized access or being overwritten.

4   An Implementation Using XML Signatures

For an exemplary implementation, XML is used to encode the tokens. The software for the generation and verification of XML signatures already exists,is easy to use and available online for free. In addition, tXML supports the integration of X.509 certificates into the signature block, which allows an exchange of certificates between client and server, if this should be necessary.

4.1   How the Process Works

For every Message that is sent the sender generates a NRO token, for every message that is received the recipient generates a NRD token. To generate a token, the hash value of the whole MMS APDU as it is stored in the memory is calculated using an appropriate (cryptographic) hash algorithm. This implementation uses SHA-256. In addition to the hash value additional information, namely the names of both sender and addressee, the timestamp, the flag indicating the type of the token and the name of the used security policy are embedded into a XML document representing the token. To sign a NRO/NRD token, the respective XML document is read by a XML parser creating a "Document Object Model" to which the signature node will be appended. The verification of a token works similar, the token is read by a XML parser and the signature node will be verified. Afterwards the hash value of the corresponding message is calculated and compared with the hash value provided by the token. If both the tests succeed, the NROT is valid and the authenticity of the message is ensured. Prior to further processing the data packet will be logged together with the corresponding tokens, ensuring that the proof of origin and receiving of the messages sent can be provided at a later time.

4.2   The Modified Communication

Figure 7 shows the communication between client and server via the MMS protocol with the additional use of tokens.

The tokens are concatenated with the messages that are sent. (Concatenations are indicated with ||). As it can be seen, the client the NROT to every message that is sent to the server. The server then processes the request and sends the response together with the generated NRD token indicating the receipt of the previous request. It also can be seen that the NRD tokens generated by the client are sent with the next MMS message, resulting in a deferred receipt by the server. However, the server gets a NRD token for all of its responses (except the conclude response), since there will be always the conclude request at the end of the association.

4.3   Example

The following example demonstrates how a data packet is provided with the corresponding NRO token. Lets assume the client sends a message resp. request

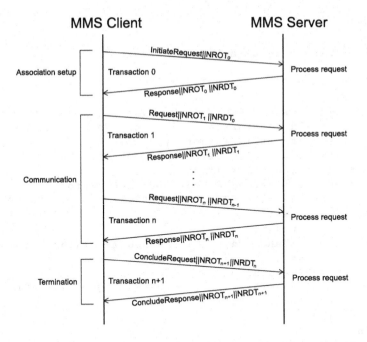

Fig. 7. The communication between client and server via MMS extended by tokens

APDU to the server. As the content of the message itself is not important for this example, it is represented as [message]. The hash value of [message] that is part of the token was generated using SHA-256 and is Base64 encoded so it can be embedded into the XML document. Note that the token also contains additional elements like <Authority> or <TimestampO> that were not mentioned before. These elements are introduced and described in the standard IEC 13888-3 [14] and can be ignored for this implementation.

The listing above shows the final message as it will be logged and sent to the server. Some parts marked with " " are omitted for reasons of clarity. It consists of the APDU [message] with the appended NROT. Note that this data packet does not contain a NRD token as it is a request to the server. Now, the server answers to the request. First, it generates the NRDT indicating the receiving of both the message and the NROT. The server generates the response message with the corresponding NROT. The server appends the NROT to the message and also appends the NRDT. Listing 1.2 only shows the NRD token that is sent together with the response message.

Note that the element <Recipient> contains the servers distinguishing name as it is the receiver of the last request message. The distinguished name of the tokens addressee is stored in the element <Origin>, since the corresponding message was received from this instance. The signature within the token was generated using ECDSA with a 256 bit key using SHA256 as hash algorithm.

```
[message]
<Token>
    <Text>This is an example token</Text>
    <Info>
        <Policy>policy_1</Policy>
        <Flag>NROT</Flag>
        <Origin>IssuerDistinguishedName</Origin>
        <Recipient>AddresseeDistinguishedName</Recipient>
        <Authority>null</Authority>
        <TimestampTG>13073622260389</TimestampTG>
        <TimestampO>null</TimestampO>
        <AdditionalInfo>SHA-256</AdditionalInfo>
        <HashValue>dAZNDGZKqaHGBJdUuV8in+J86p...</HashValue>
        <Signature>
            <SignedInfo>
                <CanonicalizationMethod Algorithm="..."/>
                <SignatureMethod Algorithm="..."/>
            </SignedInfo>
            <SignatureValue>1qGtoZoE8KyZ9UQm9huYNg4gGIhVF...
            </SignatureValue>
        </Signature>
    </Info>
</Token>
```

Listing 1.1. The NROT for a given message

```
[response message]
[NRO token for response message]
<Token>
    <Text>This is a NRDT for the previous request</Text>
    <Info>
        <Policy>policy_1</Policy>
        <Flag>NRDT</Flag>
        <Origin>SenderClientDistinguishedName</Origin>
        <Recipient>ServerdistinguishedName</Recipient>
        <Authority>null</Authority>
        <TimestampTG>13073622265765</TimestampTG>
        <TimestampO>null</TimestampO>
        <AdditionalInfo>SHA-256</AdditionalInfo>
        <HashValue>xZ3gI5Uk8BXC4Y5k+qlau4nHib...</HashValue>
        <Signature>
            <SignedInfo>
                <CanonicalizationMethod Algorithm="..."/>
                <SignatureMethod Algorithm="..."/>
            </SignedInfo>
            <SignatureValue>+8GkcTOs29KQnSKQXPTgpTqmwkgRv...
            </SignatureValue>
        </Signature>
    </Info>
</Token>
```

Listing 1.2. The response with NROT and NRDT

The NISTIR 7628 guidelines recommend using keys and curves with a security level not less than 112 bits for the use until the year 2029 and not less than 128 bits beyond the year 2030 [16]. The guidelines also names certain curves and hash algorithms that should be used.

4.4 Advantages of XML Signatures and Tokens

The XML data format and XML signatures are very flexible. It is possible to change signing algorithms or signature lengths without changing the architecture of the system. It is also possible to embed additional information and also certificates that have to be used to verify the signature.

The use of tokens to realize security services like nonrepudiation of origin and delivery is also very flexible, the main advantage is the independence of the data format of the request and response messages since only the hash value of the messages is needed to generate the tokens. Due to this flexibility the developed security solution can easily transferred to different protocols using different data formats. There are various communication protocols published or planned (also see [2,6,9]).

4.5 Possible Disadvantages of XML Signatures

The main disadvantage of XML signatures is the fact that the XML-encoded PDUs are significantly longer than PDUs encoded with different encoding rules, for example the *"Distinguished Encoding Rules"* that are used in the MMS protocol. Also, a XML parser is mandatory which may cause problems on systems with low resources. Since the tokens and the signatures do have a predefined structure, it may be possible to implement a lightweight (dedicated) XML parser for the tokens. However, tests of this implementation have shown that the main problem seems to be the additional message length which causes a heavy network utilization in some scenarios. There are possibilities to reduce the length of XML encoded messages: ITU-T X.694 describes a way to map XML schema definitions to ASN.1, which then enables different encoding rules, like the *"Distinguished Encoding Rules"* [15]. To evaluate the efficiency of this measure there are additional tests required.

5   Conclusion

The new security requirements for the MMS can only be met when the existing systems and protocols are upgraded. The solution in this paper shows a possible way to achieve end-toend authentication and non-repudiation for the MMS protocol using mechanisms described by the standard ISO 13888-3 and XML signatures. With the additional use of a logging system, all transaction that took place can be traced and attributed at a later time. Due to the simple structure reducing the required changes to a minimum, the security solution is very flexible and can also be used for different protocols. The exemplary implementation presented within this paper works fine in a simulated scenario but has to be tested in reality to check whether it is practically feasible. For this purpose, we are looking for a possibility to test the implementation in a realistic scenario.

References

1. Dournaee, B.: XML Security. RSA Press Series. Mcgraw-Hill, Osborne (2002)
2. Englert, H.: Neue Kommunikationskonzepte für den Netzbetrieb - aktuelle Entwicklungen in der IEC 61850. Smart Grids Forum, Hannover Messe (2014). https://www.vde.com/de/smart-grid/forum/beitraege/Documents/2014-04-09-neue-kommunikationskonzepte-englert.pdf. Accessed on 3 July 2015
3. Fries, S., Hof, H.-J., Dufaure, T., Seewald, M.G.: Security for the smart grid - enhancing IEC 62351 to improve security in energy automation control. Int. J. Adv. Secur. **3**(3 & 4), 169–183 (2010)
4. CEN, CENELEC, ETSI Smart Grid Coordination Group. Smart Grid Information Security, November 2012
5. CEN, CENELEC, ETSI Smart Grid Coordination Group. Smart Grid Reference Architecture, November 2012
6. IEC 61400–25: Communications for monitoring and control of wind power plants, TC 88
7. IEC 61850: Communication networks and systems in substations, TC 57
8. IEC 61850–1: Communication networks and systems in substations - Introduction and overview
9. IEC 61850-8-1: Communication networks and systems in substations - Part 8–1: Specific communication service mapping (SCSM) - Mappings to MMS (ISO 9506–1 and ISO 9506–2) and to ISO/IEC 8802–3
10. IEC 62351: Power systems management and associated information exchange - Data and communications security, TC 57
11. IEC 62351–1: Power systems management and associated information exchange - Data and communications security Part 1: Communication network and system security - introduction to security issues
12. IEC 62351–4: Power systems management and associated information exchange - Data and communications security - Part 4: Profiles including MMS
13. IEC 62351–4: Power systems management and associated information exchange - Data and communications security - Part 6: Security for IEC 61850 profiles
14. ISO/IEC 13888–3 IT Security techniques Non-repudiation - Part 3: Mechanism-susing asymmetric techniques
15. ITU-T X.694 Information technology ASN.1 encoding rules: Mapping W3C XML schema definitions into ASN.1
16. The Smart Grid Interoperability Panel - Cyber Security Working Group. NISTIR 7628 Guidelines for Smart Grid Cyber Security U.S. Department of Commerce, National Institute of Standards and Technologies, August 2010
17. Smart Grid Mandate M/490 EN: Standardization Mandate to European Standardisation Organisations (ESOs) to support European Smart Grid deployment European Commission Directorate-General for Energy, 1 March 2011
18. Systems Integration Specialists Company Inc, Overview and Introduction to the Manufacturing Message Specification (MMS) (1995). http://www.sisconet.com/downloads/mmsovrlg.pdf

19. Systems Integration Specialists Company, Inc. SISCO MMS Syntax (1994). http://www.sisconet.com/downloads/mms_abstract_syntax.txt
20. Verband der Elektrotechnik, Elektronik und Informationstechnik. VDE-Positionspapier Smart Grid Security Energieinformationsnetze und -systeme (2014). https://www.vde.com/de/InfoCenter/Studien-Reports/Seiten/Positionspapiere.aspx

Analysis of the PKCS#11 API Using the Maude-NPA Tool

Antonio González-Burgueño[1], Sonia Santiago[2], Santiago Escobar[3],
Catherine Meadows[4]($\boxtimes$), and José Meseguer[2]($\boxtimes$)

[1] University of Oslo, Oslo, Norway
antonigo@ifi.uio.no
[2] University of Illinois at Urbana-Champaign, Champaign, USA
{soniasp,meseguer}@illinois.edu
[3] DSIC-ELP, Universitat Politècnica de València, Valencia, Spain
sescobar@dsic.upv.es
[4] Naval Research Laboratory, Washington DC, USA
meadows@itd.nrl.navy.mil

Abstract. Cryptographic Application Programmer Interfaces (Crypto APIs) are designed to allow a secure interoperation between applications and cryptographic devices such as smartcards and Hardware Security Modules (HSMs). However, several Crypto APIs have been shown to be subject to attacks in which sensitive information is disclosed to an attacker, such as the RSA Laboratories Public Key Standards PKCS#11, an API widely adopted in industry. Recently, there has been a growing interest on applying automated crypto protocol analysis methods to formally analyze APIs. However, the PKCS#11 has been proven difficult to analyze using such methods since it involves non-monotonic mutable global state. In this paper we specify and analyze the PKCS#11 in Maude-NPA, a general purpose crypto protocol analysis tool.

Keywords: PKCS#11 · Cryptographic application programming interfaces (cryptographic APIs) · Symbolic cryptographic protocol analysis · Maude-NPA

1 Introduction

Standards for cryptographic protocols have long been attractive candidates for formal verification. Cryptographic protocols are tricky to design and subject to non-intuitive attacks even when the underlying cryptosystems are secure. Furthermore, when protocols that are known to be secure are implemented as standards, the modifications that are made during the standardization process may introduce new security flaws. Thus a considerable amount of work has

This work has been partially supported by NSF grant CNS 13-19109, by the EU (FEDER) and the Spanish MINECO under grant TIN 2013-45732-C4-1-P, and by Spanish Generalitat Valenciana under grant PROMETEOII/2015/013.

© Springer International Publishing Switzerland 2015
L. Chen and S. Matsuo (Eds.): SSR 2015, LNCS 9497, pp. 86–106, 2015.
DOI: 10.1007/978-3-319-27152-1_5

been done in the application of formal methods to cryptographic protocol standards [1,7,27,29]. In this work the protocols are treated *symbolically*, with the cryptosystems treated as black-box function symbols. A formal methods tool attempts to show that there is no way an attacker, by interacting with the protocol and applying the cryptographic functions symbols in any order, can break the security of the protocol. Such tools can be used both to search for attacks and to prove security with respect to the symbolic model.

Such symbolic formal analyses can be of great benefit to standards development in two ways. First of all, they offer means of verifying claims for security made by the standard, so that people can use it with more confidence. Furthermore, they can be useful in improving the standard, by discovering vulnerabilities in the protocols and attacks that can be mounted through exploiting the vulnerabilities. The explicit attacks discovered by the tools are particularly useful in that they provide the protocol designers information that can be used to help assess and repair the vulnerability.

For most analyses it has been the case that the same tool and general approach has been used to both verify the security of a protocol and to find attacks. Many of the tools employ methods that allow one to conclude that the protocol is secure if the tool terminates without finding an attack, such as heuristics that allow one to rule out redundant or useless paths (e.g. OFMC [3], Maude-NPA [14], Tamarin [30]) or abstractions (e.g. ProVerif [4]). Thus, one can use the tool first to find vulnerabilities, and then to verify security of the protocol once the vulnerabilities have been fixed. In a number of cases this has facilitated collaboration between standards developers and formal methods experts, e.g. as in [28]. This approach has worked particularly well for standards for key generation and secure communication, and these are the types of protocols that are most widely standardized, and the most well understood from the point of view of symbolic formal analysis.

However, recently another type of application has begun to attract interest: *Cryptographic Application Programming Interfaces*, or cryptographic APIs for short. A cryptographic API is a set of instructions by which a developer of an application may allow it to take advantage of the cryptographic functionality of a secure module. These APIs allow an application to perform such functions as creating keys, using keys to encrypt and decrypt data, and exporting and importing keys to and from other devices. Cryptographic APIs should also enforce security policies. In particular, no application should be able to retrieve a key in the clear.

Developers of cryptographic APIs have traditionally concentrated on providing functionality, not on guaranteeing security properties. This has resulted in a number of attacks on cryptographic APIs in which researchers have shown how many popular APIs can be led into an unsafe state (e.g. one in which a key is revealed to an untrusted application) via a series of legal steps. Indeed, much of the earliest work on formal analysis of cryptographic protocols focused on cryptographic APIs, e.g. [21,25,26]. However, the analysis of cryptographic APIs, did not become an area of research on its own until the early 2000's in particular after the attack found by Bond [5] in 2001 on IBM's CCA API.

Many more attacks on CCA and other systems, as well as new techniques for verifying the security of APIs, have followed; see for example the attacks described in Chap. 18 of Ross Anderson's *Security Engineering* [2].

One API that has attracted particularly wide attention is PKCS#11 [24]. This is a standard that provides both a set of commands that could be used by a cryptographic API and mechanisms for setting and enforcing security policies. These security policies are specified in terms of attributes on keys and other data that declare which operations using these terms are legal or illegal. However, no guidance is provided on what sort of restrictions should be put on·setting attributes on keys and other data so that undesirable states are avoided. Indeed, if no restrictions at all are put on the way attributes are set, it is possible to wind up with an application learning a key in the clear in just a very few steps, as Clulow points out in [9].

Since PKCS#11 is a widely used standard, much attention has been focused on correcting these deficiencies, in particular on developing means for formally verifying that a policy rules out undesirable states. But because PKCS#11 is intended to be applied to a wide variety of platforms, the problem is harder than verifying the security of an API such as IBM's CCA [20], which was only intended to be used for applications running on certain IBM systems. For one thing, the set of attributes forms a *mutable global state* which must be accounted for. Secondly, any formal verification system must be capable of verifying not just one or two policies specified by the developers of the API, as was the case of CCA, but any of a large class of policies that could be specified by a user.

Because of the complexity of the problem, researchers have tended to narrow their focus when applying cryptographic protocol analysis tools to PKCS#11. Most use of tools for the analysis of PKCS#11 makes some restriction on the policies analyzed, usually with an appeal to practicality or common use cases. They may also develop tools specifically designed for PKCS#11 analysis, or at the very least prove additional results specific to PKCS#11 that allow them to limit the size of the search space. Finally, they may concentrate primarily on either proving security or finding attacks, but not both.

In this paper we investigate the applicability of cryptographic protocol verification tools, in particular the Maude-NRL Protocol Analyzer (Maude-NPA) [14], to the analysis of cryptographic APIs such as PKCS#11, primarily concentrating on assessing its ability to find attacks. In order to perform the verification of PKCS#11, we make use of the results in [18] which show that, for a large class of reasonable policies, it is sufficient to assume that attributes never change; that is, that policies are *static*. We then show how, assuming static policies, it is possible to specify policies that put restrictions on what combination of attributes can be set to true for a particular key using Maude-NPA *never patterns*, a feature that allows the user to specify what events Maude-NPA should avoid in generating an attack. Since most policies proposed to date for PKCS#11 are expressed in this form, this leaves us in a good position to express both the API and policies in Maude-NPA. We then use Maude-NPA to reproduce the attacks found by Delaune et al. in [12]. Finally, we discuss the performance of Maude-NPA and compare it with other applications of cryptographic protocol analysis tools to the analysis of PKCS#11, in particular the use of the AVISPA tool in [32] and the use of Tamarin in [23].

The contributions of this paper are twofold. First, we advance investigation of the verification of PKCS#11 by performing the analysis of this API in a model more general than those in other works in the literature, namely a fully-unbounded session model with no abstraction nor approximation of fresh values, and making no other restrictions on policies other than that they are static. Second, we provide a new example of applicability of Maude-NPA to the analysis of cryptographic APIs. This paper extends the work presented in [19] on the analysis of IBM CCA by showing that Maude-NPA does not only support the specification of these APIs, but also the specification of policies restricting their behavior.

The rest of the paper is organized as follows. In Sect. 2 we give a high-level summary of Maude-NPA and its use of never patterns. In Sect. 3 we give an overview of PKCS#11 along with previous attacks and formal analyses. In Sect. 4 we describe how we specify the PKCS#11 API and policies in Maude-NPA. In Sect. 5 we describe the experiments we conducted and explain the results obtained. In Sect. 6 we discuss related work. Finally, in Sect. 7 we conclude the paper and discuss future work.

2   Maude-NPA

In this section we give a high-level summary of Maude-NPA. For further information, please see [14].

2.1   Preliminaries on Unification and Narrowing

We assume an order-sorted signature $\Sigma = (S, \leq, \Sigma)$ with a poset of sorts $(S, \leq)$ and an S-sorted family $\mathcal{X} = \{\mathcal{X}_s\}_{s \in S}$ of disjoint variable sets with each $\mathcal{X}_s$ countably infinite. $\mathcal{T}_\Sigma(\mathcal{X})_s$ is the set of terms of sort s, and $\mathcal{T}_{\Sigma,s}$ is the set of ground terms of sort s. We write $\mathcal{T}_\Sigma(\mathcal{X})$ and $\mathcal{T}_\Sigma$ for the corresponding order-sorted term algebras. For a term t, $Var(t)$ denotes the set of variables in t.

Positions are represented by sequences of natural numbers denoting an access path in the term when viewed as a tree. The top or root position is denoted by the empty sequence ϵ. The subterm of t at position p is $t|_p$ and $t[u]_p$ is the term t where $t|_p$ is replaced by u.

A *substitution* $\sigma \in \mathcal{S}ubst(\Sigma, \mathcal{X})$ is a sorted mapping from a finite subset of $\mathcal{X}$ to $\mathcal{T}_\Sigma(\mathcal{X})$. Substitutions are written as $\sigma = \{X_1 \mapsto t_1, \ldots, X_n \mapsto t_n\}$ where the domain of σ is $Dom(\sigma) = \{X_1, \ldots, X_n\}$ and the set of variables introduced by terms $t_1, \ldots, t_n$ is written $Ran(\sigma)$. The identity substitution is id. Substitutions are homomorphically extended to $\mathcal{T}_\Sigma(\mathcal{X})$. The application of a substitution σ to a term t is denoted by $t\sigma$.

A Σ-*equation* is an unoriented pair $t = t'$, where $t, t' \in \mathcal{T}_\Sigma(\mathcal{X})_s$ for some sort $s \in S$. Σ and a set E of Σ-equations, The E-equivalence class of a term

t is denoted by $[t]_E$ and $\mathcal{T}_{\Sigma/E}(\mathcal{X})$ and $\mathcal{T}_{\Sigma/E}$ denote the corresponding order-sorted term algebras modulo E. An *equational theory* (Σ, E) is a pair with Σ an order-sorted signature and E a set of Σ-equations.

An *E-unifier* for a Σ-equation $t = t'$ is a substitution σ such that $t\sigma =_E t'\sigma$. For $\mathcal{V}ar(t) \cup \mathcal{V}ar(t') \subseteq W$, a set of substitutions $CSU_E^W(t = t')$ is said to be a *complete* set of unifiers for the equality $t = t'$ modulo E away from W iff: (i) each $\sigma \in CSU_E^W(t = t')$ is an E-unifier of $t = t'$; (ii) for any E-unifier ρ of $t = t'$ there is a $\sigma \in CSU_E^W(t = t')$ such that $\sigma|_W \sqsupseteq_E \rho|_W$ (i.e., there is a substitution η such that $(\sigma \circ \eta)|_W =_E \rho|_W$); and (iii) for all $\sigma \in CSU_E^W(t = t')$, $Dom(\sigma) \subseteq (\mathcal{V}ar(t) \cup \mathcal{V}ar(t'))$ and $Ran(\sigma) \cap W = \emptyset$.

A *rewrite rule* is an oriented pair $l \to r$, where $l \notin \mathcal{X}$ and $l, r \in \mathcal{T}_\Sigma(\mathcal{X})_s$ for some sort $s \in S$. An *(unconditional) order-sorted rewrite theory* is a triple (Σ, E, R) with Σ an order-sorted signature, E a set of Σ-equations, and R a set of rewrite rules. The (R, E) rewriting relation $\to_{R,E}$ on $\mathcal{T}_\Sigma(\mathcal{X})$ is defined as: $t \to_{p,R,E} t'$ iff there exist $p \in Pos_\Sigma(t)$, a rule $l \to r$ in R, and a substitution σ such that $t|_p =_E l\sigma$ and $t' = t[r\sigma]_p$.

Let t be a term and W be a set of variables such that $\mathcal{V}ar(t) \subseteq W$, the R, E-narrowing relation on $\mathcal{T}_\Sigma(\mathcal{X})$ is defined as $t \rightsquigarrow_{p,\sigma,R,E} t'$ if there is a non-variable position $p \in Pos_\Sigma(t)$, a rule $l \to r \in R$ properly renamed s.t. $(\mathcal{V}ar(l) \cup \mathcal{V}ar(r)) \cap W = \emptyset$, and a unifier $\sigma \in CSU_E^{W'}(t|_p = l)$ for $W' = W \cup \mathcal{V}ar(l)$, such that $t' = (t[r]_p)\sigma$.

2.2 Maude-NPA Syntax and Semantics

Given a protocol $\mathcal{P}$, states are modeled as elements of an initial algebra $\mathcal{T}_{\Sigma_\mathcal{P}/E_\mathcal{P}}$, where $\Sigma_\mathcal{P}$ is the signature defining the sorts and function symbols (for the cryptographic functions and for all the state constructor symbols) and $E_\mathcal{P}$ is a set of equations specifying the *algebraic properties* of the cryptographic functions and the state constructors. Therefore, a state is an $E_\mathcal{P}$-equivalence class $[t]_E \in \mathcal{T}_{\Sigma_\mathcal{P}/E_\mathcal{P}}$ with t a ground $\Sigma_\mathcal{P}$-term.

In Maude-NPA a *state pattern* for a protocol P is a term t of sort State (i.e., $t \in \mathcal{T}_{\Sigma_\mathcal{P}/E_\mathcal{P}}(\mathcal{X})_{\mathsf{State}}$) which has the form $\{S_1 \,\&\, \cdots \,\&\, S_n \,\&\, \{IK\}\}$ where $\&$ is an associative-commutative union operator with identity symbol $\emptyset$. Each element in the set is either a *strand* S_i or the *intruder knowledge* $\{IK\}$ at that state.

The *intruder knowledge* $\{IK\}$ belongs to the state and is represented as a set of facts using the comma as an associative-commutative union operator with identity element *empty*. There are two kinds of intruder facts: *positive* knowledge facts (the intruder knows m, i.e., $m \in \mathcal{I}$), and *negative* knowledge facts (the intruder *does not yet know m* but *will know it in a future state*, i.e., $m \notin \mathcal{I}$), where m is a message expression.

A *strand* [16] specifies the sequence of messages sent and received by a principal executing the protocol and is represented as a sequence of messages $[msg_1^-, msg_2^+, msg_3^-, \ldots, msg_{k-1}^-, msg_k^+]$ such that msg_i^- (also written $-msg_i$) represents an input message, msg_i^+ (also written $+msg_i$) represents an output message, and each msg_i is a term of sort Msg (i.e., $msg_i \in \mathcal{T}_{\Sigma_\mathcal{P}/E_\mathcal{P}}(\mathcal{X})_{\mathsf{Msg}}$).

Strands are used to represent both the actions of honest principals (with a strand specified for each protocol role) and the actions of an intruder (with a strand for each action an intruder is able to perform on messages). In Maude-NPA strands evolve over time; the symbol | is used to divide past and future. That is, given a strand $[\ m_1^{\pm},\ \ldots,\ m_i^{\pm}\ |\ m_{i+1}^{\pm},\ \ldots,\ m_k^{\pm}\ ]$, messages $m_1^{\pm}$, $\ldots, m_i^{\pm}$ are the *past messages*, and messages $m_{i+1}^{\pm}, \ldots, m_k^{\pm}$ are the *future messages* $(m_{i+1}^{\pm}$ is the immediate future message). A strand $[msg_1^{\pm}, \ldots, msg_k^{\pm}]$ is shorthand for $[nil\ |\ msg_1^{\pm}, \ldots, msg_k^{\pm}, nil]$. An *initial state* is a state where the bar is at the beginning for all strands in the state, and the intruder knowledge has no fact of the form $m \in \mathcal{I}$. A *final state* is a state where the bar is at the end for all strands in the state and there is no intruder fact of the form $m \notin \mathcal{I}$.

Since **Fresh** variables must be treated differently from other variables by Maude-NPA, we make them explicit by writing $:: r_1, \ldots, r_k :: [m_1^{\pm}, \ldots, m_n^{\pm}]$, where each r_i first appears in an output message $m_{j_i}^{+}$ and can later appear in any input and output message of $m_{j_i+1}^{\pm}, \ldots, m_n^{\pm}$. If there are no **Fresh** variables, we write $:: nil :: [m_1^{\pm}, \ldots, m_n^{\pm}]$.

Example 1. Let us consider a subset of the PKCS#11 API. A symmetric key generated by principal A is denoted by $skey(A, r)$, where r is a unique **Fresh** variable and A denotes who generated the key. The symmetric encryption of a message M with a key $skey(A, r)$ is denoted by $senc(M, skey(A, r))$. The intruder's ability to generate its own symmetric keys is specified in Maude-NPA by the strand:

$$:: r :: \ [skey(i, r)^{+}]$$

where i is a constant denoting the intruder's name. Note that we have made explicit that the fresh variable r is generated in this strand. In this protocol the intruder is allowed to perform the symmetric encryption of a message M with a key K, assuming that it has received both M and K. This ability is specified by the following strand:

$$:: nil :: \ [M^{-},\ K^{-},\ (senc(M, K)^{+})]$$

where M is a variable of the sort for messages and K is a variable of the sort for symmetric keys. Note that no fresh variables are generated in this strand.

Since the number of states $T_{\Sigma_{\mathcal{P}}/E_{\mathcal{P}}}$ is in general infinite, rather than exploring concrete protocol states $[t]_{E_{\mathcal{P}}} \in T_{\Sigma_{\mathcal{P}}/E_{\mathcal{P}}}$ Maude-NPA explores *symbolic state patterns* $[t(x_1, \ldots, x_n)]_{E_{\mathcal{P}}} \in T_{\Sigma_{\mathcal{P}}/E_{\mathcal{P}}}(\mathcal{X})$ on the free $(\Sigma_{\mathcal{P}}, E_{\mathcal{P}})$-algebra over a set of variables $\mathcal{X}$. In this way, a state pattern $[t(x_1, \ldots, x_n)]_{E_{\mathcal{P}}}$ represents not a single concrete state but a possibly infinite set of such states, namely all the *instances* of the pattern $[t(x_1, \ldots, x_n)]_{E_{\mathcal{P}}}$ where the variables $x_1, \ldots, x_n$ have been instantiated by concrete ground terms.

The semantics of Maude-NPA is expressed in terms of *rewrite rules* that describe how a protocol moves from one state to another via the intruder's interaction with it. One uses Maude-NPA to find an attack by specifying an insecure state pattern called an *attack pattern*. Maude-NPA attempts to find a path from an initial state to the attack pattern via backwards narrowing (narrowing using the rewrite rules with the orientation reversed).

Example 2. Let us continue Example 1. In order to analyze whether the intruder can learn an honest key we specify the following attack pattern below:

$$\{SS \,\&\, \{skey(a,r) \in \mathcal{I},\, IK\}\}$$

where SS and IK are variables of the sort for sets of strands and for the intruder's knowledge, respectively. This attack pattern represents an insecure situation in which the intruder has learnt the honest key $skey(a,r)$. Note that a is a constant of the sort for names, which is different to the intruder's name i.

The backwards narrowing sequence from an initial state to an attack state is called a *backwards path* from the attack state to the initial state. Maude-NPA attempts to find paths until it can no longer form any backwards narrowing steps, at which point it terminates. If at that point it has not found an initial state, the attack pattern is judged *unreachable*. Note that Maude-NPA puts *no bound on the number of sessions*, so reachability is undecidable in general. Note also that Maude-NPA does not perform any data abstraction such as bound number of nonces. However, the tool makes use of a number of sound and complete state space reduction techniques that help to identify unreachable and redundant states [15], and thus make termination more likely.

2.3   Never Patterns in Maude-NPA

It is often desirable to exclude certain patterns from transition paths leading to an attack state. For example, one may want to determine whether or not authentication properties have been violated, e.g., whether it is possible for a responder strand to appear without the corresponding initiator strand. For this there is an optional additional field in the attack state containing the *never patterns*. Each never pattern is itself a state pattern. When we provide an attack pattern A and some never patterns $NP_1, \ldots, NP_k$ to Maude-NPA, every time the tool produces a state S via backwards narrowing from A, it checks whether there is a substitution θ such that $NP_i\theta =_{E_\mathcal{P}} S$. If that is the case, the state is discarded[1]. We will write an attack pattern A with the never patterns $NP_1, \ldots, NP_k$ as $A \parallel \mathrm{never}(NP_1) \ldots \parallel \mathrm{never}(NP_k)$.

Although never patterns were introduced as a means for specifying authentication properties, they can also be used to reduce the search space in a not necessarily complete way (an attack could be missed). In this work we only found it necessary to use such never patterns once in analysis (see Sect. 5).

Example 3. Let us continue Example 2. In order to exclude from the backwards path from the attack pattern of Example 2 the case in which the intruder uses

---

[1] Maude-NPA also checks whether $NP_i\theta$ satisfies *irreducibility constraints*, as described in [13].

the key $skey(a, r)$ to perform the symmetric encryption of any message M, we extend the attack pattern above with a never pattern as shown below:

$\{SS \ \& \ \{skey(a, r) \in \mathcal{I}, IK\}$

$\| \ never(\{ :: nil :: [(M)^-, (skey(a, r))^-, (senc(M, skey(a, r)))^+] \ \& \ SS' \ \& \ \{IK'\}\})$

where SS' and IK' are variables of the sort for sets of strands and for the intruder's knowledge, respectively.

3   PKCS#11

RSA Laboratories originally developed the Public Key Standards (PKCS) #11 in order to define a platform-independent API "Cryptoki" for the management of cryptographic tokens. Recently (in 2012) the responsibility of the maintenance of the standard was transitioned to the OASIS standards committee [31], but the standard is still referred to as PKCS#11.

PKCS#11 is intended to protect sensitive cryptographic keys as follows [17]. Once a session is initiated, the application may access the objects stored in the token, such as keys and certificates. However, access to the objects is controlled in the API via handles (which can be thought of as pointers to, or names for, the objects). These objects have attributes, e.g. boolean flags signalling properties of the object, namely `wrap`, `unwrap`, `encrypt`, `decrypt`, `sensitive` and `extract`. These flags can be either in positive form (l) or in negative form ($\neg l$), denoting that an attribute l is set or unset, respectively. Depending on whether these attributes are set or unset, certain API commands may be enabled or disabled.

New handles can be created by calling a key generation command, or by "unwrapping" an encrypted key packet. For example, if the *encrypt* function is called with the handle for a particular key, that key must have its `encrypt` attribute set. Also, a key may be exported outside the device if it is encrypted by another key, but only if it has the attributes `sensitive` and `extract` set. It is important to know that protection of the keys essentially relies on these two attributes, `sensitive` and `extract`.

Table 1 provides an informal description of a subset of the PKCS#11 key management commands. There are two kinds of commands. First, there are commands that correspond to PKCS#11 actions: the ones for wrapping and unwrapping keys, namely "Wrap" and "Unwrap", respectively; and for symmetric and asymmetric encryption and decryption, e.g. "SEncrypt" is the command for symmetric encryption, whereas "ADecrypt" corresponds to the command for asymmetric decryption. Note that there are several possibilities for the "Wrap" and "Unwrap" commands, depending on whether they use symmetric or asymmetric keys. Second, there are commands to modify attribute values, namely the "Set" and "Unset" commands. For example, "Set-Wrap" sets to true the `wrap` attribute of a key, whereas "Unset-Wrap" sets it to false.

The behavior of each command is described in Table 1 by rules of the form $T; L \overset{new \ \tilde{n}}{\rightarrow} T'; L'$. T is the set of messages that need to be received, whereas T' denotes the set of messages that are sent as a result of the messages in T

Table 1. Subset of PKCS#11 v2.01 key management commands

Name	API Command Description
Wrap (sym-sym)	$h(n_1, k_1), h(n_2, k_2)\,;\ wrap(n_1), extract(n_2) \rightarrow senc(k_2, k_1)$
Wrap (sym-asym)	$h(n_1, priv(z)), h(n_2, k_2)\,;\ wrap(n_1), extract(n_2) \rightarrow aenc(k_2, pub(z))$
Unwrap (sym-sym)	$h(n_1, k_2), senc(k_1, k_2)\,;\ unwrap(n_1) \overset{new\ r}{\rightarrow} h(r, k_1)\,;\ extract(r), L$
Unwrap (sym-asym)	$h(n_1, priv(z)), aenc(k_1, pub(z))\,;\ unwrap(n_1) \overset{new\ r}{\longrightarrow} h(r, k_1)\,;\ extract(r), L$
SEncrypt	$h(n, k), m\,;\ encrypt(n) \rightarrow senc(m, k)$
SDecrypt	$h(n, k), senc(m, k)\,;\ decrypt(n) \rightarrow m$
AEncrypt	$h(n, priv(z)), m\,;\ encrypt(n) \rightarrow aenc(m, pub(z))$
ADecrypt	$h(n, priv(z)), aenc(m, pub(z))\,;\ decrypt(n) \rightarrow m$
Set-Wrap	$h(n, k)\,;\ \neg wrap(n) \rightarrow wrap(n)$
Set-Encrypt	$h(n, k)\,;\ \neg encrypt(n) \rightarrow encrypt(n)$
Unset-Wrap	$h(n, k)\,;\ wrap(n) \rightarrow \neg wrap(n)$
Unset-Encrypt	$h(n, k)\,;\ encrypt(n) \rightarrow \neg encrypt(n)$

$L = \neg wrap(r), \neg unwrap(r), \neg encrypt(r), \neg decrypt(r), \neg sensitive(r)$

being received. L and L' are sets of attributes. More specifically, L denotes the attributes that must be set in order to execute the command, whereas L' denotes the value of the attributes after the command is executed. L' can include attributes in negative form and attributes related to freshly generated handles not appearing in L. The expression $new\ \tilde{n}$ represents the generation of fresh data that will appear in T' and L'.

The set L' is assumed to be satisfiable, i.e., it cannot contain two literals l and $\neg l$. Any variable appearing in T' must appear in T, i.e. $Var(T') \subseteq Var(T)$ and any variable appearing in L' also appears in L, i.e. $Var(L') \subseteq Var(L)$. The only new variables that can appear in T' and L' are those indicated in $new\ \tilde{n}$.

In Table 1 public and private keys are represented by terms of the form $pub(A)$ and $priv(A)$, respectively. Handles for keys are specified as terms of the form $h(N, K)$, where N is a nonce uniquely identifying the handle and K is the key. Symmetric and asymmetric encryption of a message M with a key K is denoted by terms $senc(M, K)$ and $aenc(M, K)$ respectively. Finally the attributes are specified as terms of the form $l(N)$, where l is the attribute's name, and N is a nonce that refers to a unique handle $h(N, K)$.

Example 4. Let us consider the rule corresponding to the "Wrap (sym-sym)" command. This rule allows wrapping a symmetric key using another symmetric key. If the attacker knows the handles $h(n_1, k_1)$ and $h(n_2, k_2)$, i.e., references to symmetric keys k_1 and k_2, the handle $h(n_1, k_1)$ has the attribute wrap set, and the handle $h(n_2, k_2)$ has the attribute extract set, then the attacker can learn the symmetric encryption of k_2 with k_1, represented by the term $senc(k_2, k_1)$.

In a cryptographic API threat model we assume that the application is malicious and in league with the adversary. Thus, in particular applications should not learn keys in the clear. In [9] Clulow presented a number of attacks in which sensitive keys are compromised. One of the best-known attacks is the so-called "key separation attack". The name refers to the fact that the attributes of a key can be set and unset in such a way as to give a key conflicting roles, allowing the attacker to learn sensitive keys. Figure 1 shows an example of a key separation attack presented in [9]. In this attack the attacker learns the value of a sensitive

Initial knowledge: The intruder knows $h(n_1, k_1)$ and $h(n_2, k_2)$; n_2 has the attributes **wrap** and **decrypt** set, whereas n_1 has the attributes **sensitive** and **extract** set.

Trace:

Wrap:	$h(n_2, k_2), h(n_1, k_1)$	$\rightarrow$	$senc(k_1, k_2)$
SDecrypt:	$h(n_2, k_2), senc(k_1, k_2)$	$\rightarrow$	k_1

Fig. 1. Decrypt/Wrap attack in PKCS#11 v2.01

Initial state: The intruder knows the handles $h(n_1, k_1)$, $h(n_2, k_2)$ and the key k_3; n_1 has the attributes **sensitive** and **extract** set whereas n_2 has the attributes **unwrap** and **encrypt** set.

Trace:

SEncrypt:	$h(n_2, k_2), k_3$	$\rightarrow$	$senc(k_3, k_2)$
Unwrap:	$h(n_2, k_2), senc(k_3, k_2)$	$\overset{new\ n_3}{\rightarrow}$	$h(n_3, k_3)$
Set_wrap:	$h(n_3, k_3)$	$\rightarrow$	$wrap(n_3)$
Wrap:	$h(n_3, k_3), h(n_1, k_1)$	$\rightarrow$	$senc(k_1, k_3)$
Intruder:	$senc(k_1, k_3), k_3$	$\rightarrow$	k_1

Fig. 2. Encrypt/Unwrap attack in PKCS#11 v2.01

key by wrapping it and then decrypting the resulting cyphertext with a key that has the attributes for decryption (**decrypt**) and wrapping (**wrap**) set.

Clulow suggested that this attack could be avoided by restricting key attribute changing operations so that a stored key could not have both the **decrypt** and **wrap** attributes set. However, as described in [12], this restriction does not prevent other key separation attacks not discovered by Clulow. For example, as shown in Fig. 2, if the attacker imports its own key by first encrypting it under a key k_2 whose handle has the attributes **unwrap** and **encrypt** set, and then unwrapping it, then it can export a sensitive key k_1 under k_3 to discover its value.

Again, in [12] the authors showed that restricting key attribute changing operations to avoid the attacks shown in Figs. 1 and 2 is not enough to make the API secure. The attacker can learn a sensitive key k_1 performing the sequence of commands shown in Fig. 3 if it knows a handle $h(n_2, k_2)$ which has both the **wrap** and **unwrap** attributes set.

Furthermore, when asymmetric encryption is considered, PKCS#11 is subject to the "Trojan Wrapped Key" attack, first discovered in [9], and found later on in [12], too. In this attack the attacker has the ability to smuggle a key of his own choice onto the token. If there is a key pair $pub(s_1), priv(s_1)$ on the token such that $pub(s_1)$ can be used for encrypting data and $priv(s_1)$ for unwrapping keys then the attacker can first encrypt a known key k_3 under $pub(s_1)$ and then plant its trojan key by unwrapping k_3 into a new handle. The attacker can then use this Trojan key to export other keys from the device, which it can then decrypt and recover. Figure 4 shows the exchange of messages of this attack.

Initial state: The intruder knows the handles $h(n_1, k_1)$ and $h(n_2, k_2)$, and the key k_3; n_1 has the attributes **sensitive** and **extract** set, n_2 has the attribute **extract** set. The intruder also knows the public key pub(s1) and its associated handle $h(n_3, priv(s_1))$.

Trace:

Set_wrap:	$h(n_2, k_2)$	$\rightarrow$	$wrap(n_2)$
Wrap:	$h(n_2, k_2), h(n_2, k_2)$	$\rightarrow$	$senc(k_2, k_2)$
Set_unwrap:	$h(n_2, k_2)$	$\rightarrow$	$unwrap(n_2)$
Unwrap:	$h(n_2, k_2), senc(k_2, k_2)$	$\stackrel{new\ n_4}{\rightarrow}$	$h(n_4, k_2)$
Wrap:	$h(n_2, k_2), h(n_1, k_1)$	$\rightarrow$	$senc(k_1, k_2)$
Set_decrypt:	$h(n_4, k_2)$	$\rightarrow$	$decrypt(n_4)$
SDecrypt:	$h(n_4, k_2), senc(k_1, k_2)$	$\rightarrow$	k_1

Fig. 3. Wrap/Unwrap attack in PKCS#11 v2.01

Initial state: The intruder knows the handles $h(n_1, k_1)$, $h(n_2, k_2)$ and the key k_3; n_1 has the attributes **sensitive** and **extract** set, n_2 has the attribute **extract** set. The intruder also knows the public key pub(s1) and its associated handle $h(n_3, priv(s_1))$.

Trace:

Intruder:	$k_3, pub(s_1)$	$\rightarrow$	$aenc(k_3, pub(s_1))$
Set_unwrap:	$h(n_3, priv(s_1))$	$\rightarrow$	$unwrap(n_3)$
Unwrap:	$aenc(k_3, pub(s_1)), h(n_3, priv(s_1))$	$\stackrel{new\ n_4}{\rightarrow}$	$h(n_4, k_3)$
Set_wrap:	$h(n_4, k_3)$	$\rightarrow$	$wrap(n_4)$
Wrap:	$h(n_4, k_3), h(n_1, k_1)$	$\rightarrow$	$senc(k_1, k_3)$
Intruder:	$senc(k_1, k_3), k_3$	$\rightarrow$	k_1

Fig. 4. Trojan Wrapped Key attack in PKCS#11 v2.01

Another experiment performed in [12] shows that more recent versions of the API that include new mechanisms to improve security are still subject to the same type of attacks. For example, version 2.20 of the PKCS#11[2] standard uses two more attributes: **wrap_with_trusted** and **trusted**. Whenever the "Wrap" command is executed, it tests whether if the key to be wrapped has **wrap_with_trusted** set then the wrapping key has **trusted** set. However, this does not prevent the attack shown in Fig. 5, where the attacker first attacks the trusted wrapping key, and then obtains the sensitive key k_1.

4 Specification of PKCS#11 in Maude-NPA

In this section we explain how to specify and analyze the core key management commands of PKCS#11 in Maude-NPA. First, in Sect. 4.1 we provide a

---

[2] Attacks shown in Figs. 1-4 actually correspond to attacks of PKCS#11 version 2.01, whereas the attack shown in Fig. 5 is an attack discovered for PKCS#11 version 2.20.

Initial state: The intruder knows the handles $h(n_1, k_1), h(n_2, k_2)$ and the key k_3; n_1 has the attributes sensitive, extract and wrap_with_trusted set , whereas n_2 has the attributes extract and trusted set. The intruder also knows the public key $pub(s_1)$ and its associated handle $h(n_3, priv(s_1)); n_3$ has the attribute unwrap set.

Trace:

Intruder:	$k_3, pub(s_1)$	$\rightarrow$	$aenc(k_3, pub(s_1))$
Set_unwrap:	$h(n_3, priv(s_1))$	$\rightarrow$	$unwrap(n_3)$
Unwrap:	$aenc(k_3, pub(s_1)), h(n_3, priv(s_1))$	$\overset{new\ n_4}{\rightarrow}$	$h(n_4, k_3)$
Set_wrap:	$h(n_4, k_3)$	$\rightarrow$	$wrap(n_4)$
Wrap:	$h(n_4, k_3), h(n_2, k_2)$	$\rightarrow$	$senc(k_2, k_3)$
Intruder:	$senc(k_2, k_3), k_3$	$\rightarrow$	k_2
Set_wrap:	$h(n_2, k_2)$	$\rightarrow$	$wrap(n_2)$
Wrap:	$h(n_2, k_2), h(n_1, k_1)$	$\rightarrow$	$senc(k_1, k_2)$
Intruder:	$senc(k_1, k_2), k_2$	$\rightarrow$	k_1

Fig. 5. Wrap with trusted key attack in PKCS#11 v2.20

high-level description of how we model PKCS#11 API version 2.20 in Maude-NPA. Then, in Sect. 4.2 we show in detail the specification of the PKCS#11 commands in Maude-NPA's syntax.

4.1   Formal Model of PKCS#11 in Maude-NPA

It is well-known that specifying and verifying API protocols in protocol analysis tools is challenging. Although this is not the first time an API protocol is analyzed in Maude-NPA (see [19]), there are specific features of PKCS#11 API that make its specification and analysis in Maude-NPA a non-trivial task.

First, the number of keys and handles is infinite a priori, e.g., the "Unwrap" command allows creating new handles for an existing key. The approach taken in [12] bounds the number of keys and handles for each key that can be created. For example, to find the attack shown in Fig. 1 the authors allow a maximum of 3 symmetric keys and 2 handles for each symmetric key (see [12] Sect. 5). In [18] the authors assume a bound on the number of fresh values that are generated in the course of an attack, and prove that it is sound and complete for the static policies of the specific systems they consider. However, these bounds may not apply to other systems. By contrast, in this paper we do not impose these restrictions and perform the analyses in an *unbounded session model*, making no abstractions of fresh data, e.g., an infinite number of keys an nonces can be generated. There was one exception, the analysis where attack shown in Fig. 5, where ultimately we had to restrict the number of keys the attacker generates to prevent state space explosion. However, in this case the attacker could still generate an unbounded number of *fresh* handles by executing the "Unwrap" command. The work presented in [23] also considers an unbounded session model using no abstraction, but its goal is different from ours (see Sect. 6 for further details).

Another feature of PKCS#11 is use of a global mutable state consisting of attributes. We avoid this issue by taking advantage of previous work in this area by Fröschle and Steel [18] to simplify the types of policies that need to be analyzed. In [18] Fröschle and Steel define a construct they call an *attribute policy*, and show that, for a large class of "reasonable" attribute policies called *complete* policies, it is enough to prove security in the case of *static* policies, in which the attributes of a key are never changed after it is created. In [18] the attribute state is a function assigning attribute valuations (specifications of which attributes are set and which are not) to keys. An attribute policy is a finite directed graph whose nodes are the set of allowable states, and whose edges are allowed transitions between states. A complete policy is one in which the transition policy consists of a collection of disjoint, disconnected cliques, and for each clique C, and each pair of states $c_0, c1 \in C$, we have $c_0 \cup c_1 \in C$. This allows for certain natural behaviors for which Fröschle and Steel point out that any well-designed policy should take into account. They then show that each clique has a unique *end point* in which the attacker has the greatest power. Thus any attacker behavior allowed by the policy in which each clique is replaced by the end point is allowed by any node in the clique. Thus analysis of the end point policy will tell us whether or not the original policy was safe. Moreover, since the end point policy is static, it is enough to be able to analyze static policies.

The main difference between the Fröschle-Steel paper and the model used in the Maude-NPA analysis is in the definition of attribute state. The Maude-NPA analysis has no such notion of attribute state. However, it does keep a history that may include the setting and unsetting of attribute values, and these can be mapped to allowable states. For example, if the attribute `wrap` was set for a key, and it was not subsequently unset, we may conclude that that key currently has its attribute `wrap` set. It is thus also possible to define policies for PKCS#11 keys as they are represented in Maude-NPA by mapping the states both before and after a transition T to attribute states, and then determining whether the two states and the transition edge connecting them belong to the policy. This in particular allows us to define both complete and static policies in terms of Maude-NPA histories and thus apply the results of Fröschle-Steel to analyze only static policies with the assurance that this analysis applies to the associated complete policies as well.

The use of static policies means that we do not need to represent attribute values explicitly. Instead, we can express policies defined in terms of what attributes can be set for a given key in terms of what combinations of actions enabled by those attributes are allowed using specific keys, without compromising completeness. Thus, instead of saying that the attributes $decrypt(n)$ and $wrap(n)$ cannot both we set, we say that the attacker cannot perform both "decrypt" and "wrap" actions using the same key. This allows us, as in [18], to dispense with global state entirely.

Another issue is to what extent we can use Maude-NPA to search for attacks assuming a certain policy is being enforced. Fortunately, this is easy for static policies that require that certain pairs of conflicting attributes not be both set for the same key. This policy is enforced if and only if there is no history in which the functions associated with each of the two conflicting attributes are executed on the same key. This allows us to specify policies using *never patterns* as described in Sect. 2.3. For example, in order to enforce a policy requiring, say, that no key can have the attributes `wrap` and `decrypt` set we specify a never pattern describing a generic state in which both `wrap` and `decrypt` strands have executed using the same key. Such a policy is shown in Example 6. Note that the never pattern does not specify which strand completed first, just that each should have executed at some time in the past.

4.2   Specification of PKCS#11 in Maude-NPA's Syntax

In this section we describe the specification of the PKCS#11 v2.20 key management commands in Maude-NPA's syntax. The API's signature is specified as follows. A nonce generated by principal A is denoted by $n(A, r)$, where r is a unique variable of sort Fresh and A denotes the principal that generated the nonce. This representation makes it easier to specify and keep track of the origin of nonces. E.g., one can use the notation to specify a state in which a principal accepts a nonce as coming from A when it actually comes from some $B \neq A$. Handles are represented by terms of the form $h(n(A, r), K)$, where K can be either a symmetric or an asymmetric key. Symmetric keys are represented by terms of the form $skey(A, r')$, where A denotes a name and r' is a fresh variable. Public and private keys, and symmetric and asymmetric encryption are specified similarly as explained in Sect. 3.

Each command of the PKCS#11 API is specified in Maude-NPA as a strand. Table 2 shows the specification of the commands of Table 1 representing PKCS#11 actions in Maude-NPA's syntax as an example. More specifically, for each rule $T; L \overset{new\ \tilde{n}}{\rightarrow} T'; L'$, messages in T are represented as received messages, i.e., terms of the form $-(M)$, whereas messages in T' are represented as sent messages, i.e., terms of the form $+(M)$. The generation of fresh data denoted by *new* $\tilde{n}$ is represented by variables $r_1, \ldots, r_i$ of sort Fresh made explicit at the beginning of the strand.

Example 5. The "Unwrap (sym-sym)" command of Table 1, which generates a new fresh data r, is specified in Maude-NPA as the following strand:

$$:: r :: [-(h(N_2, K_2)), -(senc(K_1, K_2)), +(h(n(A, r), K_1))]$$

where N_2 is a variable of the sort for nonces, K_1 and K_2 are variables of the sort for symmetric keys, and r is a variable of sort Fresh used to create a new handle for K_1.

Additionally, in PKCS#11 the attacker can perform symmetric encryption and decryption and create any number of symmetric keys. We assume it knows any public key and can generate only its private key. The attacker can perform asymmetric encryption with any public or private key it knows. As explained in Sect. 2, we specify each one of these capabilities as a strand in Maude-NPA.

Table 2. PKCS#11 key management commands in Maude-NPA

Command	Specification in Maude-NPA
Wrap (sym/sym)	$:: nil :: [-(h(N_1, K_1)), -(h(N_2, K_2)), +(senc(K_2, K_1))]$
Wrap (sym/asym)	$:: nil :: [-(h(N_1, priv(A))), -(h(N_2, K_2)), +(aenc(K_2, pub(A)))]$
Unwrap (sym/sym)	$:: r :: [-(h(N_2, K_2)), -(senc(K_1, K_2)), +(h(n(A, r), K_1))]$
Unwrap (sym/asym)	$:: r :: [-(h(N_1, priv(B))), -(aenc(K_1, pub(B))), +(h(n(A, r), K_1))]$
SDecrypt	$:: nil :: [-(h(N, K)), -(senc(M, K)), +(M)]$
SEncrypt	$:: nil :: [-(h(N, K)), -(M), +(senc(M, K))]$
ADecrypt	$:: nil :: [-(h(N, priv(A))), -(aenc(M, pub(A))), +(M)]$
AEncrypt	$:: nil :: [-(h(N, priv(A))), -(M), +(aenc(M, priv(A)))]$

We specify constraints on conflicting attributes as follows. Since in our model we do not explicitly represent attributes, we express these conditions in terms of the commands enabled when the conflicting attributes are set by adding *never patterns* (see Sect. 2.3) to the attack states we use to perform the analysis in Maude-NPA. More specifically, these never patterns are specified in such a way that they discard states where these commands are executed using the same handle. Let us illustrate this idea with the example below.

Example 6. Let us consider the attack shown in Fig. 2 where wrap and decrypt are considered as conflicting attributes, i.e., a given handle cannot have both the wrap and decrypt attributes set. In order to search for this attack in Maude-NPA one can specify the attack pattern shown below, which has a never pattern that discards any state that has, at least, two strands using the same handle $(h(N, K))$: one for wrapping, and the other one for decryption.

$$\{SS \& \{skey(A, r1') \in \mathcal{I}, IK\}$$
$$|| \; never(\{:: nil :: [-(h(N, K)), -(h(N_1, K_1)), +(senc(K_1, K))] \; \&$$
$$:: nil :: [-(h(N, K)), -(senc(M, K)), +(M)]$$
$$SS' \& \{IK'\}\})$$

Note that SS and SS' are variables of the sort for sets of strands, IK and IK' are variables of the sort for the intruder knowledge, A is a variable of the sort for names, and K_1 and K are variables of the sort for keys, and M is a variable of the sort for messages. The term $skey(a, r1')$ represents a sensitive key that should not be revealed to the attacker.

5 Experiments

We have specified the PKCS#11 API and analyzed several configurations in Maude-NPA following the methodology explained in Sect. 4. More specifically, we have rediscovered the attacks shown in Figs. 1, 2, 3, 4, and 5 (see Sect. 3).

Note that, unlike the experiments performed in [12] and in [18], we have not bounded the number of keys and handles that can be generated. The protocol specifications to reproduce our experiments are available on-line at http://www.dsic.upv.es/~sescobar/Maude-NPA/pkcs.html.

Table 3. Experimental Results

Attack	Length	dec / wrap	enc/ unwrap	wrap/ unwrap
Fig. 1	4	-	-	-
Fig. 2	7	✓	-	-
Fig. 3	6	✓	✓	-
Fig. 4	7	✓	✓	✓
Fig. 5	9	✓	✓	✓

Table 3 gathers the results of our experiments. For each one of the attacks explained in Sect. 3 we specify in the second column the length of the backwards reachability analysis performed by Maude-NPA until each attack was found. In the third, fourth, and fifth columns we show the different constraints on conflicting attributes that have been considered in each experiment. For example, the attack of Fig. 1 was found by Maude-NPA after 4 reachability steps, and no restriction on conflicting attributes was considered. The attack shown in Fig. 3 was found by Maude-NPA after 6 reachability steps and we considered two restrictions, namely that decrypt and wrap, and encrypt and unwrap are conflicting attributes.

In the experiments to find the attacks shown in Figs. 2 to 5 we used never patterns to specify policies on conflicting attributes. Additionally, in the case of the attack shown in Fig. 5 in order to control the size of the state search space we added an *attack preserving* never pattern (see [19]) that forces Maude-NPA to only search for states in which the attacker generated the minimal number of keys.

we added a never pattern to discard states containing more than one instance of the initial knowledge strand to reduce the state search space. Note that this never pattern preserves the completeness of the analysis because this PKCS#11 configuration is subject to an attack. That is, if Maude-NPA finds the attack when only one instance of the initial knowledge strand is allowed, it will still find the same attack (or an equivalent one) if several instances are allowed.

6   Related Work

There is a vast amount of research on the formal analysis of cryptographic APIs, so in this related work section we will concentrate on the work that is closest to ours, namely the formal analysis of PKCS#11 and PKCS#11-like systems and the use of cryptographic protocol analysis tools to analyze APIs.

Besides the work on formalizing and verifying PKCS#11 that we have already discussed, there has been further work focused on building tools for analyzing policies for PKCS#11 and PKCS#11-like systems. In [8] Centenaro et al. design a typed-base system for reasoning about the security of PKCS#11 policies, and use it to verify the security of new classes of PKCS#11 security policies they propose. This work is even able to verify implementations of PKCS#11. In [10]

Cortier and Steel develop a generic model for PKCS#11-like systems, and an algorithm and tool for verifying policies in this model. Interestingly, they show that a number of cryptographic protocols can also be modeled using their system, and they demonstrate their tool on them as well, thus showing that the relationship between cryptographic APIs and cryptographic protocols runs both ways.

The Tookan tool [6] can be used on PKCS#11 implementations. It reverse engineers security tokens, builds a formal model similar to that of [12], whose security can be checked by a model checker, and then runs any attack trace found directly on the token to validate it. The methods used by Tookan were further developed and commercialized in the commercial tool Cryptosense [11], which includes a component Cryptosense Analyzer that analyzes PKCS#11 configurations for insecurities and then tests attack traces on the system. Thus, analysis of PKCS#11 and systems like it, although challenging, has proved achievable enough to have potential commercial application.

An approach closer to ours is the work of Künnemann presented in [23], in that it relies on a protocol analysis tool, Tamarin [30] with some features in common with Maude-NPA (in particular, it performs backwards search over an unbounded number of sessions using no abstraction). In this paper Künnemann models PKCS#11 v2.20 in the Sapic calculus presented in [22], a variant of the applied pi calculus augmented with operators for state manipulation. This high-level protocol specification is then translated to a multiset rewrite system that can be verified using Tamarin. Using this tool-chain the author provides a configuration of the API and proves that it preserves the secrecy of sensitive keys.

However, the goal in [23] is different from ours. Instead of searching for attacks, Künnemann uses Tamarin to prove security of a particular configuration of PKCS#11 v2.20. This allows him to define a more restrictive model that includes their secure configuration but rules out others, in particular many of the configurations we analyzed with Maude-NPA. It also allows him to leave out certain features, such as asymmetric encryption, which, given the restrictions of his model does not give the intruder any more capabilities than symmetric encryption, and so can be omitted as redundant.

For the case of applying cryptographic protocol analysis tools one of the biggest issues facing the analysis of cryptographic APIs, is the problem of keeping the search space of a manageable size. Solutions that work for key distribution protocols, such as bounding the number of sessions, do not apply as well to cryptographic APIs, where the number and kind of "sessions" executed is under complete control of the adversary. The earliest work on formal analysis of APIs [21,25,26] dealt with the problem by relying to an extent on user input. For example, the analysis in [25] allowed users to tell the tool when they thought a state was reachable, and the analysis in [26] relied on lemmas that were conceived of and proved by the user, with machine assistance. More recently, the AVISPA analysis of PKCS#11 reported by Tsalapati in [32] uses simplifications such as a monotonic state and allowing only one handle per key.[3] According to [12],

---

[3] The thesis in which this work is contained is not publicly available, so we are relying on the account given in [12].

Steel and Carbone were able to enrich Tsalapati's AVISPA model to include non-monotonic state; however the number of sessions still needed to be bounded for analysis to be tractable, and the bounds needed to be relatively small. The Maude-NPA analysis of IBM's CCA API described in [19] relies on the use of *never patterns*, which can be used to tell Maude-NPA ignore classes of states specified by the user.

In [23], termination of the analysis is achieved, not only by restricting the kinds of configurations considered, but also by specifying model-specific heuristics that allow a more efficient evaluation of the operations that manipulate the protocol's state. In order to reduce the state search space and speed up the analysis the author defined a number of model-specific helping lemmas to rule out some states describing impossible situations or actions that do not allow the attacker to learn more knowledge in their model.

Another issue is the type of policies that need to be considered. In IBM CCA policies are represented in terms of separation of duties, which are straightforward to model in a tool such as Maude-NPA. For PKCS#11 the problem is apparently harder, since policies are expressed in terms of a global mutable state. Our first attempt to analyze PKCS#11 included a faithful model of PKCS#11 state, and analysis with respect to this model proved to be intractable. Steel and Carbone seemed to have encountered similar problems in their AVISPA analysis.

However, closer study reveals that there are natural restrictions on policies one can enforce. Fröschle and Steel [18] in particular show that it is possible to safely assume that policies are static, if one imposes some very natural restrictions on them. This means that one can specify policies in terms of which combinations of attributes are allowable. Künnemann also argues for static policies in [23], on the grounds of practicality and safety: the ability to turn attributes on and off is not needed, and can lead to security problems. The Tookan tool [6] restricts itself to a combination of static, sticky-on (once an attribute is turned on, it can't be turned off), and sticky-off (once an attribute is turned off, it can't be turned back on) attributes, on the grounds that this is what is normally seen in real implementations. These restrictions do much to make analyses more tractable and to limit the complexity of state representation. Indeed, in our Maude-NPA analysis we found we did not need to model state explicitly at all.

7   Conclusions

In this paper we have described the analysis of some PKCS#11 configurations in Maude-NPA, a cryptographic protocol analyzer tool that operates in the unbounded session model. This allowed us to perform the analysis of this API in a fully-unbounded session model making no abstraction nor approximation of fresh values, and with no assumptions about the policies other than that they were static. This in particular allowed us to reproduce attacks on PKCS#11 configurations found by Delaune et al. in [12].

We consider our work as complementary to that of [23]. In [23] Künnemann uses the protocol analysis tool Tamarin to prove security of a configuration in a restricted model. We use the protocol analysis tool Maude-NPA to reproduce attacks in a less restrictive model. This provides evidence that these tools can be of assistance in both proving security and in finding attacks, as they are for key generation and secure communication protocols.

What remains to be seen is how generally applicable these tools are to PKCS#11 and similar APIs. In particular, we note that our PKCS#11 analysis, although it was successful at reproducing attacks, did not achieve termination, so it is likely that Maude-NPA would not be helpful in proving security within the rather general model we use without some further improvements. However, we plan to keep on investigating this to determine to what degree performance can be improved, for example via the use of state space reduction techniques specific to these types of models. For example, it would be interesting to investigate the model-specific lemmas in [23] to see if they could be used in Maude-NPA. We also plan to investigate whether lemmas appropriate to other classes of models could be formulated and proved.

References

1. Abadi, M., Blanchet, B., Fournet, C.: Just fast keying in the pi calculus. ACM Trans. Inf. Syst. Secur. 10(3) (2007). doi:10.1145/1266977.1266978. http://dblp.uni-trier.de/rec/bib/journals/tissec/AbadiBF07
2. Anderson, R.J.: Security Engineering: A Guide to Building Dependable Distributed Systems, 2nd edn. Wiley Publishing (2008). http://dblp.uni-trier.de/rec/bib/books/daglib/0020262
3. Basin, D., Mödersheim, S., Viganò, L.: OFMC: A symbolic model checker for security protocols. Int. J. Inf. Secur. 4(3), 181–208 (2005)
4. Blanchet, B.: An efficient cryptographic protocol verifier based on prolog rules. In: 14th IEEE Computer Security Foundations Workshop (CSFW-14), Cape Breton, Nova Scotia, Canada, pp. 82–96. IEEE Computer Society, June 2001
5. Bond, M.: Attacks on cryptoprocessor transaction sets. In: Koç, Ç.K., Naccache, D., Paar, C. (eds.) CHES 2001. LNCS, vol. 2162, pp. 220–234. Springer, Heidelberg (2001)
6. Bortolozzo, M., Centenaro, M., Focardi, R., Steel, G.: Attacking and fixing pkcs# 11 security tokens. In: Proceedings of the 17th ACM Conference on Computer and Communications Security, pp. 260–269. ACM (2010)
7. Butler, F., Cervesato, I., Jaggard, A.D., Scedrov, A.: A formal analysis of some properties of kerberos 5 using msr. In: CSFW, p. 175. IEEE Computer Society (2002)
8. Centenaro, M., Focardi, R., Luccio, F.L.: Type-based analysis of PKCS#11 key management. In: Degano, P., Guttman, J.D. (eds.) Principles of Security and Trust. LNCS, vol. 7215, pp. 349–368. Springer, Heidelberg (2012)
9. Clulow, J.: On the security of PKCS #11. In: Walter, C.D., Koç, Ç.K., Paar, C. (eds.) CHES 2003. LNCS, vol. 2779, pp. 411–425. Springer, Heidelberg (2003)
10. Cortier, V., Steel, G.: A generic security API for symmetric key management on cryptographic devices. In: Backes, M., Ning, P. (eds.) ESORICS 2009. LNCS, vol. 5789, pp. 605–620. Springer, Heidelberg (2009)
11. Cryptosense. Cryptosense Web Page. https://cryptosense.com/

12. Delaune, S., Kremer, S., Steel, G.: Formal analysis of pkcs#11. In: Proceedings of the 21st IEEE Computer Security Foundations Symposium, CSF 2008, 23–25 June 2008, Pittsburgh, Pennsylvania, pp. 331–344. IEEE Computer Society (2008)

13. Erbatur, S., Escobar, S., Kapur, D., Liu, Z., Lynch, C., Meadows, C., Meseguer, J., Narendran, P., Santiago, S., Sasse, R.: Effective symbolic protocol analysis via equational irreducibility conditions. In: Foresti, S., Yung, M., Martinelli, F. (eds.) ESORICS 2012. LNCS, vol. 7459, pp. 73–90. Springer, Heidelberg (2012)

14. Escobar, S., Meadows, C., Meseguer, J.: Maude-NPA: Cryptographic protocol analysis modulo equational properties. In: Aldini, A., Barthe, J., Gorrieri, R. (eds.) FOSAD 2007/2008/2009. LNCS, vol. 5705, pp. 1–50. Springer, Heidelberg (2009)

15. Escobar, S., Meadows, C., Meseguer, J., Santiago, S.: State space reduction in the Maude-NRL Protocol Analyzer. Inf. Comput. **238**, 157–186 (2014)

16. Thayer Fabrega, F.J., Herzog, J., Guttman, J.: Strand spaces: what makes a security protocol correct? J. Comput. Secur. **7**, 191–230 (1999)

17. Focardi, R., Luccio, F.L., Steel, G.: An introduction to security API analysis. In: Aldini, A., Gorrieri, R. (eds.) FOSAD 2011. LNCS, vol. 6858, pp. 35–65. Springer, Heidelberg (2011)

18. Fröschle, S., Steel, G.: Analysing PKCS#11 key management APIs with unbounded fresh data. In: Degano, P., Viganò, L. (eds.) ARSPA-WITS 2009. LNCS, vol. 5511, pp. 92–106. Springer, Heidelberg (2009)

19. González-Burgueño, A., Santiago, S., Escobar, S., Meadows, C., Meseguer, J.: Analysis of the IBM CCA security API protocols in Maude-NPA. In: Chen, L., Mitchell, C. (eds.) SSR 2014. LNCS, vol. 8893, pp. 111–130. Springer, Heidelberg (2014)

20. IBM. CCA basic services reference and guide: CCA basic services reference and guide for the IBM 4758 PCI and IBM 4764 (2008). http://www-03.ibm.com/security/cryptocards/pdfs/bs327.pdf.

21. Kemmerer, R.A.: Using formal verification techniques to analyze encryption protocols. In: IEEE Symposium on Security and Privacy, pp. 134–139. IEEE Computer Society (1987)

22. Kremer, S., Künnemann, R.: Automated analysis of security protocols with global state. In: 2014 IEEE Symposium on Security and Privacy, SP 2014, 18–21 May, 2014, Berkeley, CA, USA, pp. 163–178 (2014)

23. Künnemann, R.: Automated backward analysis of PKCS#11 v2.20. In: Focardi, R., Myers, A. (eds.) POST 2015. LNCS, vol. 9036, pp. 219–238. Springer, Heidelberg (2015)

24. RSA Laboratories. PKCS#11: Cryptographic token interface standard. https://www.emc.com/emc-plus/rsa-labs/standards-initiatives/pkcs-11-cryptographic-token-interface-standard.htm

25. Longley, D., Rigby, S.: An automatic search for security flaws in key management schemes. Comput. Secur. **11**(1), 75–89 (1992)

26. Meadows, C.: Applying formal methods to the analysis of a key management protocol. J. Comput. Secur. 1(1) (1992)

27. Meadows, C., Cervesato, I., Syverson, P.: Specification and analysis of the group domain of interpretation protocol using NPATRL and the NRL protocol analyzer. J. Comput. Secur. **12**(6), 893–932 (2004)

28. Meadows, C., Syverson, P.F., Cervesato, I.: Formal specification and analysis of the group domain of interpretation protocol using NPATRL and the NRL protocol analyzer. J. Comput. Secur. **12**(6), 893–931 (2004)

29. Meadows, C.: Analysis of the internet key exchange protocol using the NRL protocol analyzer. In: IEEE Symposium on Security and Privacy, pp 216–231. IEEE Computer Society (1999)
30. Meier, S., Schmidt, B., Cremers, C., Basin, D.: The TAMARIN prover for the symbolic analysis of security protocols. In: Sharygina, N., Veith, H. (eds.) CAV 2013. LNCS, vol. 8044, pp. 696–701. Springer, Heidelberg (2013)
31. OASIS. OASIS PKCS 11 TC. OASIS PKCS 11 TC Home Page. https://www.oasis-open.org/committees/tc_home.php?wg_abbrev=pkcs11
32. Tsalapati, E.: Analysis of PKCS#11 using AVISPA tools. Master's thesis, University of Edinburgh (2007)

Analysis on Cryptographic Algorithm

How to Manipulate Curve Standards: A White Paper for the Black Hat
http://bada55.cr.yp.to

Daniel J. Bernstein[1,2]($\boxtimes$), Tung Chou[1]($\boxtimes$), Chitchanok Chuengsatiansup[1]($\boxtimes$),
Andreas Hülsing[1]($\boxtimes$), Eran Lambooij[1]($\boxtimes$), Tanja Lange[1]($\boxtimes$),
Ruben Niederhagen[1]($\boxtimes$), and Christine van Vredendaal[1]($\boxtimes$)

[1] Department of Mathematics and Computer Science, Technische Universiteit
Eindhoven, P.O. Box 513, 5600 MB Eindhoven, Netherlands
blueprint@crypto.tw, c.chuengsatiansup@tue.nl,
andreas.huelsing@googlemail.com, e.lambooij@student.tue.nl,
tanja@hyperelliptic.org, ruben@polycephaly.org, c.v.vredendaal@tue.nl
[2] Department of Computer Science, University of Illinois at Chicago,
Chicago, IL 60607–7045, USA
djb@cr.yp.to

Abstract. This paper analyzes the cost of breaking ECC under the following assumptions: (1) ECC is using a standardized elliptic curve that was actually chosen by an attacker; (2) the attacker is aware of a vulnerability in some curves that are not publicly known to be vulnerable.

This cost includes the cost of exploiting the vulnerability, but also the initial cost of computing a curve suitable for sabotaging the standard. This initial cost depends heavily upon the acceptability criteria used by the public to decide whether to allow a curve as a standard, and (in most cases) also upon the chance of a curve being vulnerable.

This paper shows the importance of accurately modeling the actual acceptability criteria: i.e., figuring out what the public can be fooled into accepting. For example, this paper shows that plausible models of the "Brainpool acceptability criteria" allow the attacker to target a one-in-a-million vulnerability and that plausible models of the "Microsoft NUMS criteria" allow the attacker to target a one-in-a-hundred-thousand vulnerability.

This work was supported by the European Commission under contracts INFSO-ICT-284833 (PUFFIN) and H2020-ICT-645421 (ECRYPT-CSA), by the Netherlands Organisation for Scientific Research (NWO) under grant 639.073.005, and by the U.S. National Science Foundation under grant 1018836. "Any opinions, findings, and conclusions or recommendations expressed in this material are those of the author(s) and do not necessarily reflect the views of the National Science Foundation." Calculations were carried out on two GPU clusters: the Saber cluster at Technische Universiteit Eindhoven; and the K10 cluster at the University of Haifa, funded by ISF grant 1910/12. Permanent ID of this document: bada55ecd325c5bfeaf442a8fd008c54. Date: 2015.09.25. See web site: bada55.cr.yp.to.

L. Chen and S. Matsuo (Eds.): SSR 2015, LNCS 9497, pp. 109–139, 2015.
DOI: 10.1007/978-3-319-27152-1_6

Keywords: Elliptic-curve cryptography · Verifiably random curves · Verifiably pseudorandom curves · Minimal curves · Nothing-up-my-sleeve numbers · ANSI X9 · NIST · SECG · Brainpool · Microsoft NUMS

1   Introduction

1.1   Elliptic-Curve Cryptography. Elliptic-curve cryptography (ECC) has a reputation for high security and has become increasingly popular. For definiteness we consider the elliptic-curve Diffie–Hellman (ECDH) key-exchange protocol, specifically "ephemeral ECDH", which has a reputation of being the best way to achieve forward secrecy. The literature models ephemeral ECDH as the following protocol $\text{ECDH}_{E,P}$, Diffie–Hellman key exchange using a point P on an elliptic curve E:

1. Alice generates a private integer a and sends the ath multiple of P on E.
2. Bob generates a private integer b and sends bP.
3. Alice computes abP as the ath multiple of bP.
4. Bob computes abP as the bth multiple of aP.
5. Alice and Bob encrypt data using a secret key derived from abP.

There are various published attacks showing that this protocol is breakable for many elliptic curves E, no matter how strong the encryption is. See Section 2 for details. However, there are also many (E, P) for which the public literature does not indicate any security problems. Similar comments apply to, e.g., elliptic-curve signatures.

This model begs the question of where the curve (E, P) comes from. The standard answer is that a central authority generates a curve for the public (while advertising the resulting benefits for security and performance). This does not mean that the public will accept arbitrary curves; our main objective in this paper is to analyze the security consequences of various possibilities for what the public will accept. The general picture is that Alice, Bob, and the central authority Jerry are actually participating in the following three-party protocol ECDH_A, where A is a function determining the public acceptability of a standard curve:

−1. Jerry generates a curve E, a point P, auxiliary data S with $A(E, P, S) = 1$. (The "seeds" for the NIST curves are examples of S; see Section 4.)
0. Alice and Bob verify that $A(E, P, S) = 1$.
1. Alice generates a private integer a and sends aP.
2. Bob generates a private integer b and sends bP.
3. Alice computes abP as the ath multiple of bP.
4. Bob computes abP as the bth multiple of aP.
5. Alice and Bob encrypt data using a secret key derived from abP.

Our paper analyzes the consequences of Jerry cooperating with an eavesdropper Eve to break the encryption used by Alice and Bob. The central question is how Jerry can use his curve-selection flexibility to minimize the attack cost.

Obviously the cost c_A of breaking ECDH_A depends on A, the same way that the cost $c_{E,P}$ of breaking $\text{ECDH}_{E,P}$ depends on (E, P). One might think that, to evaluate c_A, one simply has to check what the public literature says about $c_{E,P}$, and then minimize $c_{E,P}$ over all (E, P, S) with $A(E, P, S) = 1$. The reality is more complicated, for three reasons:

1. There may be vulnerabilities not known to the public: curves E for which $c_{E,P}$ is smaller than indicated by the best public attacks. Our starting assumption is that Jerry and Eve are secretly aware of a vulnerability that applies to a fraction ϵ of all curves that the public believes to be secure. The obvious strategy for Jerry is to standardize a vulnerable curve.
2. Some choices of A limit the number of curves E for which there *exists* suitable auxiliary data S. If $1/\epsilon$ is much larger than this limit then Jerry cannot expect any vulnerable (E, P, S) to have $A(E, P, S) = 1$. We show that, fortunately for Jerry, this limit is much larger than the public thinks it is. See Sections 5 and 6.
3. Other choices of A do not limit the number of vulnerable E for which S *exists* but nevertheless complicate Jerry's task of *finding* a vulnerable (E, P, S) with $A(E, P, S) = 1$. See Section 4 for analysis of the cost of this computation.

If Jerry succeeds in finding a vulnerable (E, P, S) with $A(E, P, S) = 1$, then Eve simply exploits the vulnerability, obtaining access to the information that Alice and Bob have encrypted for transmission.

Of course, this could require considerable computation for Eve, depending on the details of the secret vulnerability. Our goal in this paper is not to evaluate the cost of Eve's computation, but rather to evaluate the impact of A and ϵ upon the cost of Jerry's computation.

For this evaluation it is adequate to use simplified models of secret vulnerabilities. We specify various artificial curve criteria that have no connection to vulnerabilities but that are satisfied by (E, P, S) with probability ϵ for various sizes of ϵ. We then evaluate how difficult it is for Jerry to find (E, P, S) that satisfy these criteria and that have $A(E, P, S) = 1$.

The possibilities that we analyze for A are models built from data regarding what the public will accept. Consider, for example, the following data: the public has accepted without complaint the constants $\sin(1), \sin(2), \ldots, \sin(64)$ in MD5, the constants $\sqrt{2}, \sqrt{3}, \sqrt{5}, \sqrt{10}$ in SHA-1, the constants $\sqrt[3]{2}, \sqrt[3]{3}, \sqrt[3]{5}, \sqrt[3]{7}$ in SHA-2, the constant $(1 + \sqrt{5})/2$ in RC5, the constant $e = \exp(1)$ in Brainpool, the constant $1/\pi$ in ARIA, etc. All of these constants are listed in [48] as examples of "nothing up my sleeve numbers". Extrapolating from this data, we confidently predict that the public would accept, e.g., the constant $\cos(1)$ used in our example curve BADA55-VPR-224 in Section 5. Enumerating a complete list of acceptable constants would require more systematic psychological experiments, so we have chosen a conservative acceptability function A in Section 5 that allows just 17 constants and their reciprocals.

The reader might object to our specification of ECDH_A as implicitly assuming that the party sabotaging curve choices is the same as the party issuing curve standards to be used by Alice and Bob. In reality, these two parties are different, and having the first party exercise sufficient control over the second party is often a delicate exercise in finesse. See, for example, [29,20].

1.2   Organization. Section 2 reviews the curve attacks known to the public and analyzes the probability that a curve resists these attacks; this probability has an obvious impact on the cost of generating curves. Section 3, as a warm-up, shows how to manipulate curve choices when A merely checks for these public vulnerabilities.

Section 4 shows how to manipulate "verifiably random" curve choices obtained by hashing seeds. Section 5 shows how to manipulate "verifiably pseudo-random" curve choices obtained by hashing "nothing-up-my-sleeves numbers". Section 6 shows how to manipulate "minimal" curve choices. Section 7 shows how to manipulate "the fastest curve".

1.3   Research Contributions of this Paper. We appear to be the first to formally introduce the three-party protocol ECDH_A. The general idea of Section 4 is not new, but our cost analysis is new. We are the first to implement the attack, showing how little computer power is necessary to target highly unusual curve properties. Our theoretical and experimental analysis of the percentage of secure curves (see Section 2) is also new.

The general idea of Sections 5 and 6 is new. We are the first to show that curves using so-called "nothing-up-my-sleeves numbers" can very well be manipulated to contain a secret vulnerability. We present concrete ways to gain many bits of freedom and analyze how likely a new vulnerability needs to be in order to hide in this framework. It is surprising that millions of curves can be generated by plausible variations of the Brainpool [14] curve-generation procedure, and that hundreds of thousands of curves can be generated by plausible variations of the Microsoft [13] curve-generation procedure.

In followup work to Section 5, Aumasson has posted a "Generator of 'nothing-up-my-sleeve' (NUMS) constants" that "generates close to 2 million constants, and is easily tweaked to generate many more". See [4].

2   Public Security Analyses

Standards evaluating or claiming the security of various elliptic curves include [1,28,16,17,37,2,14,39,18,19,40,41,3,38]. These standards vary in many details, and also demonstrate important variations in public acceptability criteria, an issue explored in depth later in this paper.

Some public criteria have become so widely known that all of the above standards agree upon them. Jerry's curves need to satisfy these criteria. This means not only that Jerry will be unable to use these public attacks as back doors, but also that Jerry will have to bear these criteria in mind when searching for a vulnerable curve. Perhaps the vulnerability known secretly to Jerry does not occur in curves that satisfy the public criteria; on the other hand, perhaps this vulnerability occurs *more* frequently in curves that satisfy the public criteria than in general curves. The chance ϵ of a curve being vulnerable is defined relative to the curves that the public will accept.

This section has three goals:

- Review these standard criteria for "secure" curves, along with attacks known to the public.
- Analyze the probability δ that a curve satisfies the standard security criteria. This has a direct influence on Jerry's curve-generation cost. Two particular criteria, "small cofactor" and "small twist cofactor", are satisfied by only a small fraction of curves.
- Analyze the probability that a curve is actually feasible to break by various public attacks. It turns out that there are many probabilities on different scales, showing that one should also consider a range of probabilities ϵ for Jerry's secret vulnerability. Recall that ϵ is, by definition, the probability that curves passing the public criteria are secretly vulnerable to Jerry's attack.

Each curve that Jerry tries works with probability only $\delta\epsilon$. The number of curves that Jerry can afford to try and is allowed to try depends on various optimizations and constraints analyzed later in this paper; combining this number with $\delta\epsilon$ immediately reveals Jerry's overall success chance at creating a vulnerable curve that passes the public criteria, avoiding alarms from the public researchers.

2.1 Warning: Math Begins Here. For simplicity we cover only prime fields here. If Jerry's secret vulnerability works only in binary fields then we would expect Jerry to have a harder time convincing his targets to use vulnerable curves, although of course he will try.

Let E be an elliptic curve defined over a large prime field $\mathbb{F}_p$. One can always write E in the form $y^2 = x^3 + ax + b$. Most curve standards choose $a = -3$ for efficiency reasons. Practically all curves have low-degree isogenies to curves with $a = -3$, so this choice does not affect security.

Write $|E(\mathbb{F}_p)|$ for the number of points on E defined over $\mathbb{F}_p$, and write $|E(\mathbb{F}_p)|$ as $p + 1 - t$. Hasse's theorem (see, e.g., [45]) states that $|E(\mathbb{F}_p)|$ is in the "Hasse interval" $[p + 1 - 2\sqrt{p}, p + 1 + 2\sqrt{p}]$; i.e., t is between $-2\sqrt{p}$ and $2\sqrt{p}$.

Define ℓ as the largest prime factor of $|E(\mathbb{F}_p)|$, and define the "cofactor" h as $|E(\mathbb{F}_p)|/\ell$. Let P be a point on E of order ℓ.

2.2 Review of Public ECDLP Security Criteria. Elliptic curve cryptography is based on the believed hardness of the *elliptic-curve discrete-logarithm problem* (ECDLP), i.e., the belief that it is computationally infeasible to find a scalar k satisfying $Q = kP$ given a random multiple Q of P on E. The state-of-the-art public algorithm for solving the ECDLP is Pollard's rho method (with negation), which on average requires approximately $0.886\sqrt{\ell}$ point additions. Most publications require the value ℓ to be large; for example, the SafeCurves web page [9] requires that $0.886\sqrt{\ell} > 2^{100}$.

Some standards put upper limits on the cofactor h, but the limits vary. FIPS 186-2 [37, page 24] claims that "for efficiency reasons, it is desirable to take the cofactor to be as small as possible"; the 2000 version of SEC 1 [16, page 17] required $h \leq 4$; but the 2009 version of SEC 1 [18, pages 22 and 78] claims that there are efficiency benefits to "some special curves with cofactor larger than four" and thus requires merely $h \leq 2^{\alpha/8}$ for security level 2^{α}. We analyze a few possibilities for h and later give examples with $h = 1$; many standard curves have $h = 1$.

Another security parameter is the *complex-multiplication field discriminant* (CM field discriminant) which is defined as $D = (t^2 - 4p)/s^2$ if $(t^2 - 4p)/s^2 \equiv 1$ (mod 4) or otherwise $D = 4(t^2 - 4p)/s^2$, where t is defined as $p + 1 - |E(\mathbb{F}_p)|$ and s^2 is the largest square dividing $t^2 - 4p$. One standard, Brainpool, requires $|D|$ to be large (by requiring a related quantity, the "class number", to be large). However, other standards do not constrain D, there are various ECC papers choosing curves where D is small, and the only published attacks related to the size of D are some improvements to Pollard's rho method on a few curves. If Jerry needs a curve with small D then it is likely that Jerry can convince the public to accept the curve. We do not pursue this possibility further.

All standards prohibit efficient *additive and multiplicative transfers*. An additive transfer reduces the ECDLP to an easy DLP in the additive group of $\mathbb{F}_p$; this transfer is applicable when ℓ equals p. A degree-k multiplicative transfer reduces the ECDLP to the DLP in the multiplicative group of $\mathbb{F}_{p^k}$ where the problem can be solved efficiently using index calculus if the *embedding degree k* is not too large; this transfer is applicable when ℓ divides $p^k - 1$. All standards prohibit $\ell = p$, ℓ dividing $p - 1$, ℓ dividing $p + 1$, and ℓ dividing various larger $p^k - 1$; the exact limit on k varies from one standard to another.

2.3   ECC Security vs. ECDLP Security. The most extensive public list of requirements is on the SafeCurves web page [9]. SafeCurves covers hardness of ECDLP, generally imposing more stringent constraints than the standards listed in Section 2.2; for example, SafeCurves requires the discriminant D of the CM field to satisfy $|D| > 2^{100}$ and requires the order of p modulo ℓ, i.e., the embedding degree, to be at least $(\ell - 1)/100$. Potentially more troublesome for Jerry is that SafeCurves also covers the general security of ECC, i.e., the security of ECC implementations.

For example, if an implementor of NIST P-224 ECDH uses the side-channel-protected scalar-multiplication algorithm recommended by Brier and Joye [15], reuses an ECDH key for more than a few sessions, and fails to perform a moderately expensive input validation that has no impact on normal usage, then a *twist attack* finds the user's secret key using approximately 2^{58} elliptic-curve additions. See [9] for details. SafeCurves prohibits curves with low *twist security*, such as NIST P-224.

Luckily for Jerry, the other standards listed above focus on ECDLP hardness and impose very few additional ECC security constraints. This gives Jerry the freedom to choose a non-SafeCurves-compliant curve that encourages insecure ECC implementations even if ECDLP is difficult.

From Jerry's perspective, there is some risk that twist-security and other SafeCurves criteria will be added to future standards. This paper considers the possibility that Jerry is forced to generate twist-secure curves; it is important for

Jerry to be able to sabotage curve standards even under the harshest conditions. Obviously it is also preferable for Jerry to choose a curve for which *all* implementations are insecure, rather than merely a curve that encourages insecure implementations.

Twist-security requires the twist E' of the original curve E to be secure. If $|E(\mathbb{F}_p)| = p+1-t$ then $|E'(\mathbb{F}_p)| = p+1+t$. Define ℓ' as the largest prime factor of $p + 1 + t$. SafeCurves requires $0.886\sqrt{\ell'} > 2^{100}$ to prevent Pollard's rho method; $\ell' \neq p$ to prevent additive transfers; and p having order at least $(\ell' - 1)/100$ modulo ℓ' to prevent multiplicative transfers. SafeCurves also requires various "combined attacks" to be difficult; this is automatic when cofactors are very small, i.e. when $(p + 1 - t)/\ell$ and $(p + 1 + t)/\ell'$ are very small integers.

2.4 The Probability δ of Passing Public Criteria. This subsection analyzes the probability of random curves passing the public criteria described above.

We begin by analyzing how many random curves have small cofactors. As illustrations we consider cofactors $h = 1$, $h = 2$, and $h = 4$. Note that, for primes p large enough to pass a laugh test (at least 224 bits), curves with these cofactors automatically satisfy the requirement $0.886\sqrt{\ell} > 2^{100}$; in other words, requiring a curve to have a small cofactor supersedes requiring a curve to meet minimal public requirements for security against Pollard's rho method.

Let $\pi(x)$ be the number of primes $p \leq x$, and let $\pi(S)$ be the number of primes p in a set S. The prime-number theorem states that the ratio between $\pi(x)$ and $x/\log x$ converges to 1 as $x \to \infty$, where log is the natural logarithm. Explicit bounds such as [42] are not sufficient to count the number of primes in a short interval $I = [x - y, x]$, but there is nevertheless ample experimental evidence that $\pi(I)$ is very close to $y/\log x$ when y is larger than $\sqrt{x}$.

The number of integers in I of the form $\ell, 2\ell$, or 4ℓ, where ℓ is prime, is the same as the total number of primes in the intervals $I_1 = [x - y, x]$, $I_2 = [(x - y)/2, x/2]$ and $I_4 = [(x - y)/4, x/4]$, namely $\pi(I_1) + \pi(I_2) + \pi(I_4) \approx \sum_{h \in \{1,2,4\}} (y/h)/\log(x/h)$. Take $x = p+1+2\sqrt{p}$ and $y = 4\sqrt{p}$ to see the number of such integers in the Hasse interval. The total number of integers in the Hasse interval is almost exactly $4\sqrt{p}$, so the chance of an integer in the interval having the form $\ell, 2\ell$, or 4ℓ is approximately

$$\sum_{h \in \{1,2,4\}} \frac{1}{h \log\left((p + 1 + 2\sqrt{p})/h\right)}. \tag{1}$$

This does not imply, however, that the same approximation is valid for the number of points on a random elliptic curve. It is known, for example, that the number of points on an elliptic curve is odd with probability almost exactly $1/3$, not $1/2$; this suggests that the number is prime less often than a uniformly distributed random integer in the Hasse interval would be.

A further difficulty is that we need to know not merely the probability that the cofactor h is small, but the joint probability that both h and $h' = (p+1+t)/\ell'$ are small. Even if one disregards the subtleties in the distribution of $p+1-t$, one should not expect (e.g.) the probability that $p+1-t$ is prime to be independent of the probability that $p+1+t$ is prime: for example, if one quantity is odd then the other is also odd.

Galbraith and McKee in [25, Conjecture B] formulated a precise conjecture for the probability of any particular h (called "k" there). Perhaps the techniques of [25] can be extended to formulate a conjecture for the probability of any particular pair (h, h'). However, no such conjectures appear to have been formulated yet, let alone tested.

To collect facts we performed the following experiment: take $p = 2^{224} - 2^{96} + 1$ (the NIST P-224 prime, which is also used in the following sections), and count the number of points on 1000000 curves. Specifically, we took the curves $y^2 = x^3 - 3x + 1$ through $y^2 = x^3 - 3x + 1000001$, skipping the non-elliptic curve $y^2 = x^3 - 3x + 2$. It is conceivable that the small coefficients make these curves behave nonrandomly, but the same type of nonrandomness appears naturally in Section 6, so this is a relevant experiment. Furthermore, the simple description makes the experiment easy to reproduce.

Within this sample we found probability 0.003705 of $h = 1$, probability 0.002859 of $h = 2$, and probability 0.002372 of $h = 4$, with total $0.008936 \approx 2^{-7}$. We also found, unsurprisingly, practically identical probabilities for the twist cofactor: probability 0.003748 of $h' = 1$, probability 0.002902 of $h' = 2$, and probability 0.002376 of $h' = 4$, with total 0.009026.

For comparison, Formula (1) evaluates to approximately 0.011300 (about 25 % too optimistic), built from 0.006441 for $h = 1$ (about 74 % too optimistic), 0.003235 for $h = 2$ (about 13 % too optimistic), and 0.001625 for $h = 4$ (about 32 % too pessimistic).

We found probability $0.000032 \approx 2^{-15}$ of $h = h' = 1$. Our best estimate, with the caveat of considerable error bars, is therefore that Jerry must try about 2^{15} curves before finding one with $h = h' = 1$. If Jerry is free to neglect twist security, searching only for $h = 1$, then the probability jumps by two orders of magnitude to about 2^{-8}. If Jerry is allowed to take any $h \in \{1, 2, 4\}$ then the probability is about 2^{-7}.

These probabilities are not noticeably affected by the SafeCurves requirements regarding the CM discriminant, additive transfers, and multiplicative transfers. Specifically, random curves have a large CM field discriminant, practically always meeting the SafeCurves CM criterion; none of our experiments found a CM field discriminant below 2^{100}. We also found, unsurprisingly, no curves with $\ell = p$. As for multiplicative transfers: Luca, Mireles, and Shparlinski gave a theoretical estimate [34] for the probability that for a sufficiently large prime number p and a positive integer K with $\log K = O(\log \log p)$ a randomly chosen elliptic curve $E(\mathbb{F}_p)$ has embedding degree $k \leq K$; this result shows that curves with small embedding degree are very rare. The SafeCurves bound $K = (\ell - 1)/100$ is not within the range of applicability of their theorem, but experimentally we found that about 99 % of all curves had a high embedding degree $\geq K$.

2.5   The Probabilities for Various Feasible Attacks. We now consider various feasible public attacks as models of Jerry's secret vulnerability. Specifically, for each attack, we evaluate the probability that the attack works against curves that were *not* chosen to be secure against this type of attack. Any such probability is a reasonable guess for an ϵ of interest to Jerry.

At the low end is, e.g., an additive transfer, applying only to curves having exactly p points. The probability here is roughly $p^{-1/2}$: e.g., below 2^{-100} for the NIST P-224 prime.

At the high end, most curves fail the "rho" and "twist" security criteria; see Section 2.4. But this does not mean that the curves are feasible to break, or that the breaking cost is low enough for Jerry to usefully apply to billions of targets. These security criteria are extremely cautious, staying far away from anything potentially breakable by these attacks. For example, $\ell \approx 2^{150}$ fails the SafeCurves security criteria but still requires about 2^{75} elliptic-curve operations to break by the rho attack, costing roughly 100 million watt-years of energy with current hardware, a feasible but highly nontrivial cost. A much smaller $\ell \approx 2^{120}$ would require about 2^{60} elliptic-curve operations, and breaking 2^{30} targets by standard multiple-target techniques would again require about 2^{75} elliptic-curve operations. Even smaller values of ℓ are of interest for twist attacks.

The prime-number theorem can be used to estimate the probabilities of various sizes of ℓ as in Section 2.4, but it loses applicability as ℓ drops below $\sqrt{p}$. To estimate the probability for a wider range of ℓ we use the following result by Dickman (see, e.g., [27]). Define $\Psi(x, y)$ as the number of integers $\leq x$ whose largest prime factor is at most y; these numbers are called y-smooth integers. Dickman's result is $\Psi(x, y) \sim x\rho(u)$ as $x \to \infty$ with $x = y^u$. Here ρ, the "Dickman ρ function", satisfies $\rho(u) = 1$ for $0 \leq u \leq 1$ and $-u\rho'(u) = \rho(u-1)$ for $u \geq 1$, where ρ' means the right derivative. It is not difficult to compute $\rho(u)$ to high accuracy.

We experimentally verified how well ℓ adheres to this estimate, again for the NIST P-224 prime. For each k we used the Dickman rho function to compute an estimate for the number of integers in the Hasse interval whose largest prime factor has exactly k bits. We divided this by $4\sqrt{p}$ (the size of the Hasse interval) to obtain an estimated fraction. We also experimentally computed (for a somewhat smaller sample than in Section 2.4) the fraction of curves where ℓ has k bits, and the fraction of curves where ℓ' has k bits. For $\log_2 \ell \approx 224$ these experimental fractions are below the estimated fraction for the reasons explained in Section 2.4; for smaller values of ℓ the estimate closely matches the experimental data.

About 20 % of the 224-bit curves have $\ell < 2^{100}$, producing a tolerable rho attack cost, around 2^{50} elliptic-curve operations. However, $\rho(u)$ drops rapidly as u increases (it is roughly $1/u^u$), so the chance of achieving this reasonable cost also drops rapidly as the curve size increases. For 256-bit curves the chance is $\rho(2.56) \approx 0.12 \approx 2^{-3}$. For 384-bit curves the chance is $\rho(3.84) \approx 0.0073 \approx 2^{-7}$. For 512-bit curves the chance is $\rho(5.12) \approx 0.00025 \approx 2^{-12}$.

Table 1. Estimated probability that an elliptic curve modulo p has largest twist prime at most 2^{2k} and second largest twist prime at most 2^k, i.e., that an elliptic curve modulo p is vulnerable to a twist attack using approximately 2^k operations. Estimates rely on the method of [5] to compute asymptotic semismoothness probabilities.

p	$k = 30$	$k = 40$	$k = 50$	$k = 60$	$k = 70$	$k = 80$
P-224 prime	$2^{-15.74}$	$2^{-8.382}$	$2^{-4.752}$	$2^{-2.743}$	$2^{-1.560}$	$2^{-0.8601}$
P-256 prime	$2^{-20.47}$	$2^{-11.37}$	$2^{-6.730}$	$2^{-4.132}$	$2^{-2.551}$	$2^{-1.557}$
P-384 prime	$2^{-42.10}$	$2^{-25.51}$	$2^{-16.65}$	$2^{-11.37}$	$2^{-7.977}$	$2^{-5.708}$
P-521 prime	$2^{-68.64}$	$2^{-43.34}$	$2^{-29.57}$	$2^{-21.16}$	$2^{-15.63}$	$2^{-11.81}$

We now switch from considering rho attacks against arbitrary curves to considering twist attacks against curves with cofactor 1. For a twist attack to fit into 2^{50} elliptic-curve operations, the largest prime ℓ' dividing $p+1+t$ must be below 2^{100}, but also the *second-largest* prime dividing $p + 1 + t$ must be below 2^{50}; see generally [9]. In other words, $p + 1 + t$ must be $(2^{100}, 2^{50})$-semismooth. Recall that an integer is defined to be (y, z)-semismooth if none of its prime factors is larger than y and at most one of its prime factors is larger than z. The portion of the twist attack corresponding to the second-largest prime is difficult to batch across multiple targets, so it is reasonable to consider even smaller limits for that prime.

We estimated this semismoothness probability using the same approach as for rho attacks. First, estimate the semismoothness probability for $p+1+t$ as the semismoothness probability for a uniform random integer in the Hasse interval. Second, estimate the semismoothness probability for a uniform random integer using a known two-variable generalization of ρ. Third, compute this generalization using a method of Bach and Peralta [5]. The results range from $2^{-6.730}$ for 256-bit curves down to $2^{-29.57}$ for 521-bit curves. Table 1 shows the results of similar computations for several sizes of primes and several limits on feasible attack costs.

To summarize, feasible attacks in the public literature have a broad range of success probabilities against curves not designed to resist those attacks; probabilities listed above include 2^{-4}, 2^{-8}, 2^{-11}, 2^{-16}, and 2^{-25}. It is thus reasonable to consider a similarly broad range of possibilities for ϵ, the probability that a curve passing public security criteria is vulnerable to Jerry's secret attack.

3 Manipulating Curves

Here we target users with minimal acceptability criteria: i.e., we assume that $A(E, P, S)$ checks only the public security criteria for (E, P) described in Section 2. The auxiliary data S might be used to communicate, e.g., a precomputed $|E(\mathbb{F}_p)|$ to be verified by the user, but is not used to constrain the choice of (E, P). Curves that pass the acceptability criteria are safe against known attacks, but have no protection against Jerry's secret vulnerability.

3.1   Curves Without Public Justification. Here are two examples of standard curves distributed without any justification for how they were chosen. These examples suggest that there are many ECC users who do in fact have minimal acceptability criteria.

The ANSSI FRP256V1 standard [3] is a curve of the form $y^2 = x^3 - 3x + b$ over $\mathbb{F}_p$ with a b that appears random, accompanied by a point P. The curve meets the ECDLP requirements reviewed in Section 2. Similar comments apply to the OSCCA standard curve [41,40]. The only further data provided with these curves is data that could also have been computed efficiently by users from p, b, P. Nothing in the curve documentation suggests any verification that would have further limited the choice of curves.

3.2   The Attack. The attack is straightforward. Since the only things that users check are the public security criteria, Jerry can continue generating curves for a fixed p (either randomly or not) that satisfy the public criteria until he gets one that is vulnerable to his secret attack. Alternatively, Jerry can generate curves vulnerable to his secret attack and check them against the public security criteria. Every attack (publicly) known so far allows efficient computation of vulnerable curves, so it seems likely that the same will be true for Jerry's secret vulnerability. After finding a vulnerable curve, Jerry simply publishes it.

Of course, Jerry's vulnerability must not depend on any properties excluded by the public security criteria, and there must be enough vulnerable curves. Enumerating 2^7 vulnerable curves over $\mathbb{F}_p$ is likely to be sufficient if Jerry can ignore twist-security, and enumerating 2^{15} vulnerable curves over $\mathbb{F}_p$ is likely to be sufficient even if twist-security is required. See Section 2.

Even if Jerry's curves are less frequent, Jerry can increase his chances by also varying the prime p. To simplify our analysis we do not take advantage of this flexibility in this section: we assume that Jerry is forced to reuse a particular standard prime such as a NIST prime or the ANSSI prime. We emphasize, however, that the standard security requirements do not seriously scrutinize the choice of prime, and existing standards vary in their choices of primes. Any allowed variability in p would also improve the attack in Section 5, and we do vary p in Section 6.

3.3   Implementation. We implemented the attack to demonstrate that it is really feasible in practice. In our implementation the same setting as above is adopted and even made more restrictive: the resulting curve should be of the form $y^2 = x^3 - 3x + b$ over $\mathbb{F}_p$, where p is the same as for the ANSSI curve. The public security criteria we consider are all the standard ECDLP security criteria plus twist security, and we further require that both cofactors are 1.

We use a highly structured parameter b as an artificial model of a secret vulnerability. We show that we can construct a curve with such a b that passes all the public criteria. In reality, Jerry would select a curve with a secret vulnerability rather than a curve with our artificial model of a vulnerability, and would use a trustworthy curve name such as TrustedCurve-R-256.

Our attack is implemented using the Sage computer algebra system [46]. We took Ox5AFEBADA55ECC5AFEBADA55ECC5AFEBADA55ECC5AFEBADA55ECC5AFEBADA55EC as the start value for b and incremented b until we found a curve that meets the public security criteria. This corresponds to Jerry iteratively checking whether curves that are vulnerable to the secret attack fulfill the public criteria.

As a result we found a desired curve, which we call BADA55-R-256, with b = Ox5AFEBADA55ECC5AFEBADA55ECC5AFEBADA55ECC5AFEBADA55ECC5AFEBADA5A57 after 1131 increments within 78 min on a single core of an AMD CPU. One can easily check using a computer-algebra system that the curve does meet all the public criteria. It is thus clear that users who only verify public security criteria can be very easily attacked, and Jerry has an easy time if he is working for or is trusted by ANSSI, OSCCA, or a similar organization.

4 Manipulating Seeds

Section 3 deals with the easiest case for Jerry that the users are satisfied verifying public security criteria. However some audiences might demand justifications for the curve choices. In this section, we consider users who are suspicious that the curve parameters might be maliciously chosen to enable a secret attack. Empirically many users are satisfied if they get a *hash verification routine* as justification; see, e.g., ANSI X9.62 [1], IEEE P1363 [28], SEC 2 [19], or NIST FIPS 186-2 [37]. Hash verification routines mean that Jerry cannot use a very small set of vulnerable curves, but we will show below that he has good chances to get vulnerable curves deployed if they are just somewhat more common.

4.1 Hash Verification Routine. As the name implies, a hash verification routine involves a cryptographic hash function. The inputs to the routine are the curve parameters and a seed that is published along with the curve. Usually the seed is hashed to compute a curve parameter or point coordinate. The ways of computing the parameters differ but the public justification is that these bind the curve parameters to the hash value, making them hard to manipulate since the hash function is preimage resistant. In addition the user verifies a set of public security criteria. We focus on the obstacle that Jerry faces and call curves that can be verified with such routines *verifiably hashed curves*.

Below we recall the curve verification routine for the NIST P-curves. The routine is specified in NIST FIPS 186-2 [37].

Each NIST P-curve is of the form $y^2 = x^3 - 3x + b$ over a prime field $\mathbb{F}_p$ and is published with a seed s. The hash function SHA-1 is denoted as SHA1; recall that SHA-1 produces a 160-bit hash value. The bit length of p is denoted by m. We use bin(i) to denote the 20-byte big-endian representation of some integer i and use int(j) to denote the integer with binary expansion j. For given parameters b, p, and s, the verification routine is:

1. Let $z \leftarrow \text{int}(s)$. Compute $h_i \leftarrow \text{SHA1}(s_i)$ for $0 \leq i \leq v$, where $s_i \leftarrow \text{bin}((z+i) \bmod 2^{160})$ and $v = \lfloor (m-1)/160 \rfloor$.
2. Let h be the rightmost $m - 1$ bits of $h_0 || h_1 || \cdots || h_v$. Let $c \leftarrow \text{int}(h)$.
3. Verify that $b^2 c = -27$ in $\mathbb{F}_p$.

To generate a verifiably hashed curve one starts with a seed and then follows the same steps 1 and 2 as above. Instead of step 3 one tries to solve for b given c; this succeeds for about 50 % of all choices for s. The public perception is that this process is repeated with fresh seeds until the first resulting curve satisfies all public security criteria.

4.2 Acceptability Criteria. One might think that the public acceptability criteria are defined by the NIST verification routine stated above: i.e., $A(E, P, s) = 1$ if and only if (E, P) passes the public security criteria from Section 2 and (E, s) passes the verification routine stated above with seed s and E defined as $y^2 = x^3 - 3x + b$.

However, the public acceptability criteria are not actually so strict. P1363 allows $y^2 = x^3 + ax + b$ without the requirement $a = -3$. P1363 does require $b^2 c = a^3$ where c is a hash as above, but neither P1363 nor NIST gives a justification for the relation $b^2 c = a^3$, and it is clear that the public will accept different relations. For example, the Brainpool curves (see Section 5) use the simpler relations $a = g$ and $b = h$ where g and h are separate hashes. One can equivalently view the Brainpool curves as following the P1363 procedure but using a different hash for c, namely computing c as g^3/h^2 where again g and h are separate hashes. Furthermore, even though NIST and Brainpool both use SHA-1, SHA-1 is not the only acceptable hash function; for example, Jerry can easily argue that SHA-1 is outdated and should be replaced by SHA-2 or SHA-3.

We do not claim that the public would accept *any* relation, or that the public would accept *any* choice of "hash function", allowing Jerry just as much freedom as in Section 3. The exact boundaries of public acceptability are complicated and not immediately obvious. We have determined approximations to these boundaries by extrapolating from existing data; see, e.g., Section 5.

4.3 The Attack. Jerry begins the attack by defining a public hash verification routine. As explained above, Jerry has some flexibility to modify this routine. This flexibility is not *necessary* for the rest of the attack in this section (for example, Jerry can use exactly the NIST verification routine) but a more favorable routine does improve the *efficiency* of the attack. Our cost analysis below makes a particularly efficient choice of routine.

Jerry then tries one seed after another until finding a seed for which the verifiably hashed curve (1) passes the public security criteria but (2) is subject to his secret vulnerability. Jerry publishes this seed and the resulting curve, pretending that the seed was the first random seed that passed the public security criteria.

4.4 Optimizing the Attack. Assume that the curves vulnerable to Jerry's secret attack are randomly distributed over the curves satisfying the public security criteria. Then the success probability that a seed leads to a suitable curve is the probability that a curve is vulnerable to the secret attack times the probability that a curve satisfies the public security criteria. Depending on which condition is easier to check Jerry runs many hash computations to compute candidate b's, checks them for the easier criterion and only checks the surviving choices for the other criterion. The hash computations and security checks for each seed are independent from other seeds; thus, this procedure can be parallelized with an arbitrary number of parallel computing instances.

We generated a family of curves to show the power of this method and highlight the computing power of hardware accelerators (such as GPUs or Xeon Phis). We began by defining our own curve verification routine and implementing the corresponding secret generation routine. The hash function we use is Keccak with 256-bit output instead of SHA-1. The hash value is $c = \text{int}(\text{Keccak}(s))$, and the relation is simply $b = c$ in $\mathbb{F}_p$. All choices are easily justified: Keccak is the winner of the SHA-3 competition and much more secure than SHA-1; using a hash function with a long output removes the weird order of hashed components that smells suspicious and similarly $b = c$ is as simple and unsuspicious as it can get. In reality, however, these choices greatly benefit the attack: the GPUs efficiently search through many seeds in parallel, one single computation of Keccak has a much easier data flow than in the method above, and having b computed without any expensive number-theoretic computation (such as square roots) means that the curve can be tested already on the GPUs and only the fraction that satisfies the first test is passed on to the next stage. Of course, for a real vulnerability we would have to add the cost of checking for that vulnerability, but minimizing overhead is still useful.

Except for the differences stated above, we followed closely the setting of the NIST P-curves. The target is to generate curves of the form $y^2 = x^3 - 3x + b$ over $\mathbb{F}_p$, and we consider 3 choices of p: the NIST P-224, P-256, and P-384 primes. (For P-384 we switched to Keccak with 384-bit output.) As a placeholder "vulnerability" we define E to be vulnerable if b starts with the hex-string BADA55EC. This fixes 8 hex digits, i.e., it simulates a 1-out-of-2^{32} attack. In addition we require that the curves meet the standard ECDLP criteria plus twist security and have both cofactors equal to 1.

4.5 Implementation. Our implementation uses NVIDIA's CUDA framework for parallel programming on GPUs. A high-end GPU today allows several thousand threads to run in parallel, though at a frequency slightly lower than high-end CPUs. We let each thread start with its own random seed. The threads then hash the seeds in parallel. After hashing, each thread outputs the hash value if it starts with the hex-string BADA55EC. To restart, each seed is simply increased by 1, so no new source of randomness is required. Checking whether outputs from GPUs also satisfy the public security criteria is done by running a Sage [46] script on CPUs. Since only 1 out of 2^{32} curves has the desired pattern, the CPU computation is totally hidden by GPU computation. Longer strings, corresponding to less likely vulnerabilities, make GPUs even more powerful for our attack scheme.

In the end we found 3 "vulnerable" verifiably hashed curves: BADA55-VR-224, BADA55-VR-256, and BADA55-VR-384, each corresponding to one of the three NIST P-curves. As an example, BADA55-VR-256 was found within

7 hours, using a cluster of 41 NVIDIA GTX780 GPUs. Each GPU is able to carry out 170 million 256-bit-output Keccak hashes in a second. Most of the instructions are bitwise logic instructions. On average each core performs 0.58 bitwise logic instructions per cycle while the theoretical maximum throughput is 0.83. We have two explanations for the gap: first, each thread uses many registers, which makes the number of active warps too small to fully hide the instruction latency; second, there is not quite enough instruction-level parallelism to fully utilize the cores in this GPU architecture. We also tested our implementation on K10 GPUs. Each of them carries out only 61 million hashes per second. This is likely to be caused by register spilling: the K10 GPUs have only 63 registers per thread instead of the 255 registers of the GTX780. Using a sufficient amount of computing power easily allows Jerry to deal with secret vulnerabilities that have smaller probabilities of occurrence than 2^{-32}.

5   Manipulating Nothing-up-my-sleeve Numbers

In 1999, M. Scott complained about the choice of unexplained seeds for the NIST curves [44] and concluded "Do they want to be distrusted?" In the same vein the German ECC Brainpool consortium expressed skepticism [14, Introduction] and suggested using natural constants in place of random seeds. They coined the term "verifiably pseudorandom" for this method of generating seeds. Others speak of "nothing-up-my-sleeves numbers". We comment that "nothing-up-my-sleeves numbers" also appear in other areas of cryptography and can be manipulated in similar ways, but this paper focuses on manipulation of elliptic curves.

5.1   The Brainpool Procedure. Brainpool requires that "curves shall be generated in a pseudo-random manner using seeds that are generated in a systematic and comprehensive way". Brainpool produces each curve coefficient by hashing a seed extracted from the bits of $e = \exp(1)$. This first curve cannot be expected to meet Brainpool's security criteria, so Brainpool counts systematically upwards from this initial seed until finding a curve that does meet the security criteria. Brainpool uses a similar procedure to generate primes.

For example, for 224 bits, the procedures specified in the Brainpool standard [14, Section 5] produce the following "verifiably pseudorandom" integers p, a, b defining an elliptic curve $y^2 = x^3 + ax + b$ over $\mathbb{F}_p$:

$p = $ 0xD7C134AA264366862A18302575D1D787B09F075797DA89F57EC8C0FF

$a = $ 0x2B98B906DC245F2916C03A2F953EA9AE565C3253E8AEC4BFE84C659E

$b = $ 0x68AEC4BFE84C659EBB8B81DC39355A2EBFA3870D98976FA2F17D2D8D

We have added underlines to point out an embarrassing collision of substrings, obviously quite different from what one expects in "pseudorandom" strings.

What happened here is that the Brainpool procedure generates each of a and b as truncations of concatenations of various hash outputs (since the selected hash function, SHA-1, produces only 160-bit outputs), and there was a collision in the hash inputs. Specifically, Brainpool uses the same seed-increment function for three purposes: searching for a suitable a; moving from a to b; and moving within the concatenations. The first hash used in the concatenation for a was fed through this increment function to obtain the second hash, and was fed through the same increment function to obtain the first hash used in the concatenation for b, producing the overlap visible above.

A reader who checks the Brainpool standard [14] will find that the 224-bit curve listed there does not have the same (a, b), and does not have this overlap. The reason for this is that, astonishingly, the 224-bit standard Brainpool curve was not actually produced by the standard Brainpool procedure. In fact, although the reader will find overlaps in the standard 192-bit, 256-bit, 384-bit, and 512-bit Brainpool curves, *none* of the standard Brainpool curves below 512 bits were produced by the standard Brainpool procedure. In the case of the 160-bit, 224-bit, 320-bit, and 384-bit Brainpool curves, one can immediately demonstrate this discrepancy by observing that the gap listed between "seed A" and "seed B" in [14, Section 11] is larger than 1, while the standard procedure always produces a gap of exactly 1.

A procedure that actually *does* generate the Brainpool curves appeared a few years later in the Brainpool RFC [32]. We reimplemented the two procedures in a unified framework and now explain how they differ:

- The procedure in [32] assigns seeds to an $(a^*ab^*b)^*$ pattern. It tries consecutive seeds for a until finding that $-3/a$ is a 4th power, then tries further seeds for b until finding that b is not a square, then checks whether the resulting curve meets Brainpool's security criteria. If this fails, it goes back to trying further seeds for a etc.
- The original procedure in [14] assigns seeds to an $(a^*ab)^*$ pattern. It tries consecutive seeds for a until finding that $-3/a$ is a 4th power, then uses the next seed for b, then checks whether b is a non-square and whether the curve meets Brainpool's security criteria. If this fails, it goes back to trying further seeds for a etc.

We were surprised to discover the failure of the Brainpool standard procedure to generate the Brainpool standard curves. We have not found this failure discussed, or even mentioned, anywhere in the Brainpool RFCs or on the Brainpool web pages. We have also not found any updates or errata to the Brainpool standard after [14]. One would expect that having a "verifiably pseudorandom" curve not actually produced by the specified procedure would draw more public attention, unless the public never actually tried verifying the curves, an interesting possibility for Jerry. We do not explore this line of thought further: we assume that future curves will be verified by the public, using tools that Jerry is unable to corrupt.

The Brainpool standard also includes the following statement [14, page 2]: "It is envisioned to provide additional curves on a regular basis for users who wish to change curve parameters regularly, cf. Annex H2 of [X9.62], paragraph 'Elliptic curve domain parameter cryptoperiod considerations'." However, the procedure for generating further "verifiably pseudorandom" curves is not discussed.

One possibility is to continue the original procedure past the first (a, b) pair, but this makes new curves more and more expensive to verify. Another possibility is to replace e by a different natural constant.

5.2 The BADA55-VPR-224 Procedure. We now present a new and improved verifiably pseudorandom 224-bit curve, BADA55-VPR-224. BADA55-VPR-224 uses the standard NIST P-224 prime, i.e., $p = 2^{224} - 2^{96} + 1$.

To avoid Brainpool's complications of concatenating hash outputs, we upgrade from the deprecated SHA-1 hash function to the state-of-the-art maximum-security SHA3-512 hash function. We also upgrade to requiring maximum twist security: i.e., both the cofactor and the twist cofactor are required to be 1.

Brainpool already generates seeds using $\exp(1) = e$ and generates primes using $\arctan(1) = \pi/4$, and MD5 already uses $\sin(1)$, so we use $\cos(1)$. We eliminate Brainpool's contrived, complicated search pattern for a: we simply count upwards, trying every seed for a, until finding the first secure (a, b). The full 160-bit seed for a is the 32-bit counter followed by $\cos(1)$. We complement this seed to obtain the seed for b, ensuring maximal difference between the two seeds. This procedure is simpler and more natural than the Brainpool procedure in [14, Section 5]. Here is the resulting curve:

$a =$ 0x7144BA12CE8A0C3BEFA053ED BADA555A42391AC64F052376E041C7D4AF23195E
BD8D83625321D452E8A0C3BB0A048A26115704E45DCEB346A9F4BD9741D14D49,

$b =$ 0x5C32EC7FC48CE1802D9B70DBC3FA574EAF015FCE4E99B43EBE3468D6EFB2276B
A3669AFF6FFC0F4C6AE4AE2E5D74C3C0AF97DCE17147688DDA89E734B56944A2

5.3 How BADA55-VPR-224 Was Generated: Exploring the Space of Acceptable Procedures. The surprising collision of Brainpool substrings had an easy explanation: two hashes in the Brainpool procedure were visibly given the same input. The surprising appearance of the 24-bit string BADA55 in a above has no such easy explanation. There are 128 hexadecimal digits in a, so one expects this substring to appear anywhere within a with probability $123/2^{24} \approx 2^{-17}$.

The actual explanation is as follows. We decided in advance that we would force BADA55 to appear somewhere in a as our artificial model of a "vulnerability". We then identified millions of natural-sounding "verifiably pseudorandom" procedures, and enumerated (using a few hours on our cluster) approximately 2^{20} of these procedures. The space of "verifiably pseudorandom" procedures has many dimensions analyzed below, such as the choice of hash function, the length of the input seed, the update function between seeds, and the initial constant for deriving the seed: i.e., each procedure is defined by a combination of hash function, seed length, etc. The exact number of choices available in any particular dimension is relatively unimportant; what is important is the exponential effect from combining many dimensions.

Since 2^{20} is far above 2^{17}, it is unsurprising that our "vulnerability" appeared in quite a few of these procedures. We selected one of those procedures and presented it as Section 5.2 as an example of what could be shown to the public. We could have easily chosen a more restrictive "vulnerability".

The structure of this attack means that Jerry can use the same attack to target a real vulnerability that has probability 2^{-17}, or (with reasonable success chance) even 2^{-20}, perhaps even reusing our database of curves.

In this section we do not manipulate the choice of prime, the choice of curve shape, the choice of cofactor criterion, etc. Taking advantage of this flexibility (see Section 6) would increase the number of natural-sounding Brainpool-like procedures above 2^{30}.

Our experience is that Alice and Bob, when faced with a *single* procedure such as Section 5.2 (or Section 5.1), find it extremely difficult to envision the entire space of possible procedures (they typically see just a few dimensions of flexibility), and find it inconceivable that the space could have size as large as 2^{20}, never mind 2^{30}. This is obviously a helpful phenomenon for Jerry.

5.4 Manipulating Bit-Extraction Procedures. Consider the problem of extracting a fixed-length string of bits from (e.g.) the constant $e = \exp(1) = 2.71828\ldots = (10.10110111\ldots)_2$. Plausible options for the starting bit position include the most significant bit (position 2^1); immediately after the binary point (position 2^{-1}); the most significant *nibble* (position 2^3); the most significant *byte* (position 2^7); and the byte at position 0 (also position 2^7). These options can be viewed as using different maps from real numbers x to real numbers y with $0 \le y < 1$: the first map takes x to $|x|/2^{\lfloor \log_2 |x| \rfloor}$, the second map takes x to $x - \lfloor x \rfloor$, the third map takes x to $|x|/16^{\lfloor \log_{16} |x| \rfloor}$, etc. Brainpool used the third of these options, describing it as using "the hexadecimal representation" of e. Jerry can use similarly brief descriptions for any of the options without drawing the public's attention to the existence of other options. We implemented the first, second, and fourth options; for an average constant this produced slightly more than 2 distinct possibilities for real numbers y.

Jerry can easily get away with extracting a k-bit integer from y by truncation (i.e., $\lfloor 2^k y \rfloor$) or by rounding (i.e., $\lceil 2^k y \rfloor$). Jerry can defend truncation (which has fundamentally lower accuracy) as simpler, and can defend rounding as being quite standard in mathematics and the physical sciences; but we see no reason to believe that Jerry would be challenged in the first place. We implemented both options, gaining a further factor of 1.5.

Actually, Brainpool uses the bit position indicated above only for the low-security 160-bit Brainpool curve (which Jerry can disregard as already being a non-problem for Eve). Brainpool shifts to subsequent bits of e for the 192-bit curve, then to further bits for the 224-bit curve, etc. Brainpool uses 160 bits for each curve (see below), so the seed for the 256-bit curve (which Jerry can reasonably guess would be the most commonly used curve) is shifted by 480 bits. This number 480 depends on how many lower security levels are allocated (an obvious target of manipulation), and on exactly how many bits are allocated to those seeds. A further option, pointed out in [36] by Merkle (Brainpool RFC co-author), is to reverse the order of curve sizes; the number 480 then depends on how many *higher* security levels are allocated. Yet another option is to put curve sizes in claimed order of usage. We did not implement any of the options described in this paragraph.

5.5 Manipulating Choices of Hash Functions. The latest (July 2013) revision of the NIST ECDSA standard [38, Section 6.1.1] specifically requires that "the security strength of a hash function used [for curve generation] **shall** meet or exceed the security strength associated with the bit length". The original NIST curves are exempted from this rule by [38, footnote 2], but this rule prohibits SHA-1 for (e.g.) new 224-bit curves. On the other hand, a more recent Brainpool-related curve-selection document [36] states that "For a PRNG, SHA-1 was (and still is) sufficiently secure."

Jerry has at least 10 plausible options for standard hash functions used to generate (e.g.) 256-bit curves: SHA-1, SHA-256, SHA-384, SHA-512, SHA-512/256, SHA3-256, SHA3-384, SHA3-512, SHAKE128, and SHAKE256. There are also several non-NIST hash functions with longer track records than SHA-3. Any of RIPEMD-128, RIPEMD-160, RIPEMD-256, RIPEMD-320, Tiger, Tiger/128, Tiger/160, and Whirlpool would have been easily justifiable as a choice of hash function before 2006. MD5 and all versions of Haval would have been similarly justifiable before 2004.

Since we targeted a 224-bit curve we had even more standard NIST hash-function options. For simplicity we implemented just 10 hash-function options, namely the following variants of Keccak, the SHA-3 competition winner: Keccak-224, Keccak-256, Keccak-384, Keccak-512, "default" Keccak ("capacity" $c = 576$, 128 output bytes), Keccak-128 (capacity $c = 256$, 168 output bytes), SHA3-224 (which has different input padding from Keccak-224, changing the output), SHA3-256, SHA3-384, and SHA3-512. All of these Keccak/SHA-3 choices can be implemented efficiently with a single code base and variable input parameters.

5.6 Manipulating Counter Sizes. The simplest way to obtain a 160-bit "verifiably pseudorandom" output with SHA-1 is to hash the empty string. Curve generation needs many more outputs (since most curves do not pass the public security criteria), but the simplest way to obtain 2^b "verifiably pseudorandom" outputs is to hash all b-bit inputs.

Hash-function implementations are often limited to byte-aligned inputs, so it is natural to restrict b to a multiple of 8. If each output has chance 2^{-15} of producing an acceptable curve (see Section 2) then $b = 16$ finds an acceptable curve with chance nearly 90 % ("this is retroactively justified by our successfully finding a curve, so there was no need for us to consider backup plans"); $b = 24$ fails with negligible probability ("we chose the smallest b for which the probability of failure was negligible"); $b = 32$ is easily justified by reference to 32-bit machines; $b = 64$ is easily justified by reference to 64-bit machines.

Obviously Brainpool takes a more complicated approach, using bits of some natural constant to further "randomize" its outputs. The standard way to randomize a hash is to concatenate the randomness (e.g., bits of e) with the input being hashed (the counter). Brainpool instead *adds* the randomness to the input being hashed. The Brainpool choice is not secure as a general-purpose randomized hash, although these security problems are of no relevance to curve generation. There is no evidence of public objections to Brainpool's use of addition here (and to the overall complication introduced by the extra randomization), so there is also no reason to think that the public would object to the more standard concatenation approach.

Overall there are 13 plausible possibilities here: the 4 choices of b above, with the counter on the left of the randomness; the 4 choices of b above, with the counter on the right of the randomness; the counter being added to the randomness; and 4 further possibilities in which the randomness is partitioned into an initial value for a counter (for the top bits) and the remaining seed (for the bottom bits). We implemented the first 9 of these 13 possibilities.

5.7 Manipulating Hash Input Sizes. ANSI X9.62 requires ≥ 160 input bits for its hash input. One way for Jerry to advertise a long input is that it allows many people to randomly generate curves with a low risk of collision. For example, Jerry can advertise a 160-bit or 256-bit or 384-bit input as allowing 2^{64} or 2^{64} or 2^{128} curves respectively with only a 2^{-32} or 2^{-128} or 2^{-128} chance of collision. All of these numbers sound perfectly natural. Of course, what Jerry is actually producing is a single standard for many people to use, so multiple-curve collision probabilities are of no relevance, but (in the unlikely event of being questioned) Jerry can simply say that the input length was chosen for "compatibility" with having users generate their own curves.

Jerry can advertise longer input lengths as providing "curve coverage". A 512-bit input will cover a large fraction of curves, even for primes as large as 512 bits. A 1024-bit input is practically guaranteed to cover all curves, and to produce probabilities indistinguishable from uniform. Jerry can also advertise, as input length, the "natural input block length of the hash function".

We implemented all 6 possibilities listed above. We gained a further factor of 2 by storing the seed (and counter) in big-endian format ("standard network byte order") or little-endian format ("standard CPU byte order").

5.8 Manipulating the (a, b) Hash Pattern. It should be obvious from Section 5.1 that there are many degrees of freedom in the details of how a and b are generated: how to distribute seeds between a and b; whether to require $-3/a$ to be a 4th power in $\mathbb{F}_p$; whether to require b to be a non-square in $\mathbb{F}_p$; whether to concatenate hash outputs from left to right or right to left; exactly how many bits to truncate hash outputs to (Brainpool uses one bit fewer than the prime; Jerry can argue for the same length as the prime "for coverage", or more bits "for indistinguishability"); whether to truncate to rightmost bits (as in Brainpool) or leftmost bits (as in various NIST requirements; see [38]); et al.

For simplicity we eliminated the concatenation and truncation, always using a hash function long enough for the target 224-bit prime. We also eliminated the options regarding squares etc. We implemented a total of just 8 choices here. These choices vary in (1) whether to allocate seeds primarily to a or primarily to b and (2) how to obtain the alternate seed (e.g., the seed for a) from the primary seed (e.g., the seed for b): plausible options include complement, rotate 1 byte left, rotate 1 byte right, and four standard versions of 1-bit rotations.

5.9   Manipulating Natural Constants. As noted in Section 1, the public has accepted dozens of "natural" constants in various cryptographic functions, and sometimes reciprocals of those constants, without complaint. Our implementation started with just 17 natural constants: π, e, Euler gamma, $\sqrt{2}$, $\sqrt{3}$, $\sqrt{5}$, $\sqrt{7}$, $\log(2)$, $(1+\sqrt{5})/2$, $\zeta(3)$, $\zeta(5)$, $\sin(1)$, $\sin(2)$, $\cos(1)$, $\cos(2)$, $\tan(1)$, and $\tan(2)$. We gained an extra factor of almost 2 by including reciprocals.

5.10   Implementation. Any combination of the above manipulations defines a "systematic" curve-generation procedure. This procedure outputs the first curve parameters (using the specified update function) that result in a "secure" curve according to the public security tests. However, performing all public security tests for each set of parameters considered by each procedure is very costly. Instead, we split the attack into two steps:

1. For a given procedure f_i we iterate over the seeds $s_{i,k}$ using the specific update function of f_i. We check each parameter candidate from seed $s_{i,k}$ for our secret BADA55 vulnerability. After a certain number of update steps the probability that we passed valid, secure parameters is very high; thus, we discard the procedure and start over with another one. If we find a candidate exhibiting the vulnerability, we perform the public security tests on this particular candidate. If the BADA55 candidate passes, we proceed to step 2.
2. We perform the whole public procedure f_i starting with seed $s_{i,0}$ and check whether there is any valid parameter set passing the public security checks already before the BADA55 parameters are reached. If there is such an earlier parameter set, we return to step 1 with the next procedure f_{i+1}.

The largest workload in our attack scenario is step 2, the re-checking for earlier safe curve parameters before BADA55 candidates. The public security tests are not well suited for GPU parallelization; the first step of the attack procedure is relatively cheap and a GPU parallelization of this step does not have a remarkable impact on the overall runtime. Therefore, we implemented the whole attack only for the CPUs of the cluster and left the GPUs idle.

We initially chose 8000 as the limit for the update counter to have a very good chance that the first secure twist-secure curve starting from the seed is the curve with our vulnerability. For example, BADA55-VPR-224 was found with counter just 184, and there was only a tiny risk of a smaller counter producing a secure twist-secure curve (which we checked later, in the second step). In total $\approx 2^{33}$ curves were covered by this limited computation; more than 2^{18} were secure and twist-secure. We then pushed the 8000 limit higher, performing more computation and finding more curves. This gradually increased the risk of the counter not being minimal, something that we would have had to address by the techniques of Section 6; but this issue still did not affect, e.g., BADA55-VPR2-224, which was found with counter 28025.

6   Manipulating Minimality

Instead of supporting "verifiably pseudorandom" curves as in Section 5, some researchers have advocated choosing "verifiably deterministic" curves.

Both approaches involve specifying a "systematic" procedure that outputs a curve. The difference is that in a "verifiably pseudorandom" curve the curve coefficient is the output of a hash function for the *first hash input* that meets specified curve criteria, while a "verifiably deterministic" curve uses the *first curve coefficient* that meets specified curve criteria. Typically the curve uses a "verifiably deterministic" prime, which is the *first prime* that meets specified prime criteria.

Eliminating the hash function and hash input makes life harder for Jerry: it eliminates the main techniques that we used in previous sections to manipulate curve choices. However, as we explain in detail in this section, Jerry still has many degrees of freedom. Jerry can manipulate the concept of "first curve coefficient", can manipulate the concept of "first prime", can manipulate the curve criteria, and can manipulate the prime criteria, with public justifications claiming that the selected criteria provide convenience, ease of implementation, speed of implementation, and security.

In Section 5 we did not manipulate the choice of prime: we obtained a satisfactory level of flexibility in other ways. In this section, the choice of prime is an important component of Jerry's flexibility. It should be clear to the reader that the techniques in this section to manipulate the prime, the curve criteria, etc. can be backported to the setting of Section 5, adding to the flexibility there.

We briefly review a recent proposal that fits into this category and then proceed to work out how much flexibility is left for Jerry.

6.1   NUMS Curves. In early 2014, Bos, Costello, Longa, and Naehrig [13] proposed 13 Weierstrass and 13 Edwards curves, spread over 3 different security levels. Each curve was generated following a deterministic procedure (similar to the procedure proposed in [8]). Given that there are up to 10 different procedures per security level we cannot review all of them here but [13] is a treasure trove of arguments to justify different prime and curve properties and we will use this to our benefit below.

The same authors together with Black proposed a set of 6 of these curves as an Internet-Draft [12] referring to these curves as "Nothing Up My Sleeve (NUMS) Curves". Note that this does not match the common use of "nothing up my sleeves"; see, e.g., the Wikipedia page [48]. These curves are claimed in [30] to have "independently-verifiable provenance", as if they were not subject to any possible manipulation; and are claimed in [11] to be selected "without any hidden parameters, reliance on randomness or any other processes offering opportunities for manipulation of the resulting curves". What we analyze in this section is the extent to which Jerry can manipulate the resulting curves.

6.2   Choice of Security Level. Jerry may propose curves aiming for multiple security levels. To quote the Brainpool-curves RFC [32] "The level of security provided by symmetric ciphers and hash functions used in conjunction with the elliptic curve domain parameters specified in this RFC should roughly match or exceed the level provided by the domain parameters." Table 1 in that document justifies security levels of 80, 96, 112, 128, 160, 192, and 256 bits. We consider the highest five to be easy sells. For the smaller ones Jerry will need to be more creative and, e.g., evoke the high cost of energy for small devices.

6.3   Choice of Prime. There are several parts to choosing a prime once the security level is fixed.

Choice of Prime Size. For a fixed security level α it should take about 2^α operations to break the DLP. The definition of "operation" leaves some flexibility. Plausible options for the bit length r of the prime include not just 2α, $2\alpha - 1$, and $2\alpha - 2$ used in [13] but also various smaller and larger values, accounting in various ways for the number of bit operations in an elliptic-curve operation; the $\sqrt{\pi/4}$ in the Pollard-rho complexity, and the impact of multi-target attacks. We simplify by counting just 8 options here.

Choice of Prime Shape. The choices for the prime shape are:

- A random prime. This might seem somewhat hard to justify outside the scope of the previous section because arithmetic in $\mathbb{F}_p$ becomes slower, but members of the ECC Brainpool working group published several helpful arguments [33]. The most useful one is that random primes mean that the blinding factor in randomizing scalars against differential side-channel attacks can be chosen smaller.
- A pseudo-Mersenne prime, i.e. a prime of the shape $2^r \pm c$. The most common choice is to take c to be the smallest integer for a given r which leads to a prime because this makes reduction modulo the prime faster. (To reduce modulo $2^r \pm c$, divide by 2^r and add $\mp c$ times the dividend to the remainder.) See, e.g., [13]. Once r is fixed there are two choices for the two signs.
- A Solinas prime, i.e. a prime of the form $2^r \pm 2^v \pm 1$ as chosen for the Suite B curves [39]. Also for these primes speed of modular reduction is the common argument. The difference $r - v$ is commonly chosen to be a multiple of the word size. Jerry can easily argue for multiples of 32 and 64. We skip this option in our count because it is partially subsumed in the following one.
- A "Montgomery-friendly" prime, i.e. a prime of the form $2^{r-v}(2^v - c) \pm 1$. These curves speed up reductions if elements in $\mathbb{F}_p$ are represented in Montgomery representation, $r - v$ is a multiple of the word size and c less than the word size. Common word sizes are 32 and 64, giving two choices here. We ignore the flexibility of the $\pm$ because that determines p modulo 4, which is considered separately.

There are of course infinitely many random primes; in order to keep the number of options reasonable we take 4 as an approximation of how many prime shapes can be easily justified, making this a total of 8 options.

Choice of Prime Congruence. Jerry can get an additional bit of freedom by choosing whether to require $p \equiv 1 \pmod 4$ or to require $p \equiv 3 \pmod 4$. A common justification for the latter is that computations of square roots are particularly fast which could be useful for compression of points, see, e.g., [14,13]. (In fact one can also compute square roots efficiently for $p \equiv 1 \pmod 4$, in particular for $p \equiv 5 \pmod 8$, but Jerry does not need to admit this.) To instead justify $p \equiv 1 \pmod 4$, Jerry can point to various benefits of having $\sqrt{-1}$ in the field: for example, twisted Edwards curves are fastest when $a = -1$, but completeness for $a = -1$ requires $p \equiv 1 \pmod 4$.

If Jerry chooses twisted Hessian curves he can justify restricting to $p \equiv 1 \pmod 3$ to obtain complete curve arithmetic.

6.4 Choice of Ordering of Field Elements. The following curve shapes each have one free parameter. It is easy to justify choosing this parameter as the smallest parameter under some side conditions. Here smallest can be chosen to mean smallest in $\mathbb{N}$ or as the smallest power of some fixed generator g of $\mathbb{F}_p^*$. The second option is used in, e.g., a recent ANSSI curve-selection document [24, Section 2.6.2]: "we define ... g as the smallest generator of the multiplicative group ... We then iterate over ... $b = g^n$ for $n = 1, \ldots$, until a suitable curve is found." Each choice below can be filled with these two options.

6.5 Choice of Curve Shape and Cofactor Requirement. Jerry can justify the following curve shapes:

1. Weierstrass curves, the most general curve shape. The usual choice is $y^2 = x^3 - 3x + b$, leaving one parameter b free. For simplicity we do not discuss the possibility of choosing values other than -3.
2. Edwards curves, the speed leader in fixed-base scalar multiplication offering complete addition laws. The usual choices are $ax^2 + y^2 = 1 + dx^2y^2$, for $a \in \{\pm 1\}$, leaving one parameter d free. The group order of an Edwards curve is divisible by 4.
3. Montgomery curves, the speed leader for variable-base scalar multiplication and the simplest to implement correctly. The usual choices are $y^2 = x^3 + Ax^2 + x$, leaving one parameter A free. The group order of a Montgomery curve is divisible by 4.
4. Hessian curves, a cubic curve shape with complete addition laws (for twisted Hessian). The usual choices are $ax^3 + y^3 + 1 = dxy$, where a is a small non-cube, leaving one parameter d free. The group order of a Hessian curve is divisible by 3, making twisted Hessian curves the curves with the smallest cofactor while having complete addition.

Weierstrass Curves. Most standards expect the point format to be (x, y) on Weierstrass curves. Even when computations want to use the faster Edwards and Hessian formulas, Jerry can easily justify specifying the curve in Weierstrass form. This also ensures backwards compatibility with existing implementations that can only use the Weierstrass form.

The following are examples of justifiable choices for the cofactor h of the curve: exactly 1, as in Brainpool; exactly 2, minimum for [8]; exactly 3, minimum for Hessian arithmetic; exactly 4, minimum for Edwards arithmetic; exactly 12, minimum for both Hessian and Edwards arithmetic; the first curve having cofactor below $2^{\alpha/8}$, as in [19] and [38]; below $2^{\alpha/8}$ and a multiple of 3; below $2^{\alpha/8}$ and a multiple of 4; below $2^{\alpha/8}$ and a multiple of 12; the SafeCurves requirement of a largest prime factor above 2^{200}. On average these choices produce slightly more than 8 options; the last few options sometimes coincide.

The curve is defined as $y^2 = x^3 - 3x + b$ where b is minimal under the chosen criterion. Changing from positive b to negative b changes from a curve to its twist if $p \equiv 3 \pmod 4$, and (as illustrated by additive transfers) this change does not necessarily preserve security. However, this option makes only a small difference in our final total, so for simplicity we skip it.

Hessian Curves. A curve given in Hessian form (and chosen minimal there) can be required to have minimal cofactor, minimal cofactor while being compatible with Edwards form, cofactor smaller than $2^{\alpha/8}$, or largest prime factor larger than 2^u. This leads to 8 options considering positive and negative values of d. Of course other restrictions on the cofactor are possible.

Edwards Curves. For Edwards curves we need to split up the consideration further:

Edwards curves with $p \equiv 3 \pmod 4$. Curves with $a = -1$ are attractive for speed but are not complete in this case. Nevertheless [13] argues for this option, so we have additionally the choice between aiming for a complete or an $a = -1$ curve.

A curve given in (twisted) Edwards form (and chosen minimal there) can be required to have minimal cofactor, minimal cofactor while being compatible with Hessian form, cofactor smaller than $2^{\alpha/8}$, or largest prime factor larger than 2^u (and the latter in combination with Hessian if desired). This leads to at least 8 choices considering completeness; for minimal cofactors [13] shows that minimal choices for positive and negative values of d are not independent. To stay on the safe side we count these as 8 options only.

Edwards curves with $p \equiv 1 \pmod 4$. The curves $x^2 + y^2 = 1 + dx^2y^2$ and $-x^2 + y^2 = 1 - dx^2y^2$ are isomorphic because -1 is a square, hence taking the smallest positive value for d finds the same curve as taking the smallest negative value for the other sign of a. Jerry can however insist or not insist on completeness. Justifying non-completeness if the smallest option is complete however seems a hard sell.

Because $2p + 2 \equiv 4 \pmod 8$ one of the curve and its twist will have order divisible by 8 while the other one has remainder 4 modulo 8. Jerry can require cofactor 4, as the minimal cofactor, or cofactor 8 if he chooses the twist with minimal cofactor as well and is concerned that protocols will only multiply by the cofactor of the curve rather than by that of the twist. The other options are the same as above. Again, to stay on the safe side, we count this as 8 options only.

Montgomery Curves. There is a significant overlap between choosing the smallest Edwards curve and the smallest Montgomery curve. In order to ease counting and avoid overcounting we omit further Montgomery options.

Summary of Curve Choice. We have shown that Jerry can argue for $8 + 8 + 8 = 24$ options.

6.6   Choice of Twist Security. We assume, as discussed in Section 2, that future standards will be required to include twist security. However, Jerry can play twist security to his advantage in changing the details of the twist-security requirements. Here are three obvious choices:

– Choose the cofactor of the twist as small as possible. Justification: This offers maximal protection.
– Choose the cofactor of the twist to be secure under the SEC recommendation, i.e. $h' < 2^{\alpha/8}$. Justification: This is considered secure enough for the main curve, so it is certainly enough for the twist.
– Choose the curve such that the curve passes the SafeCurves requirement of 2^{100} security against twist attacks. Justification: Attacks on the twist cannot use Pollard rho but need to do a brute-force search in the subgroups. The SafeCurves requirement captures the actual hardness of the attack.

Jerry can easily justify changes to the bound of 2^{100} by pointing to a higher security level or reducing it because the computations in the brute-force part are more expensive. We do not use this flexibility in the counting.

6.7   Choice of Global vs. Local Curves. Jerry can take the first prime (satisfying some criteria), and then, for that prime, take the first curve coefficients (satisfying some criteria). Alternatively, Jerry can take the first possible curve coefficients, and then, for those curve coefficients, take the first prime. These two options are practically guaranteed to produce different curves. For example, in the Weierstrass case, Jerry can take the curve $y^2 = x^3 - 3x + 1$, and then search for the first prime p so that this curve over $\mathbb{F}_p$ satisfies the requirements on cofactor and twist security. If Jerry instead takes $y^2 = x^3 - 3x + g$ as in [24, Section 2.6.2], p must also meet the requirement that g be primitive in $\mathbb{F}_p$.

In mathematical terminology, the second option specifies a curve over a "global field" such as the rationals $\mathbb{Q}$, and then reduces the curve modulo suitable primes. This approach is particularly attractive when presented as a family of curves, all derived from the same global curve.

6.8   More Choices. Brainpool [14] requires that the number of points on the curve is less than p but also presents an argument for the opposite choice:

> To avoid overruns in implementations we require that $\#E(GF(p)) < p$. In connection with digital signature schemes some authors propose to use $q > p$ for security reasons, but the attacks described e.g. in [BRS] appear infeasible in a thoroughly designed PKI.

So Jerry can choose to insist on $p < |E(\mathbb{F}_p)|$ or on $p > |E(\mathbb{F}_p)|$.

6.9   Overall Count. We have shown that Jerry can easily argue for 4 (security level) ·8 (prime size) ·8 (prime shape) ·2 (congruence) ·2 (definition of first) ·24 (curve choice) ·3 (twist conditions) ·2 (global/local) ·2 ($p \lessgtr |E(\mathbb{F}_p)|$) = 294912 choices.

7   Manipulating Security Criteria

A recent trend is to introduce top performance as a selection requirement. This means that Alice and Bob accept only *the fastest curve*, as demonstrated by benchmarks across a range of platforms. The most widely known example of this approach is Bernstein's Curve25519, the curve $y^2 = x^3 + 486662x^2 + x$ modulo the particularly efficient prime $2^{255} - 19$, which over the past ten years has set speed records for conservative ECC on many different platforms, using implementations from 23 authors. See [6,26,22,7,10,31,35,43,21,23,47].

The annoyance for Jerry in this scenario is that, in order to make a case for his curve, he needs to present implementations of the curve arithmetic on a variety of devices, showing that his curve is fastest across platforms. Jerry could try to falsify his speed reports, but it is increasingly common for the public to demand verifiable benchmarks using open-source software.

Jerry can hope that some platforms will favor one curve while other platforms will favor another curve; Jerry can then use arguments for a "reasonable" weighting of platforms as a mechanism to choose one curve or the other. However, it seems difficult to outperform Curve25519 even on *one* platform. The prime $2^{255} - 19$ is particularly efficient, as is the Montgomery curve shape $y^2 = x^3 + 486662x^2 + x$. The same curve is also expressible as a complete Edwards curve, allowing fast additions without the overhead of checking for exceptional cases. Twist security removes the overhead of checking for invalid inputs. Replacing 486662 with a larger curve coefficient produces identical performance on many platforms but loses a measurable amount of performance on some platforms, violating the "top performance" requirement.

In Section 6, Jerry was free to, e.g., claim that $p \equiv 3 \pmod 4$ provides "simple square-root computations" and thus replace $2^{255} - 19$ with $2^{255} - 765$; claim that "compatibility" requires curves of the form $y^2 = x^3 - 3x + b$; etc. The new difficulty in this section is that Jerry is facing "top performance" advocates who reject $2^{255} - 765$ as not providing top performance; who reject $y^2 = x^3 - 3x + b$ as not providing top performance; etc.

Jerry still has some flexibility in defining what security requirements to take into account. Taking "the fastest curve" actually means taking the fastest curve *meeting specified security requirements*, and the list of security requirements is a target of manipulation.

Most importantly, Jerry can argue for any size of ℓ. However, if there is a faster curve with a larger ℓ satisfying the same criteria, then Jerry's curve will be rejected. Furthermore, if Jerry's curve is only marginally larger than a significantly faster curve, then Jerry will have to argue that a tiny difference in security levels (e.g., one curve broken with 0.7× or 0.5× as much effort as another) is meaningful, or else the top-performance advocates will insist on the significantly faster curve.

The choice of prime has the biggest impact on speed and closely rules the size of ℓ. For pseudo-Mersenne primes larger than 2^{224} the only possibly competitive ones are: $2^{226} - 5, 2^{228} + 3, 2^{233} - 3, 2^{235} - 15, 2^{243} - 9, 2^{251} - 9, 2^{255} - 19, 2^{263} + 9, 2^{266} - 3, 2^{273} + 5, 2^{285} - 9, 2^{291} - 19, 2^{292} + 13, 2^{295} + 9, 2^{301} + 27, 2^{308} + 27, 2^{310} + 15, 2^{317} + 9, 2^{319} + 9, 2^{320} + 27, 2^{321} - 9, 2^{327} + 9, 2^{328} + 15, 2^{336} - 3, 2^{341} + 5, 2^{342} + 15, 2^{359} + 23, 2^{369} - 25, 2^{379} - 19, 2^{390} + 3, 2^{395} + 29, 2^{401} - 31, 2^{409} + 29, 2^{414} - 17, 2^{438} + 25, 2^{444} - 17, 2^{452} - 3, 2^{456} + 21, 2^{465} + 29, 2^{468} - 17, 2^{488} - 17, 2^{489} - 21, 2^{492} + 21, 2^{495} - 31, 2^{508} + 15, 2^{521} - 1$. Preliminary implementation work shows that the Mersenne prime $2^{521} - 1$ has such efficient reduction that it outperforms, e.g., the prime $2^{512} - 569$ from [13]; perhaps it even outperforms primes below 2^{500}. We would expect implementation work to also show, e.g., that $2^{319} + 9$ is significantly faster than $2^{320} + 27$, and Jerry will have a hard time arguing for $2^{320} + 27$ on security grounds. Considering other classes of primes, such as Montgomery-friendly primes, might identify as many as 100 possibly competitive primes, but it is safe to estimate that fewer than 80 of these primes will satisfy the top-performance devotees, and further implementation work is likely to reduce the list even more. Note that in this section, unlike other sections, we take a count that is optimistic for Jerry.

Beyond the choice of prime, Jerry can use different choices of security criteria. However, most of the flexibility in Section 6 consists of speed claims, compatibility claims, etc., few of which can be sold as security criteria. Jerry *can* use the different twist conditions, the choice whether $p < |E(\mathbb{F}_p)|$ or $p > |E(\mathbb{F}_p)|$, and possibly two choices of cofactor requirements. Jerry can also choose to require completeness as a security criterion, but this does not affect curve choice in this section: the complete formulas for twisted Hessian and Edwards curves are *faster* than the incomplete formulas for Weierstrass curves. The bottom line is that multiplying fewer than 80 primes by 12 choices of security criteria produces fewer than 960 curves. The main difficulty in pinpointing an exact number is carrying out detailed implementation work for each prime; we leave this to future work.

References

1. Accredited Standards Committee X9: American national standard X9.62-1999, public key cryptography for the financial services industry: the elliptic curve digital signature algorithm (ECDSA) (1999)
2. Accredited Standards Committee X9: American national standard X9.63-2001, public key cryptography for the financial services industry: key agreement and key transport using elliptic curve cryptography (2001)
3. Agence nationale de la sécurité des systèmes d'information: Publication d'un paramétrage de courbe elliptique visant des applications de passeport électronique et de l'administration électronique française (2011)
4. Aumasson, J.P.: Generator of "nothing-up-my-sleeve" (NUMS) constants (2015). https://github.com/veorq/numsgen/blob/master/numsgen.py
5. Bach, E., Peralta, R.: Asymptotic semismoothness probabilities. Math. Comput. **65**(216), 1701–1715 (1996)
6. Bernstein, D.J.: Curve25519: New Diffie-Hellman speed records. In: Yung, M., Dodis, Y., Kiayias, A., Malkin, T. (eds.) PKC 2006. LNCS, vol. 3958, pp. 207–228. Springer, Heidelberg (2006)
7. Bernstein, D.J., Duif, N., Lange, T., Schwabe, P., Yang, B.Y.: High-speed high-security signatures. J. Crypt. Eng. **2**, 77–89 (2012)
8. Bernstein, D.J., Hamburg, M., Krasnova, A., Lange, T.: Elligator: elliptic-curve points indistinguishable from uniform random strings. In: Sadeghi, A., Gligor, V.D., Yung, M. (eds.) ACM CCS 2013, pp. 967–980. ACM (2013)
9. Bernstein, D.J., Lange, T.: SafeCurves: choosing safe curves for elliptic-curve cryptography (2015). http://safecurves.cr.yp.to. Accessed 21 May 2015
10. Bernstein, D.J., Schwabe, P.: NEON Crypto. In: Prouff, E., Schaumont, P. (eds.) CHES 2012. LNCS, vol. 7428, pp. 320–339. Springer, Heidelberg (2012). http://dx.doi.org/10.1007/9783642330278
11. Black, B., Bos, J.W., Costello, C., Langley, A., Longa, P., Naehrig, M.: Rigid parameter generation for elliptic curve cryptography (2015). https://tools.ietf.org/html/draft-black-rpgecc-01
12. Black, B., Bos, J.W., Costello, C., Longa, P., Naehrig, M.: Elliptic curve cryptography (ECC) nothing up my sleeve (NUMS) curves and curve generation (2014). https://tools.ietf.org/html/draft-black-numscurves-00
13. Bos, J.W., Costello, C., Longa, P., Naehrig, M.: Selecting elliptic curves for cryptography: an efficiency and security analysis. J. Cryptographic Eng. 1–28 (2015). doi:10.1007/s13389-015-0097-y
14. ECC Brainpool: ECC Brainpool standard curves and curve generation (2005). http://www.ecc-brainpool.org/download/Domain-parameters.pdf
15. Brier, E., Joye, M.: Weierstraß elliptic curves and side-channel attacks. In: Naccache, D., Paillier, P. (eds.) PKC 2002. LNCS, vol. 2274, pp. 335–345. Springer, Heidelberg (2002)
16. Certicom Research: SEC 1: Elliptic curve cryptography, version 1.0 (2000)
17. Certicom Research: SEC 2: Recommended elliptic curve domain parameters, version 1.0 (2000)
18. Certicom Research: SEC 1: Elliptic curve cryptography, version 2.0 (2009)
19. Certicom Research: SEC 2: Recommended elliptic curve domain parameters, version 2.0 (2010)

20. Checkoway, S., Fredrikson, M., Niederhagen, R., Everspaugh, A., Green, M., Lange, T., Ristenpart, T., Bernstein, D.J., Maskiewicz, J., Shacham, H.: On the practical exploitability of Dual EC in TLS implementations. In: 23rd USENIX Security Symposium (USENIX Security 2014). USENIX Association, San Diego (2014)
21. Chou, T.: Sandy2x: fastest Curve25519 implementation ever (2015). http://csrc. nist.gov/groups/ST/ecc-workshop-2015/presentations/session6-chou-tung.pdf
22. Costigan, N., Schwabe, P.: Fast elliptic-curve cryptography on the Cell Broadband engine. In: Preneel, B. (ed.) AFRICACRYPT 2009. LNCS, vol. 5580, pp. 368–385. Springer, Heidelberg (2009)
23. Düll, M., Haase, B., Hinterwälder, G., Hutter, M., Paar, C., Sánchez, A.H., Schwabe, P.: High-speed Curve25519 on 8-bit, 16-bit, and 32-bit microcontrollers. Designs, Codes and Cryptography (to appear, 2015). https://cryptojedi.org/ papers/mu25519-20150417.pdf
24. Flori, J.P., Plût, J., Reinhard, J.R., Ekerå, M.: Diversity and transparency for ECC (2015). http://csrc.nist.gov/groups/ST/ecc-workshop-2015/ papers/session4-flori-jean-pierre.pdf
25. Galbraith, S.D., McKee, J.: The probability that the number of points on an elliptic curve over a finite field is prime. J. London Math. Soc. **62**, 671–684 (2000)
26. Gaudry, P., Thomé, E.: The mpFq library and implementing curve-based key exchanges. In: SPEED: Software Performance Enhancement for Encryption and Decryption, pp. 49–64 (2007). http://www.loria.fr/gaudry/papers.en.html
27. Granville, A.: Smooth Numbers: Computational Number Theory and Beyond, pp. 267–323. Cambridge University Press (2008). http://en.scientificcommons.org/ 43534098, http://www.math.leidenuniv.nl/psh/ANTproc/09andrew.pdf
28. Institute of Electrical and Electronics Engineers: IEEE 1363–2000: Standard specifications for public key cryptography (2000)
29. Kelsey, J.: Choosing a DRBG algorithm (2003?). https://github.com/ matthewdgreen/nistfoia/blob/master/6.4.2014
30. LaMacchia, B., Costello, C.: Deterministic generation of elliptic curves (a.k.a. "NUMS" curves) (2014). https://www.ietf.org/proceedings/90/slides/ slides-90-cfrg-5.pdf
31. Langley, A., Moon, A.: Implementations of a fast elliptic-curve digital signature algorithm (2013). https://github.com/floodyberry/ed25519-donna
32. Lochter, M., Merkle, J.: RFC 5639: Elliptic curve cryptography (ECC) Brainpool standard curves and curve generation (2010)
33. Lochter, M., Merkle, J., Schmidt, J.M., Schütze, T.: Requirements for standard elliptic curves (2014), position Paper of the ECC Brainpool. http://www. ecc-brainpool.org/20141001_ECCBrainpool_PositionPaper.pdf
34. Luca, F., Mireles, D.J., Shparlinski, I.E.: MOV attack in various subgroups on elliptic curves. Illinois J. Math. **48**(3), 1041–1052 (2004)
35. Mahé, E.M., Chauvet, J.M.: Fast GPGPU-based elliptic curve scalar multiplication (2014). https://eprint.iacr.org/2014/198.pdf
36. Merkle, J.: Re: [Cfrg] ECC reboot (Was: When's the decision?) (2014). https:// www.ietf.org/mail-archive/web/cfrg/current/msg05353.html
37. National Institute for Standards and Technology: FIPS PUB 186–2: Digital signature standard (2000)
38. National Institute for Standards and Technology: FIPS PUB 186–4: Digital signature standard (DSS) (2013)
39. National Security Agency: Suite B cryptography / cryptographic interoperability (2005). https://web.archive.org/web/20150724150910/www.nsa.gov/ia/programs/ suiteb_cryptography/

40. State Commercial Cryptography Administration (OSCCA), China: Public key cryptographic algorithm SM2 based on elliptic curves, December 2010. http://www.oscca.gov.cn/UpFile/2010122214822692.pdf
41. State Commercial Cryptography Administration (OSCCA), China: Recommanded curve parameters for public key cryptographic algorithm SM2 based on elliptic curves, December 2010. http://www.oscca.gov.cn/UpFile/2010122214836668.pdf
42. Rosser, J.B., Schoenfeld, L.: Approximate formulas for some functions of prime numbers. Illinois J. Math. **6**, 64–94 (1962)
43. Sasdrich, P., Güneysu, T.: Efficient elliptic-curve cryptography using Curve25519 on reconfigurable devices. In: Goehringer, D., Santambrogio, M.D., Cardoso, J.M.P., Bertels, K. (eds.) ARC 2014. LNCS, vol. 8405, pp. 25–36. Springer, Heidelberg (2014)
44. Scott, M.: Re: NIST announces set of Elliptic Curves (1999). https://groups.google.com/forum/message/raw?msg=sci.crypt/mFMukSsORmI/FpbHDQ6hM_MJ
45. Silverman, J.H.: The Arithmetic of Elliptic Curves. Graduate Texts in Mathematics, vol. 106. Springer, New York (2009)
46. Stein, W., et al.: Sage Mathematics Software (Version 6.1.1). The Sage Development Team (2014). http://www.sagemath.org
47. Hutter, M., Schilling, J., Schwabe, P., Wieser, W.: NaCl's crypto_box in hardware. In: Güneysu, T., Handschuh, H. (eds.) CHES 2015. LNCS, vol. 9293, pp. 81–101. Springer, Heidelberg (2015)
48. Wikipedia: Nothing up my sleeve number (2015). http://www.en.wikipedia.org/wiki/Nothing_up_my_sleeve_number. Accessed 20 May 2015

Security of the SM2 Signature Scheme Against Generalized Key Substitution Attacks

Zhenfeng Zhang[1], Kang Yang[1 (✉)], Jiang Zhang[2], and Cheng Chen[1]

[1] Laboratory of Trusted Computing and Information Assurance,
Institute of Software, Chinese Academy of Sciences, Beijing, China
{zfzhang,yangkang,chencheng}@tca.iscas.ac.cn
[2] State Key Laboratory of Cryptology, Beijing, China
jiangzhang09@gmail.com

Abstract. Though existential unforgeability under adaptively chosen-message attacks is well-accepted for the security of digital signature schemes, the security against key substitution attacks is also of interest, and has been considered for several practical digital signature schemes such as DSA and ECDSA. In this paper, we consider *generalized* key substitution attacks where the base element is considered as a part of the public key and can be substituted. We first show that the general framework of certificate-based signature schemes defined in ISO/IEC 14888-3 is vulnerable to a *generalized* key substitution attack. We then prove that the Chinese standard SM2 signature scheme is existentially unforgeable against adaptively chosen-message attacks in the generic group model if the underlying hash function h is uniform and collision-resistant and the underlying conversion function f is almost-invertible, and the SM2 digital signature scheme is secure against the *generalized* key substitution attacks if the underlying hash functions H and h are modeled as *non-programmable* random oracles (NPROs).

Keywords: Digital signatures · Key substitution attacks · Provable security

1 Introduction

The well-known security notion for digital signature schemes is existential unforgeability under adaptively chosen-message attacks (EUF-CMA) introduced by Goldwasser et al. [8,9], which states that an adversary given any signatures for messages of its choice is unable to create a valid signature for a new message. However, as a de facto standard security notion for signature schemes, the security notion of EUF-CMA fails to capture the duplicate-signature key selection attacks introduced by Blake-Wilson and Menezes [3]. Later, this type of attacks is called key substitution (KS) attacks by Menezes and Smart [12],

The work was supported by National Basic Research Program of China (No. 2013CB338003), and National Natural Science Foundation of China (No. 61170278).

© Springer International Publishing Switzerland 2015
L. Chen and S. Matsuo (Eds.): SSR 2015, LNCS 9497, pp. 140–153, 2015.
DOI: 10.1007/978-3-319-27152-1_7

when they investigated the security of signature schemes in a multi-user setting. The KS attacks for some EUF-CMA secure signature schemes can be found in [3,7,12,15]. Informally, a KS-adversary is given a public key pk and a valid message-signature pair (m, σ) under the public key pk, and attempts to produce another public key pk' such that the message-signature pair (m, σ) is still valid under the different public key $pk' \neq pk$. In [3], Blake-Wilson and Menezes showed that if the underlying signature scheme suffers from the KS attacks, the station-to-station (STS) key agreement protocol [6] using a message authentication code (MAC) algorithm to provide key confirmation is vulnerable to unknown key-share (UKS) attacks. This gives a direct evidence that KS attacks might be harmful in practice. Two types of key substitution attacks are considered by Menezes and Smart [12]: if the KS-adversary is further required to output the private key sk' corresponding to pk', then this kind of KS attacks are called the weak-key substitution (WKS) attacks, else this type of attacks are referred to as the strong-key substitution (SKS) attacks. Obviously, a signature scheme which is secure against SKS attacks is also secure against WKS attacks. Afterwards, Bohli, Rohrich and Steinwandt [4] explored the security of some practical signature schemes against key substitution attacks in the presence of a *malicious* signer, where an adversary is given a set of domain parameters *params*, and aims at outputting two different public keys pk and pk' and a message-signature pair (m, σ) such that (1) both public keys pk and pk' are valid under the same set of domain parameters *params*, and (2) the pair (m, σ) is valid under both pk and pk' (also with respect to the same set of domain parameters *params*).

Besides, a related notion–domain parameter substitution attacks[1] are considered in [16,17]. In this kind of attacks, an adversary is given a set of domain parameters *params*, a public key pk and a signing oracle. The goal of the adversary is to output a new set of domain parameters *params'* and a signature σ on an un-queried message m such that (1) *params'* passes the test for the domain parameters verification algorithm, and (2) the pair (m, σ) is valid under the same public key pk but with respect to the different set of domain parameters *params'*.

SM2 Digital Signature Scheme. The SM2 digital signature scheme [2] was issued by the State Cryptography Administration Office of Security Commercial Code Administration in 2010, and had become the Chinese cryptographic public key algorithm standard GM/T 0003.2-2012. Later, it was adopted by Trusted Computing Group (TCG) in the TPM 2.0 specification [10] which will be published as the international standard ISO/IEC 11889:2015 [11].

1.1   Our Contributions

In this paper, we consider *generalized* key substitution attacks where the base element is regarded as a part of the public key and can be substituted. In detail,

---

[1] In [17], Vaudenay referred to this type of attacks presented in [16,17] as domain parameter shifting attacks. Later, this kind of attacks are called domain parameter substitution attacks in [4].

given a public key (g, pk) that g is a basis and pk is the other part of the public key and a valid message-signature pair (m, σ) under (g, pk), the goal of an adversary is to output another public key (g', pk') such that $(g, pk) \neq (g', pk')$ and (m, σ) is also valid under (g', pk'). This is possible when the domain parameters are generated by a signer, or the domain parameters are not properly validated, and has been considered by Blake-Wilson and Menezes [3] when examining the security against key substitution attacks on DSA and ECDSA signature schemes.

We first examine the security of a general framework of certificate-based signature schemes specified by ISO/IEC CD 14888-3 [1], and show that it is vulnerable to *generalized* key substitution attacks in the *weak* sense that the adversary knows the private key corresponding to the substituted public key.

Then, we analyze the security of the SM2 signature scheme [2] against chosen-message attack and generalized KS attacks respectively. Concretely, we not only show that SM2 signature scheme satisfies the EUF-CMA notion in the generic group model [14] provided that the underlying hash function h is uniform and collision-resistant and the underlying conversion function f is almost-invertible, but also give a formal proof that SM2 is secure against the *generalized* key substitution attacks in the *strong* sense (i.e., the adversary is not required to output the private key corresponding to the substituted public key) if the underlying hash functions H and h are both modeled as *non-programmable* random oracles [13].

2 Preliminaries

Notation. Throughout this paper, κ denotes the security parameter. We denote by $s \xleftarrow{\$} S$ the fact that s is picked uniformly at random from a finite set S. We write $(y_1, y_2, \dots) \leftarrow A(x_1, x_2, \dots)$ as the process that runs a randomized algorithm A on input $(x_1, x_2, \dots)$ and obtains its output $(y_1, y_2, \dots)$. The notation $[n]$ denotes the set $\{1, \dots, n\}$ for some positive integer n. We use $\mathbb{F}_q$ and $\mathbb{A}_n$ to denote the set $\{0, 1, \dots, q-1\}$ and a group with the order n respectively. We say that a function $f : \mathbb{N} \to [0, 1]$ is negligible if for every positive c and sufficiently large κ we have $f(\kappa) < 1/\kappa^c$, and is overwhelming if $1 - f$ is a negligible function.

2.1 Collision-Resistant Hash Functions

A hash function $h : \{0, 1\}^* \mapsto \mathcal{R}$ is said to be collision-resistant if for any probabilistic polynomial time (PPT) adversary $\mathcal{A}$, there exists a negligible function $\nu(\cdot)$ such that

$$\Pr[(x, y) \leftarrow \mathcal{A}(1^\kappa, h) : x \neq y \wedge h(x) = h(y)] \leq \nu(\kappa),$$

where $\mathcal{R}$ denotes the range of h.

2.2 Uniform (Smooth) Hash Functions

Following the definition [5], the uniformity (or smoothness) of a hash function h is described as below. Let $h : \{0, 1\}^* \mapsto \mathcal{R}$ be a hash function. Let $\mathcal{D} \subseteq \{0, 1\}^*$ such that

1. For $x \xleftarrow{\$} \mathcal{D}$, $y = h(x)$ can be efficiently generated.
2. For each $y \in \mathcal{R}$, the set $S_y = h^{-1}(y) \cap \mathcal{D}$ is sufficiently large so that the probability $1/|S_y|$ is sufficiently small (negligible) to make guessing a randomly picked secret element of S_y infeasible.

We say that h is uniform for $\mathcal{D}$ if for any PPT adversary $\mathcal{A}$, there exists a negligible function $\nu(\cdot)$ such that

$$\left| \Pr[x \xleftarrow{\$} \mathcal{D} : \mathcal{A}(1^\kappa, h, h(x)) = 1] - \Pr[y \xleftarrow{\$} \mathcal{R} : \mathcal{A}(1^\kappa, h, y) = 1] \right| \le \nu(\kappa)$$

2.3 Almost-Invertibility of Conversion Functions

SM2 uses a conversion function $f : \mathbb{A}_n \mapsto \mathbb{F}_n$ which could be efficiently computed. Almost-invertibility of the conversion function is associated with the EUF-CMA security of SM2, and is defined in [5]. Concretely, a conversion function f is almost-invertible if an almost-inverse of f is efficiently computed. An almost-inverse of f is a probabilistic polynomial time (PPT) algorithm $f^{-1} : \mathbb{F}_n \mapsto \mathbb{A}_n$ which on input $x \in \mathbb{F}_n$ produces a $Q \in \mathbb{A}_n \cup \{\texttt{Invalid}\}$ such that:

- The probability $Q \ne \texttt{Invalid}$ is at least $1/10$ over random choices of x and the almost-inverse f^{-1}.
- If $Q \ne \texttt{Invalid}$, $f(Q) = x$.
- If independently random inputs $x \xleftarrow{\$} \mathbb{F}_n$ are repeatedly input to the algorithm f^{-1} until the output $Q \ne \texttt{Invalid}$, the probability distribution of the resulting Q is computationally indistinguishable from the distribution of a random element $Q \in \mathbb{A}_n$.

3 Definitions

Following the definitions in [12], we present the syntax and security notions of a signature scheme in the multi-user setting. Concretely, the syntax is described in Definition 1, the security model for existential unforgeability under adaptively chosen-message attacks (EUF-CMA) [9] is formalized in Definition 2, and the security notion for (generalized) strong key substitution (SKS) attacks is defined in Definition 3.

Definition 1 (Syntax). *A signature scheme in the multi-user setting consists of the following algorithms.*

- Setup(1^κ). On input a security parameter κ, the setup algorithm returns the domain parameters *params*.
- Keygen(*params*). On input the domain parameters *params*, the key generation algorithm returns a public-private key pair (PK, SK). Recall that PK contains the base element.
- Sign(*params*, SK, m). On input the domain parameters *params*, the private key SK and a message m, the signing algorithm returns a signature σ on the message m.

Experiment $\mathbf{Exp}_{SIG,\mathcal{F}}^{\text{EUF-CMA}}(\kappa)$

$\quad \mathcal{Q} := \varnothing;\ params \leftarrow \mathsf{Setup}(1^\kappa);\ (\text{PK}, \text{SK}) \leftarrow \mathsf{Keygen}(params);$

$\quad (m^*, \sigma^*) \leftarrow \mathcal{F}^{\mathsf{Sign}(params, \text{SK}, \cdot)}(params, \text{PK});$

$\quad$ If $m^* \notin \mathcal{Q} \wedge \mathsf{Verify}(params, \text{PK}, \sigma^*, m^*) = 1$, return 1.

$\quad$ Otherwise, return 0.

Signing oracle $\mathsf{Sign}(params, \text{SK}, m)$

$\quad \sigma \leftarrow \mathsf{Sign}(params, \text{SK}, m);$

$\quad \mathcal{Q} := \mathcal{Q} \cup \{m\};$

$\quad$ Return σ.

Fig. 1. Experiment for EUF-CMA security

- $\mathsf{Verify}(params, \text{PK}, \sigma, m)$. On input the domain parameters $params$, the public key PK and a candidate signature σ on a message m, the deterministic verification algorithm returns 1 if σ is valid on m under PK and 0 otherwise.

Besides, we define an additional checking algorithm Check to check the validity of a public key PK. Specifically, given the domain parameters $params$ and a candidate public key PK, the checking algorithm $\mathsf{Check}(params, \text{PK})$ returns 1 if and only if the public key PK is valid under the domain parameters $params$.

Definition 2 (EUF-CMA). *We say that a signature scheme is existentially unforgeable under adaptively chosen-message attacks (EUF-CMA) if for any probabilistic polynomial time (PPT) adversary $\mathcal{F}$, there exists a negligible function $\nu(\cdot)$ such that*

$$\mathbf{Adv}_{SIG,\mathcal{F}}^{\text{EUF-CMA}}(\kappa) := \Pr[\mathbf{Exp}_{SIG,\mathcal{F}}^{\text{EUF-CMA}}(\kappa) = 1] \leq \nu(\kappa),$$

where $\mathbf{Exp}_{SIG,\mathcal{F}}^{\text{EUF-CMA}}(\kappa)$ is defined in **Fig. 1.**

Definition 3 (SKS). *We say that a signature scheme is secure against (generalized) strong key substitution attacks if for any PPT adversary $\mathcal{F}$, there exists a negligible function $\nu(\cdot)$ such that*

$$\mathbf{Adv}_{SIG,\mathcal{F}}^{\text{SKS}}(\kappa) := \Pr[\mathbf{Exp}_{SIG,\mathcal{F}}^{\text{SKS}}(\kappa) = 1] \leq \nu(\kappa),$$

where $\mathbf{Exp}_{SIG,\mathcal{F}}^{\text{SKS}}(\kappa)$ is defined in **Fig. 2.**

Experiment $\mathbf{Exp}^{\mathrm{SKS}}_{SIG,\mathcal{F}}(\kappa)$

$\quad \mathcal{Q} := \varnothing;\ params \leftarrow \mathsf{Setup}(1^{\kappa});\ (\mathrm{PK}, \mathrm{SK}) \leftarrow \mathsf{Keygen}(params);$

$\quad (m^{*}, \sigma^{*}, \mathrm{PK}^{*}) \leftarrow \mathcal{F}^{\mathsf{Sign}(params,\mathrm{SK},\cdot)}(params, \mathrm{PK});$

$\quad$ If $\mathrm{PK}^{*} \neq \mathrm{PK}\ \wedge\ \mathsf{Check}(params, \mathrm{PK}^{*}) = 1\ \wedge\ (m^{*}, \sigma^{*}) \in \mathcal{Q}$

$\qquad \wedge\ \mathsf{Verify}(params, \mathrm{PK}^{*}, \sigma^{*}, m^{*}) = 1$, return 1.

$\quad$ Otherwise, return 0.

Signing oracle $\mathsf{Sign}(params, \mathrm{SK}, m)$

$\quad \sigma \leftarrow \mathsf{Sign}(params, \mathrm{SK}, m);$

$\quad \mathcal{Q} := \mathcal{Q} \cup \{(m, \sigma)\};$

$\quad$ Return σ.

Fig. 2. Experiment for (generalized) strong key substitution attacks

We could easily modify the definition for (generalized) strong key substitution (SKS) attacks to capture the notion for (generalized) weak key substitution (WKS) attacks. The experiment for (generalized) WKS attacks is the same as $\mathbf{Exp}^{\mathrm{SKS}}_{SIG,\mathcal{F}}(\kappa)$ defined in **Fig. 2**, except that an adversary against (generalized) WKS attacks is further required to output the private key SK^{*} of the substituted public key PK^{*}.

4   Generalized WKS Attacks Against a General Framework of ISO/IEC CD 14888-3

In this section, we first review the general framework of certificate-based mechanisms of ISO/IEC CD 14888-3 [1] in the setting that a signer chooses the base element as a part of its public key, and then show that the general framework is vulnerable to the *generalized* weak key substitution (WKS) attacks.

General Framework. The general framework of certificate-based mechanisms specified in ISO/IEC CD 14888-3 [1] is presented as follows.

- $\mathsf{Setup}(1^{\kappa})$. Given a security parameter κ, pick a finite commutative group $\mathbf{E}$ where multiplicative notation is used, and a prime divisor q of the cardinality of $\mathbf{E}$, and choose an element $G \in \mathbf{E}$ of order q. Return $params := (\mathbf{E}, q, G)$ as a set of domain parameters.
- $\mathsf{Keygen}(params)$. Given the set of domain parameters $params = (\mathbf{E}, q, G)$, choose $X \xleftarrow{\$} \mathbb{F}_{q} \backslash \{0\}$. Then, compute $Y := G^{X}$. Actually, in the certificate-based mechanisms of ISO/IEC CD 14888-3, Y is equal to either G^{X} or $G^{X^{-1}}$

relying on the specific mechanism. Without loss of generality, we only consider the case that $Y = G^X$. Finally, output PK $= (G, Y)$ and SK $= X$ as the public key and the private key respectively.

- Sign($params$, SK, M). Given the set of domain parameters $params = (\mathbf{E}, q, G)$, the private key SK $= X$ and a message M, the signing process is executed as follows:

  1. (*Producing the randomizer*) Choose $K \overset{\$}{\leftarrow} \mathbb{F}_q \backslash \{0\}$.
  2. (*Producing the pre-signature*) Compute $\Pi := G^K$.
  3. (*Preparing the message for signing*) Depending on the particular mechanism, one of M_1 and M_2 is set as M, and the other is set as empty.
  4. (*Computing the witness (the first part of the signature)*) The values of Π and M_1 are taken as inputs to the witness function which is specified in the concrete mechanism. The output of the witness function is the witness R.
  5. (*Computing the assignment*) The witness R, M_2 and (optionally) Y are taken as input to the assignment function which is defined in the particular mechanism. Then, the assignment function outputs assignment $T = (T_1, T_2)$ where T_1 and T_2 are integers such that $0 < |T_1| < q$ and $0 < |T_2| < q$.
  6. (*Computing the second part of the signature*) Let S be the second part of the signature and (A, B, C) is a permutation of three elements (S, T_1, T_2) depending on the particular mechanism. Solve the following signature equation for S where $S \in \mathbb{F}_q \backslash \{0\}$:

  $$AK + BX + C \equiv 0 \ (\mathrm{mod} \ q).$$

  7. Output $\sigma := (R, S)$ as the signature.

- Verify($params$, PK, σ, M). Given the set of domain parameters $params$, the public key PK $= (G, Y)$ and a candidate signature $\sigma = (R, S)$ on a message M, the verification process is executed as below:

  1. (*Preparing message for verification*) Divide the message M into two parts M_1 and M_2.
  2. (*Retrieving the assignment*) Recompute the assignment $T = (T_1, T_2)$ using the assignment function with the inputs R, M_2 and (optionally) Y.
  3. (*Recomputing the pre-signature*) Set (A, B, C) as (S, T_1, T_2) according to the order specified in the signature algorithm. Recompute the pre-signature $\Pi' := Y^m G^n$ where $m = -A^{-1}B \ \mathrm{mod} \ q$ and $n = -A^{-1}C \ \mathrm{mod} \ q$.
  4. (*Recomputing the witness*) Recompute the witness R' via executing the witness function with the inputs Π' and M_1.
  5. (*Verifying the witness*) If $R = R'$, then return 1, else return 0.

Generalized WKS Attacks. Recall that given a valid message-signature pair $(M, (R, S))$ under the public key PK $= (G, Y)$ of some legitimate user, the goal of a *generalized* WKS adversary $\mathcal{A}$ is to produce a public-private key pair (PK$' = (G', Y' = (G')^{X'}), X'$) such that PK$' \neq$ PK, but the message-signature pair

$(M, (R, S))$ is still valid under the public key PK′. A *generalized* WKS adversary $\mathcal{A}$ for the general framework of certificate-based mechanisms of ISO/IEC CD 14888-3 [1] is described as follows.

The adversary $\mathcal{A}$ first computes m and n with $(M, (R, S))$ and (optionally) Y following the verification process. Then, $\mathcal{A}$ computes $\Pi = Y^m G^n$. In the following, the attack manner of $\mathcal{A}$ is divided into the following two cases depending on whether or not Y is used to generate (T_1, T_2):

- If Y is not used to generate (T_1, T_2), then the values of m and n, which are created with $(M, (R, S))$, remain unchanged according to the verification process. Choose $X' \xleftarrow{\$} \mathbb{F}_q \backslash \{0\}$, and then compute $G' := \Pi^{1/(mX'+n)}$ and $Y' := (G')^{X'}$. Finally, output the new public-private key pair (PK′ $= (G', Y'), X')$. It is easy to see that $\Pi' = (Y')^m (G')^n = (G')^{mX'+n} = \Pi$.

- If Y is used to generate (T_1, T_2), choose $t \xleftarrow{\$} \mathbb{F}_q \backslash \{0\}$ and compute $Y' := \Pi^t$. Then, compute m' and n' with $(M, (R, S))$ and Y' following the verification process. Finally, compute $X' = (1 - tm')/tn' \mod q$ and $G' = \Pi^{tX'}$, and then output the new public-private key pair (PK′ $= (G', Y'), 1/X')$. Again, one can easily verify that $\Pi' = (Y')^{m'} (G')^{n'} = \Pi^{tm'+tn'X'} = \Pi$.

Since in both cases we always have that Π' is equal to Π, the message-signature pair $(M, (R, S))$ is valid under the new public key PK′ according to the verification process. This shows that the above constructed adversary $\mathcal{A}$ will break the security against *generalized* WKS attacks on the general framework with probability 1.

5   Security of the SM2 Signature Scheme

In this section, we first recall the description of SM2 digital signature scheme, and then we present the formal security proofs showing that SM2 satisfies both EUF-CMA security and the security against *generalized* strong key substitution attacks.

5.1   SM2 Digital Signature Scheme

The Chinese digital signature standard SM2 [2] is based on elliptic curve which has a formal of $y^2 + xy = x^3 + ax^2 + b$ over $\mathbb{F}_q$ for some integer $q = 2^m$, and $y^2 = x^3 + ax + b$ over $\mathbb{F}_q$ for some large prime q. In other words, the curve is parameterized by q and (a, b). Denote $E(\mathbb{F}_q)$ as the additive finite group which consists of all the integer points (including the infinity point 0) on the elliptic curve. In the following, we give the formal description of the four algorithms of the SM2 signature scheme.

- Setup(1^κ): Given a security parameter κ, generate the elliptic curve parameters (q, a, b, n) such that n is a prime divisor of the cardinality of $E(\mathbb{F}_q)$ and $|n| \geq 2\kappa$, where q, a, b is the curve parameter. Choose a (random) generator $G \in E(\mathbb{F}_q)$ of order n. Output a set of domain parameters $params := (q, a, b, n, G)$.

Let $h : \{0,1\}^* \mapsto \mathbb{F}_n$ and $H : \{0,1\}^* \mapsto \{0,1\}^{256}$ be two cryptographic hash functions. Let $\mathbb{A}_n \subseteq E(\mathbb{F}_q)$ be the cyclic group generated by G. The conversion function $f : \mathbb{A}_n \mapsto \mathbb{F}_n$ is defined as $f(Q) = x_Q \mod n$, where x_Q is an integer representation of the x-coordinate of the elliptic curve point $Q \in \mathbb{A}_n$.

- Keygen($params$): Given the domain parameters $params = (q, a, b, n, G)$, pick $d \xleftarrow{\$} \mathbb{F}_n \backslash \{0, n-1\}$ and compute $Y = dG$. Output the public-private key pair $\left(\text{PK} = (G, Y), \text{SK} = d\right)$.

- Sign($params$, PK, SK, m): Given the set of domain parameters $params = (q, a, b, n, G)$, the private key SK $= d$, the public key PK $= (G, Y)$, and a message m, let $Z = H(ENTL\|ID\|a\|b\|G\|Y)$ where $ENTL$ denotes the length of ID and ID is the identity of the owner of PK and do the following:
  1. Let $\bar{m} = Z\|m$.
  2. Compute $e := h(\bar{m})$.
  3. Choose $k \xleftarrow{\$} \mathbb{F}_n \backslash \{0\}$.
  4. Compute $x_1 := f(kG)$.
  5. Compute $r := (e + x_1) \mod n$. If $r = 0$ or $r + k = n$, go back to step 3.
  6. Compute $s := (k - rd)/(1 + d) \mod n$.
  7. The signature on m is $\sigma := (r, s)$.

- Verify($params$, PK, σ', m'): Given the set of domain parameters $params = (q, a, b, n, G)$, the public key PK $= (G, Y)$, and a signature $\sigma' = (r', s')$ on a message m', let $Z = H(ENTL\|ID\|a\|b\|G\|Y)$ and do the following:
  1. If $r' \notin [1, n-1]$, output 0 and exit.
  2. If $s' \notin [1, n-1]$, output 0 and exit.
  3. Let $\bar{m}' = Z\|m'$.
  4. Compute $e' := h(\bar{m}')$.
  5. Compute $t := (r' + s') \mod n$. If $t = 0$, output 0 and exit.
  6. Compute $x_1' := f(s'G + tY)$.
  7. Compute $R := (e' + x_1') \mod n$.
  8. If $R = r'$, then output 1, else output 0.

The conversion function $f : \mathbb{A}_n \mapsto \mathbb{F}_n$ of SM2 is exactly the same as that of ECDSA, and has been shown to be almost-invertible in [5].

5.2   EUF-CMA Security of SM2

Now, we proceed to give a formal security proof showing the EUF-CMA security of SM2. Formally, we have the following theorem.

Theorem 1. *If h is a uniform and collision-resistant hash function, and the conversion function f is almost-invertible, SM2 is existentially unforgeable under adaptively chosen-message attacks in the generic group model.*

Note that in the generic group model, an adversary is not given direct access to the group, but rather only receives "handles" representing group elements. More concretely, the adversary must interact with an oracle to perform the group operations (including scalar-multiplication and addition) and obtain handles for new elements. In particular, it is assumed that the adversary can only use handles previously received from its environment. Back to our case, in addition to directly to get group element handles from group operation queries, the adversary can also obtain handles from the public key and the signatures from signing oracle queries. Actually, the adversary can use the handles in the public key and the signatures as the "bases" to perform further groups operations. More formally, let (G, Y) be the group element handles in the public key, and let $(V_1, \ldots, V_{q_s})$ be the group element handles created in the signing queries, where q_s is the number of signing queries made by the adversary. Then, by the assumption that all the group elements that the adversary want to compute have a form of $z_1 G + z_2 Y + z_3 V_1 + \ldots + z_{q_s+2} V_{q_s}$, where $z_1, \ldots, z_{q_s+2}$ are known integers chosen by the adversary. Thus, we can unify all the group operation queries by the coefficient vector $(z_1, \ldots, z_{q_s+2})^2$. For example, multiplying the base element G by an integer z can be expressed as a group operation query $(z, 0, \ldots, 0)$.

Proof. In the following, for any PPT forger $\mathcal{F}$, we show there exists a challenger $\mathcal{C}$ to simulate the attack environment for $\mathcal{F}$ such that the advantage of $\mathcal{F}$ is negligible. In order to answer the group operation queries from $\mathcal{F}$, the challenger $\mathcal{C}$ will keep a table L to record the information generated in the group operation queries. Formally, $\mathcal{C}$ first generates the handle G of the base element by choosing $G \overset{\$}{\leftarrow} \mathbb{A}_n$, and adds $(1, G, -, -)$ into the table L, where $\mathbb{A}_n$ is a set supporting efficient sampling and representing the underlying group. Then, $\mathcal{C}$ chooses an integer $d \overset{\$}{\leftarrow} \mathbb{F}_n \backslash \{0, n-1\}$, $Y \overset{\$}{\leftarrow} \mathbb{A}_n$ as the handle of multiplying G by d, and adds $(d, Y, -, -)$ into the table L. Let q_s be the number of signing queries made by $\mathcal{F}$, q_c be the current number of signing queries during the interaction between $\mathcal{C}$ and $\mathcal{F}$, and denote $V_1, \ldots, V_{q_s}$ as the group element handles that will be generated in the signing queries. $\mathcal{C}$ answers $\mathcal{F}$'s group operation queries and signing queries as follows.

- For a group operation query with input $(z_1, \ldots, z_{q_s+2})$ (i.e., $\mathcal{F}$ wants to compute $z_1 G + z_2 Y + z_3 V_1 + \ldots + z_{q_s+2} V_{q_s}$), $\mathcal{C}$ does the following:
  1. Let j be the maximum index such that $z_j \neq 0$.
  2. If $j > q_c + 2$, then return $\bot$ and exit.
  3. Otherwise, for each $i \in \{1, \ldots, j\}$, retrieve k_i from the entry $(k_i, V_i, -, -)$ in table L and compute $z' = z_1 + z_2 d + z_3 k_1 + \ldots + z_j k_j \mod n$.
  4. If there exists an entry $(z', V', -, -)$ in table L, $\mathcal{C}$ directly returns V' to $\mathcal{F}$. Otherwise, it distinguishes the following two cases:
     - Case 1: If $z_2 = 0$, choose $V' \overset{\$}{\leftarrow} \mathbb{A}_n$, add $(z', V', (z_1, \ldots, z_{q_s+2}), -)$ into table L. Finally, return V' to $\mathcal{F}$.
     - Case 2: If $z_2 \neq 0$, randomly choose $Z' \overset{\$}{\leftarrow} \{0, 1\}^{256}, m' \overset{\$}{\leftarrow} \mathcal{M}^3$, and compute

$$V' = f^{-1}(z_2 - z_1 - z_3 k_1 - \ldots - z_{q_s+2} k_{q_s} - h(Z' \| m'))$$

---

[2] In this case, if some z_i is equal to 0, it means that the corresponding group element is not involved in the computation.

[3] We use $\mathcal{M}$ to denote the efficiently sampling message space of SM2.

until $V' \in \mathbb{A}_n$. Then, add $(z', V', (z_1, \ldots, z_{q_s+2}), Z'\|m')$ into table L, and return V' to $\mathcal{F}$.

- For a signing query on some message m, $\mathcal{C}$ chooses $k \xleftarrow{\$} \mathbb{F}_n\setminus\{0\}$, and makes a group operation query $(k, 0, \ldots, 0)$ by itself to obtain a handle V. Then, it computes $x = f(V), r = h(Z\|m)+x \mod n$, and $s = (k-rd)/(1+d) \mod n$, where Z is the other information as determined in the signing algorithm SM2. Finally, $\mathcal{C}$ returns (r, s) as the signature on m to $\mathcal{F}$.

After making polynomial time queries of the above two types, the adversary $\mathcal{F}$ will output a forged signature (r^*, s^*) for $m^* \notin \{m_i\}_{i\in[q_s]}$. Below, we prove the probability that the forged signature is valid to be negligible under the assumption that h is uniform and collision-resistant.

Analysis. Note that $\mathcal{C}$ honestly generates the public key and the signatures, if $\mathcal{C}$ also perfectly answers the group operation queries, then we have that $\mathcal{C}$ almost simulates a perfect attack environment for $\mathcal{F}$. Actually, it is easy to check that all the group element handles are uniformly chosen at random except in Case 2 of the group operation query. Now, we argue that the handle V' generated in Case 2 is also uniformly distributed. In fact, since Z' and m' are uniformly chosen at random, and h is a uniform function, we have that the input of f^{-1} in Case 2 is uniformly distributed. By the fact that f^{-1} is almost-invertible, we have V' is uniformly distributed. In addition, by the Schwartz-Zippel Lemma, the probability that there exist two entries $(z', V', -, -)$ and $(z'', V'', -, -)$ in table L such that $z' \neq z''$ but $V' = V''$ (i.e., $\mathcal{C}$ fails to simulate the generic group model due to the inconsistency) is bounded by $O\left(\frac{(q_G+q_s)^2}{n}\right)$ which is negligible, where q_G denotes the total number of group operation queries made by $\mathcal{F}$. This finally shows that $\mathcal{C}$ almost perfectly simulates the attack environment for $\mathcal{F}$.

In order to finish the proof, we only have to show that the probability that (r^*, s^*) is a valid signature on m^* is negligible. Before continuing, we note that the secret key d is perfectly hidden from the adversary $\mathcal{F}$. This is because in the generic group model, d is chosen independently from the group element handle Y in the public key, and d is perfectly hidden from the signature (r, s) in the signing query (due to the randomly choices of k and V). Let $k^* = s^* + (s^*+r^*)d$, then (r^*, s^*) is a valid signature on m^* if and only if there exists an entry $(k^*, V^*, -, -)$ in table L such that $r^* - h(Z\|m^*) = f(V^*)$. We first claim that $V^* \notin \{G, Y\}$ holds with overwhelming probability. Otherwise, the adversary can deterministically compute d from (r^*, s^*) by using the fact that $s^* \neq 0$ and $s^* + r^* \neq 0$, which contradicts to the fact that d is perfectly hidden from the adversary $\mathcal{F}$. In other words, V^* can only be created either in answering the group operation query or in answering the signature queries. We distinguish the following two cases:

- If $V^* \in \{V_1, \ldots, V_{q_s}\}$, then let $V^* = V_i$ for some i, and let (r_i, s_i) be the signature on some message m_i and auxiliary information Z_i in the i-th signing query. In other words, we have $s^* + (s^*+r^*)d = s_i + (s_i+r_i)d$. By the fact that d is perfectly hidden from the adversary $\mathcal{F}$, this can only happen with non-negligible probability when both $s^* = s_i$ and $s^* + r^* = s_i + r_i$ hold. In this case, (r^*, s^*) is a valid signature on m^* if and only if the equation $h(Z_i\|m_i) = r_i - f(V_i) = r^* - f(V^*) = h(Z\|m^*)$ holds. Since $m^* \neq m_i$, this means that $\mathcal{F}$ has to find a collision $(Z_i\|m_i, Z\|m^*)$ of the hash function h. Under the assumption that h is collision-resistant, this can only happen with negligible probability.

– Else, V^* is created by a group operation query with input $(z_1^*, \ldots, z_{q_s+2}^*)$. In this case, we have $k^* = s^* + (s^* + r^*)d = z_1^* + z_2^*d + z_3^*k_1 + \ldots + z_{q_s+2}^*k_{q_s}$. Again, by the fact that d is perfectly hidden from the adversary $\mathcal{F}$, this can only happen with non-negligible probability when both $s^* = z_1^* + z_3^*k_1 + \ldots + z_{q_s+2}^*k_{q_s}$ and $s^* + r^* = z_2^*$ hold. By a simple computation, we have $r^* = z_2^* - z_1^* - z_3^*k_1 - \ldots - z_{q_s+2}^*k_{q_s}$. Besides, according to the strategy of $\mathcal{C}$ (in Case 2), there exists a pair (Z', m') chosen by $\mathcal{C}$ such that $f(V^*) = z_2^* - z_1^* - z_3^*k_1 - \ldots - z_{q_s+2}^*k_{q_s} - h(Z'\|m')$. In other words, (r^*, s^*) is a valid signature if and only if $h(Z\|m^*) = r^* - f(V^*) = h(Z'\|m')$. However, under the assumption that h is collision-resistant, the probability that $\mathcal{F}$ outputs a pair $(Z\|m^*)$ such that $h(Z\|m^*) = h(Z'\|m')$ is negligible.

In all, we have shown that under the assumption that h is uniform and collision-resistant, the probability that (r^*, s^*) is a valid signature on m^* is negligible, which completes the proof. $\qquad\square$

5.3   Security of SM2 Against Generalized SKS Attacks

In this subsection, we show that SM2 is secure against *generalized* SKS attacks. Formally, we have the following theorem.

Theorem 2. *If both H and h are modeled as non-programmable random oracles (NPROs), then SM2 is secure against generalized strong key substitution attacks.*

Proof. In the following, we will show that the advantage of any PPT adversary $\mathcal{F}$ against the *generalized* SKS security of SM2 is negligible. Formally, in order to simulate the attack environment for $\mathcal{F}$, the challenger $\mathcal{C}$ only has to generate the domain parameters $params$ and (PK, SK), and answers the signing queries honestly. More concretely, $\mathcal{C}$ first runs the Setup and Keygen algorithms to obtain $params = (q, a, b, n, G)$ and $(\text{PK}, \text{SK}) = ((G, Y), d)$ where $Y = dG$. Then, let ID be the identity of the owner of PK, and send $(params, \text{PK})$ to $\mathcal{F}$. Recall that in our model h and H are modeled as NPROs, both the challenger $\mathcal{C}$ and the adversary $\mathcal{F}$ have to access the external random oracles h and H to realize the functionality of SM2.

After receiving the i-th signing query on a message m_i, $\mathcal{C}$ honestly computes $(r_i, s_i) \leftarrow \text{Sign}(params, \text{SK}, m_i)$ by making appropriate random oracle queries to h and H, and returns $\sigma_i = (r_i, s_i)$ as the signature on m_i to $\mathcal{F}$. Let q_s be the number of the signing queries issued by $\mathcal{F}$.

Finally, $\mathcal{F}$ will return $(ID^*, m^*, (r^*, s^*), \mathrm{PK}^* = (G^*, Y^*))$ as its output[4], such that (1) $(G^*, Y^*) \neq (G, Y)$, (2) (G^*, Y^*) is valid[5], and (3) $(m^*, r^*, s^*) = (m_j, r_j, s_j)$ for some $1 \leq j \leq q_s$.

Analysis. Now, we will show that $\mathcal{F}$ can only win the SKS game with negligible probability. Specifically, the probability that the message-signature pair (m_j, r_j, s_j) is valid under PK^* is negligible in the security parameter κ. Note that (m_j, r_j, s_j) is valid under $\mathrm{PK}^* = (G^*, Y^*)$ if and only if

$$r_j = e_j + f\left(s_j G + (r_j + s_j)Y\right) = e^* + f\left(s_j G^* + (r_j + s_j)Y^*\right) \mod n, \quad (1)$$

where $e^* = h(H(ENTL^*\|ID^*\|a\|b\|G^*\|Y^*)\|m_j)$ and $e_j = h(H(ENTL\|ID \|a\|b\|G\|Y)\|m_j)$. Since $\mathrm{PK} \neq \mathrm{PK}^*$, we have that the distribution of e_j is independent from that of e^*, and that $e_j \neq e^*$ holds with overwhelming probability. This means that the distribution of $\sigma_j = (r_j, s_j)$ is independent from e^* according to the signing algorithm. In other words, the distribution of e^* is still uniform conditioned on the equation (1) holds by the assumption that both h and H are NPROs. Note that $\mathcal{F}$ must first fix $\mathrm{PK}^* = (G^*, Y^*)$ to make the appropriate H query, and that the outputs of both h and H are uniformly distributed, the probability that $e^* = h(H(ENTL^*\|ID^*\|a\|b\|G^*\|Y^*)\|m_j)$ satisfying a prior fixed equation $e^* = r_j - f\left(s_j G^* + (r_j + s_j)Y^*\right) \mod n$ is negligible, which shows that the equation (1) can only hold with negligible probability. This completes the proof of Theorem 2. $\square$

Acknowledgements. We would like to thank Hui Guo and the anonymous reviewers for their helpful comments.

References

1. ISO/IEC 1st CD 14888-3 - Information technology - Security techniques - Digital signatures with appendix - Part 3: Discrete logarithm based mechanisms
2. GM/T 0003.2-2012, Public Key Cryptographic Algorithm SM2 based on Elliptic Curves - Part 2: Digital Signature Algorithm (2010). http://www.oscca.gov.cn/UpFile/2010122214822692.pdf
3. Blake-Wilson, S., Menezes, A.: Unknown key-share attacks on the Station-to-Station (STS) Protocol. In: Imai, H., Zheng, Y. (eds.) PKC 1999. LNCS, vol. 1560, pp. 154–170. Springer, Heidelberg (1999)
4. Bohli, J.-M., Röhrich, S., Steinwandt, R.: Key substitution attacks revisited: taking into account malicious signers. Int. J. Inf. Secur. **5**(1), 30–36 (2006)
5. Brown, D.R.L.: Generic groups, collision resistance, and ECDSA. Des. Codes Crypt. **35**(1), 119–152 (2005)

---

[4] We also allow the adversary to output the identity ID^* of the owner of PK^*. This is only because both the signing and verification algorithms of SM2 have an identity input. We do not have any additional restriction on ID^*.

[5] To verify the validity of (G^*, Y^*), the following conditions need to be satisfied: (1) $G^* \in E(\mathbb{F}_q)$, (2) the order of G^* is n, and 3) $Y^* \in <G^*> \setminus \{0\}$.

6. Diffie, W., Van Oorschot, P.C., Wiener, M.J.: Authentication and authenticated key exchanges. Des. Codes Crypt. **2**(2), 107–125 (1992)
7. Geiselmann, W., Steinwandt, R.: A Key Substitution Attack on SFLASH^{v3}. Cryptology ePrint Archive, Report 2003/245 (2003). http://eprint.iacr.org/
8. Goldwasser, S., Micali, S., Rivest, R.L.: A paradoxical solution to the signature problem. In: Proceedings of the IEEE 25th Annual Symposium on Foundations of Computer Science, pp. 441–448. IEEE (1984)
9. Goldwasser, S., Micali, S., Rivest, R.L.: A digital signature scheme secure against adaptive chosen-message attacks. SIAM J. Comput. **17**(2), 281–308 (1988)
10. Trusted Computing Group. TCG TPM specification 2.0. (2013) http://www. trustedcomputinggroup.org/resources/tpm_library_specification
11. ISO/IEC 11889:2015. Information technology - Trusted Platform Module Library (2015)
12. Menezes, A., Smart, N.: Security of signature schemes in a multi-user setting. Des. Codes Crypt. **33**(3), 261–274 (2004)
13. Nielsen, J.B.: Separating random oracle proofs from complexity theoretic proofs: the non-committing encryption case. In: Yung, M. (ed.) CRYPTO 2002. LNCS, vol. 2442, pp. 111–126. Springer, Heidelberg (2002)
14. Shoup, V.: Lower bounds for discrete logarithms and related problems. In: Fumy, W. (ed.) EUROCRYPT 1997. LNCS, vol. 1233, pp. 256–266. Springer, Heidelberg (1997)
15. Tan, C.H.: Key substitution attacks on some provably secure signature schemes. IEICE Trans. Fundam. Electron. Commun. Comput. Sci. **E87-A**(1), 226–227 (2004)
16. Vaudenay, S.: The security of DSA and ECDSA. In: Desmedt, Y.G. (ed.) PKC 2003. LNCS, vol. 2567, pp. 309–323. Springer, Heidelberg (2002)
17. Vaudenay, S.: Digital signature schemes with domain parameters. In: Wang, H., Pieprzyk, J., Varadharajan, V. (eds.) ACISP 2004. LNCS, vol. 3108, pp. 188–199. Springer, Heidelberg (2004)

Side Channel Cryptanalysis of Streebog

Gautham Sekar[(✉)]

Indian Statistical Institute, Chennai Centre,
SETS Campus, MGR Knowledge City,
Taramani, Chennai 600113, India
sgautham@isichennai.res.in

Abstract. Streebog is the cryptographic hash function standard of the Russian Federation. It comprises two hash functions corresponding to two digest sizes, 256 bits and 512 bits. This paper presents a side channel attack that uses processor flag information to speed up message recovery by a factor of 2. Success is nearly guaranteed if the flag is set; the probability is 0.668 otherwise.

Keywords: Cryptographic hash function · Streebog · Side channel cryptanalysis · Carry flag · Message recovery · HMAC

1 Introduction

A hash function F takes an arbitrarily long bit string m as input and outputs a fixed length bit string H (called *hash value* or *digest*). A cryptographic hash function is meant to satisfy certain security properties, the most important of which are the following.

- **(First) preimage resistance**: given H, it is computationally infeasible to find an m' such that $F(m') = H$.
- **Second preimage resistance**: given an m and $F(m)$, it is computationally infeasible to find an $m' \neq m$ such that $F(m') = F(m)$.
- **Collision resistance**: it is computationally infeasible to find an m and an $m' \neq m$ such that $F(m) = F(m')$.

The general model for cryptographic hash functions involves what is called a compression function. The function transforms a bit string of a fixed length into a shorter string of a fixed length. The arbitrarily long message is partitioned into blocks after a process called padding (described later in the context of Streebog). The blocks are then sequentially processed, with the compression function acting on every block until all the blocks are processed. The final output is the hash value. The general model is described in good detail in [9, Sect. 2.4.1].

Streebog is a set of two hash functions and a Russian cryptographic standard (GOST R 34.10–2012) [5]. It was developed by the Center for Information Protection and Special Communications of the Federal Security Service of the Russian Federation, with participation of the Open Joint-Stock Company

© Springer International Publishing Switzerland 2015
L. Chen and S. Matsuo (Eds.): SSR 2015, LNCS 9497, pp. 154–162, 2015.
DOI: 10.1007/978-3-319-27152-1_8

"Information Technologies and Communication Systems" (JSC "InfoTeCS") [5], following a demand for "a hash function to meet modern requirements for cryptographic strength" [5]. In 2012, Streebog replaced GOST R 34.11–94 as the national standard.

The hash functions comprising Streebog have 256 bits and 512 bits as their digest lengths. We shall call the hash functions "Streebog-256" and "Streebog-512", respectively. The compression function, common to both the versions, operates on 512-bit blocks in the Miyaguchi-Preneel mode, has 13 rounds, is based on a substitution-permutation network and uses a linear transformation.

In 2010–2011, open research workshops were organised by the Chinese Academy of Sciences to discuss cryptographic algorithms proposed for inclusion in the LTE/4G mobile standards. In a seemingly similar fashion, between 2013 and 2015, the Russian Technical Committee for Standardization "Cryptography and Security Mechanisms" (TC 26), with the participation of the Academy of Cryptography of the Russian Federation and support from the JSC InfoTeCS, held an open research competition for the analysis of Streebog. In this period, several results were reported, notably in [1–3,6,10].

In [1,10], the rebound attack is used to find (semi-free-start) collisions for reduced versions of the Streebog compression function; [2] presents integral distinguishers on up to 7 rounds of the compression function; [3] reports preimages for 6-round Streebog; and [6] describes second preimage attacks on the full Streebog-512. The drawback of the attacks in [6] is that they work well only with long messages. For instance, if the length of the message is at least 2^{188} bits, then 2^{342} compression function evaluations are required. The time complexity can be brought down to as low as $O(2^{266})$ provided that the message is at least 2^{268} bits in length. For shorter messages, of bit-length $\gamma < 2^{188}$ (but greater than 512 bits), the number of compression function evaluations is estimated at $(\log_2 \gamma - 9) \cdot 2^{522 - \log_2 \gamma}$. We present in this paper the first side channel attack on the full Streebog. We also discuss the implications of our attack on the security of Streebog-based keyed-hash message authentication code (HMAC).

Processors have registers that store information on operations performed by their ALUs. For example, in the Intel IA-32 architecture, the *status flags* of the EFLAGS register indicate the result of arithmetic instructions such as ADD and DIV (divide) [7]. One of these flags, known as the *carry flag*, is a single bit that indicates an overflow in unsigned integer arithmetic. For instance, when two unsigned integers are added, the carry flag is set (to 1) if a carry is generated by the addition at the most significant bit position (we shall call this an *end carry*) and the flag is cleared (i.e., 0) otherwise. This may be exploited by an attacker as in [8] where Kelsey *et al.* use carry flag information to attack the block cipher RC5. In our side channel attack too we use the state of the carry flag. Our attack recovers a message block in about 2^{511} time with 99.9 % success rate (number of successful recoveries per 100 messages uniformly distributed at random) if the carry flag is set and 66.8 % otherwise. The only other attack known on the full Streebog is due to Guo *et al.* [6].

Table 1. Notation and conventions

Symbol/notation	Meaning
$\lvert W \rvert$	length of W in bits
$\Gamma_i(W)$	ith 64-bit word of W; $i = 0$ denotes the least significant word
$W_{(i)}$	ith bit of W; $i = 0$ denotes the least significant bit
$\parallel$	concatenation
$\oplus$	exclusive OR
fg, where f and g are functions	$f \circ g$ (composition of f and g)
LSB	least significant bit
MSB	most significant bit

The paper is organised as follows. Section 2 describes Streebog and Sect. 3 details our meesage recovery attack. We propose countermeasures to our attack in Sect. 4 and conclude in Sect. 5.

2 Description of Streebog

Table 1 lists the notation and conventions followed in the rest of this paper.

Streebog is a simple design that uses only a few elementary arithmetic operators such as XOR and modular addition, and simple functions such as substitution, permutation and linear transformation. The hash function accepts any message M of length less than 2^{512} bits and returns a digest of length 256 bits or 512 bits. The round function or compression function has 13 iterations, the first twelve of which involve a substitution-permutation layer. If $512 \nmid \lvert M \rvert$, then padding prefixes M with a bit string $pad := \{0\}^{511 - (\lvert M \rvert \bmod 512)} \parallel 1$. The padded message is then partitioned into $(k + 1)$ 512-bit blocks $M_k, M_{k-1}, \ldots, M_0$; i.e., $pad \parallel M = M_k \parallel M_{k-1} \parallel \cdots \parallel M_0$. The compression function g that processes the message block M_i takes as additional inputs the chaining value H_i (of size 512 bits) and a length counter N_i, and outputs H_{i+1}. Algorithm 1 describes the working of Streebog. The IV in the algorithm is the initial value H_0 (Streebog-256 and Streebog-512 use different 512-bit IVs).

The substitution-permutation layer includes the following components.

- Substitution function S: The input, a 512-bit string, is first partitioned into bytes. Every byte is then substituted by a byte from a set π', which is a permutation of $\{0, 1, \ldots, 255\}$, and concatenated.
- Permutation function P: Partitions its 512-bit input into bytes, permutes the bytes (i.e., shuffles their positions) and concatenates them.
- Linear transformation L: This is also a 512-bit-to-512-bit mapping. If the input is W, then $L(W) = \ell(\Gamma_7(W)) \parallel \ell(\Gamma_6(W)) \parallel \cdots \parallel \ell(\Gamma_0(W))$, where ℓ is a 64-bit-to-64-bit linear transformation that outputs the right multiplication of its input with a constant matrix $\mathbf{A}$ over $GF(2)$.

Algorithm 1. The Streebog algorithm

Require: The message M, $|M| < 2^{512}$
Ensure: A 256-bit or a 512-bit digest
1: $M \to pad \| M \to M_k \| M_{k-1} \| \cdots \| M_0$;
2: $H_0 = IV$;
3: $N_0 = 0$;
4: **for** $i = 0$ to $(k-1)$ **do**
5:    $H_{i+1} = g(H_i, M_i, N_i)$;
6:    $N_{i+1} = N_i + 512 \mod 2^{512}$;
7:    $\Sigma \leftarrow \Sigma + M_i \mod 2^{512}$;
8: $H_{k+1} = g(H_k, M_k, N_k)$;
9: $N_{k+1} = N_k + \alpha \mod 2^{512}$, where $\alpha = 512 - |pad|$;
10: $\Sigma \leftarrow \Sigma + M_k \mod 2^{512}$;
11: $H_{k+2} = g(H_{k+1}, N_{k+1}, 0)$;
12: $H = g(H_{k+2}, \Sigma, 0)$;
13: Output H if Streebog-512, else output $H \gg 256$;

– The function $X[\cdot]$: If K and W are 512-bit strings, then $X[K](W) = K \oplus W$.

The compression function g is now given by:

$$g(H_i, M_i, N_i) = E(L(P(S(H_i \oplus N_i))), M_i) \oplus H_i \oplus M_i, \qquad (1)$$

where

$$E(L(P(S(H_i \oplus N_i))), M_i) = X[K_{13}]LPSX[K_{12}]LPSX[K_{11}] \ldots$$
$$LPSX[K_1](M_i), \qquad (2)$$

(recall from Table 1 that $fg = f \circ g$) and

$$K_0 = LPS(H_i \oplus N_i), \qquad (3)$$
$$K_{j+1} = LPS(K_j \oplus C_j), \quad \text{for } j = 0, 1, \ldots, 12, \text{ and constants } C_j. \qquad (4)$$

The subkeys $K_1, K_2, \ldots, k_{13}$ are the round keys; in deriving them, K_0 is used as an initial value.

3 The Message Recovery Attack

The functions S and P do not involve modular addition or multiplication. The function X is a simple XOR operation. The linear transformation ℓ works as follows. Denoting its 64-bit input by $\beta := \beta_{(63)} \| \beta_{(62)} \| \cdots \| \beta_{(0)}$, we have:

$$\ell(\beta) = \bigoplus_{i=0}^{63} \beta_{(63-i)} \odot A[i],$$

where the product $\odot$ is defined as follows:

$$\beta_{(63-i)} \odot A[i] = \begin{cases} \{0\}^{64} & \beta_{(63-i)} = 0\,; \\ A[i] & \beta_{(63-i)} = 1\,. \end{cases}$$

Hence, from (1)–(4), it immediately follows that Streebog compression does not involve any operation, such as addition modulo 2^{512}, that can alter the state of the carry flag. This means that only steps 6, 7, 9 and 10 of Algorithm 1 can potentially affect the carry flag.[1]

Now, the maximum length of M is $2^{512} - 1$. Given a message of this length, the number of blocks will be $\lceil (2^{512} - 1)/512 \rceil = 2^{503}$.[2] If $k + 1 < 2^{503}$ (to simply calculations, this can be considered a sure event as it happens with a probability that is very close to 1 if $|M|$ is uniformly distributed at random over $\{0, 1, \ldots, 2^{512} - 1\}$), then $N_k = 512k$, $512k < N_{k+1} \leq 512(k+1)$, and the carry flag will be unaffected by steps 6 and 9. This leaves us with steps 7 and 10. Now,

$$\Sigma = \left(\sum_{i=0}^{k-1} M_i \right) \mod 2^{512} + M_k \mod 2^{512}. \tag{5}$$

$$= T_{k-1} + M_k \mod 2^{512}, \text{ say.} \tag{6}$$

Let $C := [C_{(511)} C_{(510)} \cdots C_{(0)}]$ denote the vector of carries generated in (6) such that $C_{(0)}$ is the carry at the LSB position. When $k \geq 1$ (this can also be considered a sure event), we have the following attack.

Scenario 1: Suppose that the carry flag is set at the end of Algorithm 1. If $|pad| \geq 2 \Rightarrow M_{k(511)} = 0$, or $|pad| = 0$ and $M_{k(511)} = 0$, then $T_{k-1(511)} = C_{(511)} = 1$. If the attacker knows $M_0, M_1, \ldots, M_{k-2}$, and all but the MSB of M_{k-1}, then she can recover $M_{k-1(511)}$ from $T_{k-1(511)} = 1$ performing $k - 1 < 2^{503} - 2$ additions (recall (5) and (6)).

If $|pad| = 0$ and $M_{k(511)} = 1$, or $|pad| = 1 \Rightarrow M_{k(511)} = 1$, then there are three possibilities: (i) $T_{k-1(511)} = C_{(511)} = 1$, (ii) $T_{k-1(511)} = 0$ and $C_{(511)} = 1$, (iii) $T_{k-1(511)} = 1$ and $C_{(511)} = 0$. Assuming these cases to be equally likely,[3] the attacker can assume with $2/3$ probability that $T_{k-1(511)} = 1$, and recover $M_{k-1(511)}$.

Table 2 lists the above cases and their probabilities assuming that (i) $|M_k|$ is uniformly distributed at random over $\{0, 1, \ldots, 511\}$, and (ii) every message

---

[1] The for-loop of Algorithm 1 is implemented differently in [5]. To obtain M_0, the least significant 512-bit word of the padded message is extracted. The leftover message replaces the padded message and its 512 LSBs are extracted as M_1. This process is repeated until all the message blocks have been extracted. The carry flag is evidently unaffected by the process.

[2] Therefore, even if we go with the for-loop implementation (Algorithm 1), it will have no bearing on the carry flag.

[3] Since the distribution of $|M_k|$ is uniform, given the padding scheme employed, the distribution of M_k is not uniform. This makes it tedious to compute the distribution of the carry vector C. Hence the assumption.

block other than M_k is uniformly distributed at random over $\{0, 1, \ldots, 2^{512} - 1\}$. The attack methodology is as follows. The attacker, knowing $M_0, M_1, \ldots, M_{k-2}$ and M_k, makes a guess for the 511 LSBs of M_{k-1}, obtains a value for the MSB of M_{k-1} (assuming that $T_{k-1(511)} = 1$), hashes $M_k \| M_{k-1} \| \cdots \| M_0$, and compares the digest with the given hash value. If the values do not agree, the guess is incorrect and the attacker makes another guess. The process is repeated until the hash values agree. The sum σ of $M_0, M_1, \ldots, M_{k-2}$ modulo 2^{512} can be precomputed (cost is $k - 2$); $\sigma + M_{k-1} \mod 2^{512}$ can be performed at each guess and, in doing so, can be avoided while computing the digest (i.e., $\sigma + M_{k-1} \mod 2^{512}$ can be stored and reused). To minimise memory usage, the storage element can be rewritten at the next guess. The probability of success is the probability that $T_{k-1(511)} = 1$ holds true. From Table 2, this probability is simply $510/512 + 1/768 + 1/1024 + 1/1536 \approx 0.999$. The attack requires 2^{511} hash function evaluations plus a precomputation cost of $k - 2 < 2^{503} - 3$. Memory requirements are negligible.

Table 2. Computing $Pr(T_{k-1(511)} = 1)$ when the carry flag is 1; the probability q is given the condition on $|pad|$ and r is given the conditions on $|pad|$ and $M_{k(511)}$

$|pad|$	Pr. (p)	$M_{k(511)}$	Cond. pr. (q)	$T_{k-1(511)}$	Cond. pr. (r)	Overall pr. (pqr)
≥ 2	510/512	0	1	1	1	510/512
1	1/512	1	1	1	2/3	1/768
0	1/512	0	1/2	1	1	1/1024
0	1/512	1	1/2	1	2/3	1/1536

Scenario 2: Suppose that the carry flag is 0 at the end of Algorithm 1. If $|pad| \geq 2 \Rightarrow M_{k(511)} = 0$, or $|pad| = 0$ and $M_{k(511)} = 0$, then at least one of $T_{k-1(511)}$ and $C_{(511)}$ is 0. Knowing $M_0, M_1, \ldots, M_{k-2}$, and all but the MSB of M_{k-1}, the attacker can recover $M_{k-1(511)}$ assuming that $T_{k-1(511)} = 0$. The assumption is valid in two out of the three possible cases: (i) $T_{k-1(511)} = C_{(511)} = 0$, (ii) $T_{k-1(511)} = 0$ and $C_{(511)} = 1$, (iii) $T_{k-1(511)} = 1$ and $C_{(511)} = 0$. Assuming that these cases are equally likely, $Pr(T_{k-1(511)} = 0) = 2/3$.

When $|pad| = 0$ and $M_{k(511)} = 1$ or when $|pad| = 1 \Rightarrow M_{k(511)} = 1$, then $T_{k-1(511)} = C_{(511)} = 0$.

Table 3 lists the above cases and their probabilities under the assumption that (i) $|M_k|$ is uniformly distributed at random over $\{0, 1, \ldots, 511\}$, and (ii) every message block other than M_k is uniformly distributed at random over $\{0, 1, \ldots, 2^{512} - 1\}$. The attack methodology is identical to that described under Scenario 1, except that the attacker here assumes that $T_{k-1(511)} = 0$. The probability of success is the probability that $T_{k-1(511)} = 0$ holds true. From Table 3, this probability is $170/256 + 1/512 + 1/1536 + 1/1024 \approx 0.668$. The time complexity and memory requirements are the same as that in Scenario 1.

Note: The probability that $T_{k-1} = 0$ given that the carry flag is 0 and $M_{k(511)} = 0$ is at least 1/2 since $Pr(\text{case } (i) \text{ or case } (ii)) = Pr(T_{k-1}) = 1/2$ (given the assumption that the message blocks other than M_k are uniformly distributed). Even if the conditional probability is 1/2, the success probability will be $255/512 + 1/512 + 1/2048 + 1/1024 > 1/2$ (see Table 3). The success probability calculated from Table 2 changes negligibly when 2/3 is replaced by 1/2. □

Table 3. Computing $Pr(T_{k-1(511)} = 0)$ when the carry flag is 0; the probability q is given the condition on $|pad|$ and r is given the conditions on $|pad|$ and $M_{k(511)}$

$|pad|$	Pr. (p)	$M_{k(511)}$	Cond. pr. (q)	$T_{k-1(511)}$	Cond. pr. (r)	Overall pr. (pqr)
≥ 2	510/512	0	1	0	2/3	170/256
1	1/512	1	1	0	1	1/512
0	1/512	0	1/2	0	2/3	1/1536
0	1/512	1	1/2	0	1	1/1024

In summary, by simply guessing $T_{k-1(511)}$ to be equal to the carry flag, the attacker is able to recover M_{k-1} with 2^{511} hash function evaluations and $k-2$ precomputations. The number of precomputations can be negligible in comparison to 2^{511} and even the maximum number of precomputations ($2^{503} - 4$) is considerably smaller than 2^{511}. Moreover, each precomputation is only an addition of two 512-bit integers. Consequently, the precomputation cost can be ignored. The success probability is 0.668 if the carry flag is 0 and 0.999 otherwise. Arriving at a single value for the probability is involved given the difficulty in determining the distribution of the carry vector C. It is easy to see that the attack works for any $i \in \{0, 1, \ldots, k-2\}$ in place of $k-1$. In the ideal case, either 2^{512} hash function evaluations are required or the success probability is 1/2 for 2^{511} evaluations.[4] Since the compression functions of Streebog-256 and Streebog-512 are identical, our attack applies to both the hash functions.

3.1 Implications of Our Attack

Our attack may be particularly relevant to HMACs. Proposed by Bellare *et al.* [4] as a message integrity checking mechanism, a HMAC employs a hash function h in conjunction with a secret key K and generates a MAC value as follows:

$$HMAC(K, m) = h((K_0 \oplus opad)\|h((K_0 \oplus ipad)\|m)),$$

where m is the message, *opad* and *ipad* are public constants, and K_0 is the secret key or a function of K. The lengths of K_0, *opad* and *ipad* equal the length of a

---

[4] This does not apply to M_k unless $|pad| = 0$. Knowing $|pad|$ and $M_0, M_1, \ldots, M_{k-1}$, the attacker can recover M_k in $2^{512-|pad|}$ time. Our attack is not intended to recover M_k.

message block. Given the HMAC value and $h((K_0 \oplus ipad)\|m)$, in certain cases, our attack appears to speed up the recovery of K_0 by a factor of 2. This is being further investigated.

4   Countermeasures

A simple way to preclude our attack is to introduce a low-cost arithmetic operation, after step 12 of Algorithm 1, that permanently sets or clears the carry flag. However, the approach fails if the attack model assumes that the attacker can determine the status of the carry flag after step 12.[5]

A faster and safer countermeasure is to implement the checksum using XOR; i.e., replace the addition modulo 2^{512} in steps 7 and 10 of Algorithm 1 with XOR.

5   Conclusions

In this paper, we have presented the first known side channel attack on Streebog. The attack speeds up message recovery by a factor of 2 with a probability that lies in $[0.668, 0.999]$. The attack is conjectured to be applicable to Streebog-based HMAC. We have also proposed some countermeasures.

It may be possible to improve the attack by recovering bits other than the MSB, but calculating the success probabilities is involved and beyond the scope of this paper. We leave it as a problem for future work. Use of other processor flags such as the parity flag is also worth investigating.

References

1. AlTawy, R., Kircanski, A., Youssef, A.M.: Rebound attacks on Stribog. In: Lee, H.-S., Han, D.-G. (eds.) ICISC 2013. LNCS, vol. 8565, pp. 175–188. Springer, Heidelberg (2014)
2. AlTawy, R., Youssef, A.M.: Integral distinguishers for reduced-round Stribog. Inf. Process. Lett. **114**(8), 426–431 (2014)
3. AlTawy, R., Youssef, A.M.: Preimage attacks on reduced-round Stribog. In: Pointcheval, D., Vergnaud, D. (eds.) AFRICACRYPT 2014. LNCS, vol. 8469, pp. 109–125. Springer, Heidelberg (2014)
4. Bellare, M., Canetti, R., Krawczyk, H.: Keying hash functions for message authentication. In: Koblitz, N. (ed.) CRYPTO 1996. LNCS, vol. 1109, pp. 1–15. Springer, Heidelberg (1996)
5. Federal Agency on Technical Regulation and Metrology, "NATIONAL STANDARD OF THE RUSSIAN FEDERATION GOST R 34.11-2012" (English Version), 1 January 2013
6. Guo, J., Jean, J., Leurent, G., Peyrin, T., Wang, L.: The usage of counter revisited: second-preimage attack on new Russian standardized hash function. In: Joux, A., Youssef, A. (eds.) SAC 2014. LNCS, vol. 8781, pp. 195–211. Springer, Heidelberg (2014)

---

[5] A similar assumption is made in [8].

7. Intel, "IA-32 Intel Architecture Software Developer's Manual", vol. 1 (Basic Architecture), p. 426 (2003). http://flint.cs.yale.edu/cs422/doc/24547012.pdf
8. Kelsey, J., Schneier, B., Wagner, D., Hall, C.: Side channel cryptanalysis of product ciphers. J. Comput. Secur. **8**, 141–158 (2000)
9. Preneel, B.: Analysis and Design of Cryptographic Hash Functions, PhD thesis, Katholieke Universiteit Leuven (1993)
10. Wang, Z., Yu, H., Wang, X.: Cryptanalysis of GOST R hash function. Inf. Process. Lett. **114**(12), 655–662 (2014)

Privacy

Improving Air Interface User Privacy
in Mobile Telephony

Mohammed Shafiul Alam Khan$^{(\boxtimes)}$ and Chris J. Mitchell

Information Security Group, Royal Holloway, University of London,
Egham, Surrey TW20 0EX, UK
shafiulalam@gmail.com, me@chrismitchell.net

Abstract. Although the security properties of 3G and 4G mobile networks have significantly improved by comparison with 2G (GSM), significant shortcomings remain with respect to user privacy. A number of possible modifications to 2G, 3G and 4G protocols have been proposed designed to provide greater user privacy; however, they all require significant alterations to the existing deployed infrastructures, which are almost certainly impractical to achieve in practice. In this article we propose an approach which does not require any changes to the existing deployed network infrastructures, i.e. to the serving networks or the mobile devices, but offers improved user identity protection over the air interface. The proposed scheme makes use of multiple IMSIs for an individual USIM to offer a degree of pseudonymity for a user. The only changes required are to the operation of the authentication centre in the home network and to the USIM, both owned by a single entity in the mobile telephony system. The scheme could be deployed immediately since it is completely transparent to the existing mobile telephony infrastructure. We present two different approaches to the use and management of multiple IMSIs, and report on experiments to validate its deployability.

Keywords: Multiple IMSIs USIM · Pseudonymity · Mobile telephony · User privacy

1 Introduction

While the first generation (1G) mobile telephony systems did not provide any security features, security has been an integral part of such systems since the second generation (2G). For example, GSM, perhaps the best known 2G system, provides a range of security features, including authentication of the mobile user to the network, data confidentiality across the air interface, and a degree of user pseudonymity through the use of temporary identities. Third and fourth generation (3G and 4G) systems, such as UMTS/3GPP and Long-Term Evolution (LTE), have enhanced these security features, notably by providing mutual

M.S.A. Khan—The author is a Commonwealth Scholar, funded by the UK government.

L. Chen and S. Matsuo (Eds.): SSR 2015, LNCS 9497, pp. 165–184, 2015.
DOI: 10.1007/978-3-319-27152-1_9

authentication between network and phone, and integrity protection for signalling commands sent across the air interface. However, user privacy protection has remained largely unchanged, relying in all cases on the use of temporary identities [18,29], and it has long been known that the existing measures do not provide complete protection for the user identity [11,17]. The discussion below applies equally to 2G, 3G and 4G systems, although we use 3G terminology throughout.

The problem of user identity privacy in mobile networks is more than two decades old. Samfat et al. [25] first addressed the conflicting requirements of untraceability and disclosure of identity during authentication in a mobile network. The user privacy issue has been discussed extensively in the literature [6–9,20–23,31,32], and many modifications to existing protocols have been proposed to avoid the problem [6,8,9,20,22,32]. All these proposals involve making major modifications to the air interface protocol, which would require changes to the operation of all the serving networks as well as all the deployed phones. It seems likely that making the necessary major modifications to the operation of the air interface after deployment is essentially infeasible. Many of the proposed schemes also involve the use of public key cryptography [6,9,32], which has a high computational cost, although there do exist schemes which only use symmetric cryptography [8,20]. It would therefore be extremely valuable if a scheme offering greater user privacy could be devised which did not involve making significant changes to the existing mobile telecommunications infrastructures, and had minimal computational cost. This motivates the work described in this paper. In sum, the contributions of the paper are as follows.

- We propose a new approach to the use and management of multiple IMSIs in a USIM to enhance user pseudonymity in mobile telephony systems.
- We have implemented key parts of the scheme, verifying its feasibility.
- We provide a privacy and functional analysis of the scheme.

The remainder of the paper is structured as follows. In Sect. 2 key terminology for and features of mobile telephony systems are briefly reviewed. Threats to user privacy addressed in this paper are then summarised in Sect. 3. In Sect. 4 our threat model is presented. Section 5 outlines a novel approach to improving air interface user privacy using multiple IMSIs. Sections 6 and 7 provide descriptions of two proposed approaches to the use and management of multiple IMSIs in a USIM. Results from our experimental evaluation are presented in Sect. 8. An analysis of the proposed approaches is presented in Sect. 9. Section 10 provides a brief discussion of related work. Finally, conclusions are drawn and possible directions for future work are considered in Sect. 11.

2 Background

2.1 Mobile Telephony Systems

We start by providing a brief overview of key terminology for mobile systems. A complete mobile phone is referred to as a *user equipment (UE)*, where the term encapsulates not only the *mobile equipment (ME)*, i.e. the phone, but also the *user subscriber identity module (USIM)* within it [4], where the USIM takes the form of a cut-down smart card. The USIM embodies the relationship between the human user and the issuing *home network*, including the *International Mobile Subscriber Identity (IMSI)*, the telephone number of the UE, and other user (subscriber) data, together with a secret key shared with the issuing network which forms the basis for all the air interface security features. The USIM data storage capabilities are specified in Sect. 10.1 of 3GPP TS 121 111 [13]. Information held within the USIM is stored in files, which can be divided into the following categories: *application dedicated files (ADFs)*, *dedicated files (DFs)* and *elementary files (EFs)* [14]. Most of the subscriber information is stored in EFs, which are the files we focus on in this paper.

To attach to a mobile network, a UE connects via its radio interface to a radio tower. Several radio towers are controlled by a single *radio network controller (RNC)* which is connected to one *mobile switching center/visitor location register (MSC/VLR)*. The MSC/VLR is responsible for controlling call setup and routing. Each MSC/VLR is also connected to the carrier network's *home location register (HLR)* where corresponding subscriber details can be found. The HLR is associated with an *authentication center (AuC)* that stores cryptographic credentials required for communicating with the USIM. The RNC and the MSC/VLR are part of the *visiting/serving network* whereas the HLR and the AuC are the *home network* component (see Fig. 1(a)).

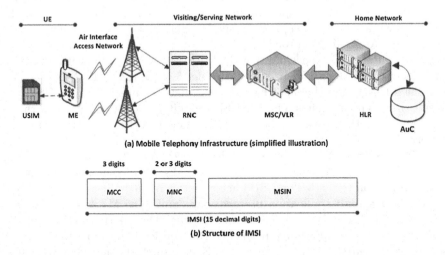

(a) Mobile Telephony Infrastructure (simplified illustration)

(b) Structure of IMSI

Fig. 1. Mobile telephony systems

To access mobile network services, a UE needs to complete mutual authentication as soon as it attaches to a network. Mutual authentication is performed using the *Authentication and Key Agreement (AKA)* protocol, described in detail below. If mutual authentication is successful, the MSC informs the HLR, which associates the UE's IMSI with the address of the MSC. The MSC also assigns a *temporary mobile subscriber identity (TMSI)* and sends the TMSI to the UE in encrypted form. The TMSI is unique to the location area in which the subscriber is currently located. Accordingly, whenever the subscriber visits a new location area, the MSC must update the TMSI value.

An IMSI is a 15-digit decimal number (see Fig. 1(b)). Of the 15 digits, the first three form the *mobile country code (MCC)*. The next two or three digits identify the network operator, and are known as the *mobile network code (MNC)*. The length of the MNC, i.e. whether it contains two or three digits, is a national matter. The remaining nine or ten digits, known as the *mobile subscriber identification number (MSIN)*, are administered by the relevant operator in accordance with the national policy [3, 28]. IMSIs therefore have geographical significance, and their use is typically managed by the network operator in blocks. The combination of the MCC and the MNC can be used to uniquely identify the home network of the IMSI. The MSIN is used by the operator to identify the subscriber for billing and other operational purposes. Each IMSI uniquely identifies the mobile user, as well as the user's home network and home country. The IMSI is stored in the USIM and is normally fixed. The elementary file EF_{IMSI} contains the value of the IMSI.

2.2   Proactive UICC

Proactive UICC is a service operating across the USIM-ME interface that provides a mechanism for a USIM to initiate an action to be taken by the ME. It forms part of the USIM application toolkit [15]. The 2G predecessor of the USIM, known as the SIM, supports a similar feature, known as *proactive SIM*, part of the SIM application toolkit.

ETSI TS 102 221 [14] specifies that the ME must communicate with the USIM using either the T=0 or T=1 protocol, specified in ISO/IEC 7816-3 [19]. In both cases the ME is always the *master* and thus initiates commands to the USIM; as a result there is no mechanism for the USIM to initiate communications with the ME. This limits the possibility of introducing new USIM features requiring the support of the ME, as the ME needs to know in advance what actions it should take. The proactive UICC service provides a mechanism that allows the USIM to indicate to the ME, using a response to an ME-issued command, that it has some information to send. The USIM achieves this by including a special status byte in the response application protocol data unit. The ME is then required to issue the *FETCH* command to find out what the information is [16]. To avoid cross-phase compatibility problems, this service is only permitted to be used between a proactive UICC and an ME that supports the proactive UICC feature. The fact that an ME supports proactive UICC is revealed when it sends a *TERMINAL PROFILE* command during UICC initialization.

The USIM can make a variety of requests using the proactive UICC service. Examples include: requesting the ME to display USIM-provided text, notifying the ME of changes to EF(s), and providing local information from the ME to the USIM [16]. The command of interest here is *REFRESH*. The *REFRESH* command requests the ME to carry out an initialisation procedure, or advises the ME that the contents of EF(s) have been changed. The command also makes it possible to restart the session by performing a reset [16].

2.3   The AKA Protocol

The AKA protocol is at the core of mobile telephony air interface security, and is regularly performed between the visited network and the UE. The involved parties are the home network (that issued the USIM), the serving network, and the UE. The AuC of the home network generates authentication vectors (used by the serving network in AKA) and sends them to the serving network. In the schemes proposed in Sects. 6 and 7, we use the $RAND$ value in an authentication vector in a novel way for management of multiple IMSIs.

The AKA protocol starts with the serving network sending a *user authentication request* to the UE. The UE checks the validity of this request (thereby authenticating the network), and then sends a *user authentication response*. The serving network checks this response to authenticate the UE. As a result, if successful, the UE and the network have authenticated each other, and at the same time they establish two shared secret keys.

In order to participate in the protocol, the UE, in fact the USIM installed inside the UE, must possess two values:

- a long term secret key K, known only to the USIM and to the USIM's home network, and
- a sequence number SQN, maintained by both the USIM and the home network.

The key K never leaves the USIM, and the values of K and SQN are protected by the USIM's physical security features.

The 48-bit sequence number SQN enables the UE to verify the 'freshness' of the user authentication request. More specifically, the request message contains two values: $RAND$ and $AUTN$, where $RAND$ is a 128-bit random number generated by the home network, and the 128-bit $AUTN$ consists of the concatenation of three values: $SQN \oplus AK$ (48 bits), AMF (16 bits), and MAC (64 bits). The MAC is a *message authentication code* (or *tag*) computed as a function of $RAND$, SQN, AMF, and the long term secret key K, using a MAC algorithm known as $f1$. The value AK is computed as a function of K and $RAND$, using a cipher mask generating function known as $f5$. The AK functions as a means of encrypting SQN; this is necessary since, if sent in cleartext, the SQN value would potentially compromise user identity confidentiality, given that the value of SQN is USIM-specific.

On receipt of these two values, the USIM uses the received $RAND$, along with its stored value of K, to regenerate the value of AK, which it can then use to recover SQN. It next uses its stored key K, together with the received values of $RAND$, SQN, and AMF, in function $f1$ to regenerate MAC (see Fig. 2); if the newly computed value agrees with the value received in $AUTN$ then the first stage of authentication has succeeded. The USIM next checks that SQN is

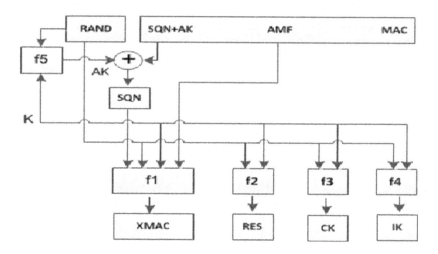

Fig. 2. Computation of AKA key values at the USIM

a 'new' value; if so it updates its stored SQN value and the network has been authenticated.

If authentication succeeds, the USIM computes another message authentication code, called RES, from K and $RAND$ using a distinct MAC function $f2$, and sends it to the network as part of the user authentication response. If this RES agrees with the value expected by the network then the UE is deemed authenticated.

3   User Privacy Threats

In a mobile telephony context, a user identity can be any of the mobile number or the IMSI of a USIM, or the international mobile station equipment identity (IMEI) of an ME. Of these various identities, the IMSI is used to identify the subscriber for authentication and access provision; limiting the degree to which its use compromises user privacy is the main focus of this paper. When a subscriber is roaming, i.e. accessing service from a network other than its home network, the IMSI is sent from the UE via the visited network to the home network. Since the IMSI is a permanent user identity, the air interface protocols are designed to minimise the number of circumstances in which it is sent across the air interface.

Clearly, providing user privacy requires that the permanent user identity cannot be intercepted when sent across the radio link. A level of identity confidentiality is provided by use of the TMSI instead of the IMSI. However, on certain occasions a UE needs to send its IMSI across the air interface in clear-text. One such case is when a UE is switched on and wishes to connect to a new network, and hence will not have an assigned TMSI [23]. Another case is where the serving network is unable to identify the IMSI from the TMSI [17].

An active adversary can intentionally simulate one of these scenarios to force a UE to transfer its IMSI in cleartext. Moreover, several further scenarios have been identified [6,7,11,31] in which user identity privacy is at significant risk. In this paper, we address this privacy threat.

4   Threat Model

The schemes we propose in this paper are designed to address real-life threats to user privacy in 3G networks. In particular we have already observed that there are circumstances in which an adversary can cause a UE to send its IMSI across the network in plaintext. This is the threat we aim to mitigate by reducing the impact of IMSI compromise. That is, although the possibility of IMSI compromise remains unchanged, we propose making the IMSI a short term identity and hence prevent the compromise of a long-term user identity. In doing so we must also ensure that two different IMSIs for the same UE are not linkable, at least via the network protocol. This issue is examined further in Sect. 9.1.

In designing our schemes we make the underlying assumption that AKA is sound, and provides mutually authenticated key establishment. We also implicitly assume that the USIM and the network have not been compromised by other means. Of course, if these assumptions are false, then very serious threats exist to both user privacy and security. Since we assume that AKA is secure, and no changes are made to its operation, we do not need to re-examine its security for the schemes discussed here. The main risk introduced by use of the multiple-IMSI schemes we propose is the possibility of loss of IMSI synchronisation between UE and home network, and this issue is addressed in Sect. 9.2.

It is important to note that two of the three schemes we describe rely on using the *RAND* sent as part of the AKA protocol as a means of signalling from the home network to the USIM. From our assumption regarding the security of AKA, we can assume that this provides an authenticated channel with replay detection. This is fundamental to the schemes presented in Sects. 6.2 and 7.

5   A Pseudonymity Approach

We consider here the possible use of multiple IMSIs for a single account to provide a form of pseudonymity on the air interface, even when it is necessary to send the IMSI in cleartext. The use of multiple IMSIs is described here using 3G terminology, but a precisely analogous approach would apply equally to both GSM and LTE systems. However, while all the techniques for IMSI distribution specified in Sects. 6 and 7 would also work for LTE, only the scheme described in Sect. 6.1 would work for GSM, since the other two schemes rely on UE authentication of the network which is not provided in GSM.

At present, a USIM holds one IMSI along with other subscription and network parameters. We propose that a USIM and the home network support the use of varying IMSIs for a single user account, in such a way that no modification is required to the operation of any intermediate entities, notably the visited (serving) network and the ME itself. This allows the provision of a more robust form of pseudonymity without making any changes to the air interface protocol. In this section we consider how a change of IMSI can be made.

The following issues need to be addressed to allow use of multiple IMSIs.

– *Transferring IMSIs.* Clearly, before a USIM switches to a new IMSI, it must be present in the USIM and in the database of the home network. Also, new IMSIs must always be chosen by the home network to avoid the same IMSI being assigned to two different USIMs. This requires a direct means of communication between the home network and the USIM (which must be transparent to the serving network and the ME, since our objective is to enable changing of an IMSI without making any changes to existing deployed equipments). In Sects. 6 and 7 we describe in detail two different strategies for transferring IMSIs from the home network to a USIM.

– *Initiating an IMSI change.* Clearly the IMSI needs to be changed in such a way that both the home network and the USIM know at all times which IMSI is being used, and the home network always knows the correspondence between the IMSI being used by the USIM and the user account. An IMSI change can be triggered either by the USIM or by the home network, as we describe below. However, use of a new IMSI is always implemented by the USIM, since it is the appearance of a mobile device in a network using a particular IMSI which causes a request to be sent by the serving network for authentication information for use in the AKA protocol. That is, when the ME sends an IMSI to the serving network, it is forwarded to the home network. Once the home network sees the 'new' IMSI it knows that an IMSI change has occurred and can act accordingly.

This requires that the home network knows that both the previously used IMSI and the 'new' IMSI belong to the same account. This will require some minor changes to the operation of the home network's account database, i.e. to allow more than one IMSI to point to a single account. However, this does not seem likely be a major problem in practice.

– *Triggering an IMSI change.* Whether the USIM or the home network is responsible for initiating a change of IMSI, logic needs to be implemented to cause such a change to take place. Regardless of whether the USIM or the home network makes the decision, logic needs to be in place in the USIM either to make the decision or to receive the instruction to make the change from the home network; for convenience we refer to this logic as an application, although this is not intended to constrain how it is implemented. The decision-making logic could take account of external factors, including, for example, the elapsed time or the number of AKA interactions since the last change; indeed, if the ME included an appropriate user-facing application, then it might also be possible to allow user-initiated changes. Of course, if the home network is responsible for triggering the change of IMSI, then it needs a means of communicating its decision to the USIM that is transparent to the existing infrastructure, including the serving network and the ME. This issue is addressed in Sects. 6 and 7.

– *Rate of change of IMSI.* The rate of change of IMSI will clearly be decided by the USIM-issuing network (which equips the USIM with the IMSI-changing application). We observe in this context that Sect. 4.2.2 of ETSI TS 131 102 [12] recommends that IMSI updates should not occur frequently. The rate of change of an IMSI could be determined by the customer contract with the issuing network; for example, a USIM which changes its IMSI frequently might cost more than a fixed-IMSI USIM (or one that only changes its IMSI occasionally), and could be marketed as a special 'high-privacy' service.

- *Implementing an IMSI change.* A mechanism will be required for the USIM to indicate to the ME that the IMSI has changed. We propose that this should be achieved by the following steps.

  1. As noted in Sect. 2.1, the IMSI is contained in the elementary file EF_{IMSI}. When the USIM wishes to change the IMSI, it first updates this file accordingly.

  2. At the first opportunity, the USIM uses the proactive UICC status byte to indicate to the ME that it wishes to issue a command.

  3. When the ME responds with a *FETCH* command, the USIM sends a *REFRESH* command to the ME.

  4. The *REFRESH* command causes the ME to read the EF_{IMSI} file, allowing it to discover the new IMSI. The next time that the ME needs to send its IMSI to the serving network, it will send the new value.

As noted above, using multiple IMSIs requires a direct and transparent means of communication between the home network and the USIM. The *Unstructured Supplementary Service Data (USSD)* protocol appears at first sight to be a possible channel for such communications. However, the protocol end points are the home network and the ME, rather than the USIM. As such, it could only be used for our purposes if the ME was aware of the multiple IMSI scheme, i.e. the ME would need to be modified — contradicting our design objectives. As a result we do not consider the use of USSD further here, although we note that it might be possible to deploy a smart phone application which could provide the necessary additional phone functionality, a possible avenue for future research.

6   Predefined Multiple IMSIs

Our first means of deploying multiple IMSIs involves a USIM being pre-equipped with a number of IMSIs. These IMSIs are all associated with a single account in the home network's account database. Initially, one of the IMSIs is stored in EF_{IMSI}. We propose below two ways of initiating an IMSI change in this case.

6.1   USIM-Initiated IMSI Change

This is the simpler of the two approaches. We suppose the USIM has an application that decides when to initiate an IMSI change. The new IMSI will clearly need to be selected from the predefined list. How the list is used is a matter for the issuing network. For example, the IMSIs could be used in cyclic order or at random (or, more probably, pseudo-randomly). The USIM changes the IMSI to the 'new' IMSI using the procedure described in Sect. 5.

6.2   Network-Initiated IMSI Change

In this case, the home network decides when to trigger an IMSI change. The home network will have a richer set of information to use to decide when to change IMSI than the USIM. For example, the home network could change the IMSI whenever the UE changes serving network or after a fixed number of calls.

As discussed in Sect. 5, when the home network decides to trigger an IMSI change, it must, by some means, send an instruction to the USIM. We propose to use the AKA protocol as the communications channel for this instruction. More specifically, we propose using the value $RAND$ of AKA to carry the signal. The IMSI change procedure operates as follows (see also Fig. 3).

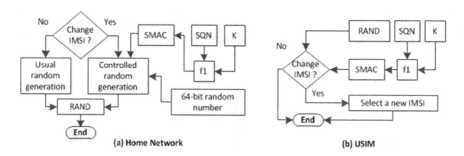

Fig. 3. IMSI change procedure for predefined multiple IMSIs

1. When the logic in the home network decides that an IMSI change is necessary, a flag is set for the appropriate user account in the AuC database of the home network.
2. Whenever the AuC generates authentication vectors for use in AKA, it checks this flag to see if an IMSI change signal is to be embedded in the $RAND$ value. If so it resets the flag and executes the following steps (as in Fig. 3(a)).
   (a) The AuC uses the MAC function $f1^1$ to generate a 64-bit MAC on the subscriber's current sequence number SQN using the subscriber's long term key K. We refer to this as the *sequence-MAC* or *SMAC*.
   (b) The AuC generates a 64-bit random number R using the same process as normally used to generate 128-bit $RAND$ values.
   (c) The AuC sets $RAND$ to be the concatenation of the R and $SMAC$.
   If an IMSI change signal is not required, the AuC generates $RAND$ in the normal way.
3. The AuC follows the standard steps to generate the authentication vector from $RAND$, and sends the vector (including $RAND$) to the serving network.

Whenever the USIM receives an authentication request, it follows the usual AKA steps. If the AKA procedure completes successfully, the USIM checks the $RAND$ in the following way (as shown in Fig. 3(b)).

---

[1] For cryptographic cleanliness it should be ensured that the data string input for this additional use of $f1$ can never be the same as the data string input to $f1$ for its other uses; alternatively, a slight variant of $f1$ could be employed here.

1. The USIM uses the received SQN and its stored key K to regenerate $SMAC$.
2. It compares the computed $SMAC$ with the appropriate part of $RAND$.
3. If they do not agree then the USIM terminates the checking process. However, if they agree then the USIM performs the next step.
4. The USIM selects a 'new' IMSI value from the stored list, and later changes the IMSI accordingly using the procedure described in Sect. 5.

We next consider how IMSI changes will work in practice. There are two cases to consider. If the home network is also the serving network then it could potentially force an instance of AKA to occur at will, i.e. making the IMSI change happen almost immediately. However, if the serving network is distinct from the home network, then the home network can only send new authentication vectors when requested by the serving network. Moreover, the serving network may delay before using the supplied authentication vector in AKA. That is, there may be a significant delay between the decision being made to change an IMSI and the signal being sent to the USIM. In either case the phone may be switched off or temporarily out of range of a base station, in which case there will inevitably be some delay. However, regardless of the length of the delay in the signal reaching the USIM (or even if it never reaches the USIM) there is no danger of loss of IMSI synchronisation between the USIM and the home network, since the home network will always keep the complete list of IMSIs allocated to the USIM.

We observe that there is always the chance that a randomly chosen $RAND$ will contain the 'correct' $SMAC$, leading to an unscheduled IMSI change by the USIM. However, the probability of this occurring is 2^{-64}, which is vanishingly small. In any case, the occurrence of such an event would not have an adverse impact, since the home network would always be aware of the link between the new IMSI and the particular USIM.

Finally, an active interceptor could introduce its own $RAND$ into the channel to try to force an IMSI change. However, given that K is not compromised and $f1$ has the properties required of a good MAC function, then no strategy better than generating a random $RAND$ will be available. Replays of old $RAND$ values will be detected and rejected as a normal part of AKA, at least for 3G and 4G networks, which enable the USIM to check the freshness of an authentication request. Finally, assuming the $SMAC$ value is indistinguishable from a random value, a standard assumption for MAC functions, then an eavesdropper will be unable to determine when an IMSI change is being requested.

7   Modifiable Multiple IMSIs

The second proposed means of deploying multiple IMSIs involves distributing new IMSI values from the home network to the USIM after its initial deployment, where the home network will choose each new IMSI from its pool of unused values. Such an approach clearly requires a means of communicating from the home network directly to the USIM, and, analogously to the scheme proposed in Sect. 6.2, we describe how the AKA protocol, and specifically the $RAND$ value, can be used for this purpose. Before describing the details of the IMSI transfer procedure, we describe some relatively minor changes which are required to the operation of the home network in order to support the scheme.

- The home network must maintain a pool of unused IMSIs, enabling the AuC to dynamically assign a new IMSI to an existing subscriber.
- For each subscriber account in its database, the home network must maintain an *IMSI-change* flag indicating whether an IMSI change is under way. The database must also hold up to two IMSIs for each subscriber; it will always hold the current IMSI (with status *allocated*) and, if the *IMSI-change* flag is set, it will also hold the new IMSI (with status *in transit*), where the possible status values for an IMSI are discussed below. If use of the new IMSI is observed then IMSI status changes are triggered (see below).
- The home network must manage the use of IMSIs so that no IMSI is assigned to more than one subscriber at any one time. This can be achieved by maintaining the status of each IMSI as one of *allocated*, *free*, or *in transit*. The set of IMSIs with status *free* corresponds to the pool of available IMSIs, as above. The status of an IMSI can be updated in the following ways.
  - When the home network selects an available IMSI from the pool to allocate to a USIM, the status is changed from *free* to *in transit*.
  - When the home network receives implicit acknowledgement (in the form of a request for authentication vectors for that IMSI from a network) of a successful IMSI change, the home network changes the status of the IMSI from *in transit* to *allocated*, and the status of the previously used IMSI for that subscriber from *allocated* to *free*. In addition, the current IMSI for the subscriber will be set equal to the new IMSI, the new IMSI will be set to null, and the *IMSI-change* flag will be reset.
  - A third case also needs to be considered, that is when an IMSI change instruction never reaches the USIM. If this case is not addressed then future IMSI changes for that USIM will be blocked. On the other hand, making a decision to abandon an IMSI change could be disastrous, i.e. if a USIM makes an IMSI change after the home network has terminated this change (and changed the status of the 'new' IMSI back to *free*), then the USIM could be rendered unusable. As a result we propose never to abandon an IMSI change, and instead to resend the new IMSI as many times as necessary until the change is accepted by the USIM. How this works should be clear from the description below.
- If the home network is required to do so by its regulatory environment, e.g. to support lawful interception, it can maintain a log of all the IMSIs assigned to a particular subscriber for however long is required. It is in any case likely to be necessary to retain this information for a period to enable processing of billing records received from visited networks.

The details of the IMSI transfer procedure are as follows (see also Fig. 4).

1. When the logic in the home network decides that an IMSI transfer is necessary for a particular subscriber, it must set the *IMSI-change* flag for that

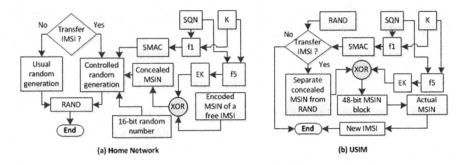

Fig. 4. IMSI change procedure for modifiable multiple IMSIs

subscriber. Observe that if an IMSI change is already under way then the flag will already be set; in this case the flag is left as it is.

2. Whenever the AuC needs to generate authentication vectors for use in AKA, it checks this flag to see if an IMSI transfer signal and a new IMSI are to be embedded in the $RAND$ value. If so it performs the following steps (as shown in Fig. 4(a)). Note that this means that, once an IMSI change has been initiated, the new IMSI will be embedded in all $RAND$ values until evidence of the successful changeover by the USIM has been observed.

   (a) The AuC uses the MAC function $f1$ to generate a 64-bit MAC on the subscriber's current sequence number SQN using the subscriber's long term cryptographic key K. The generated MAC is referred to as the $SMAC$.

   (b) The AuC generates a 48-bit encryption key EK using the key generation function $f5$. The function takes SQN as the data input and K as the key input. Note that observations regarding cryptographic cleanliness and the use here of functions $f1$ and $f5$, analogous to those given in Sect. 6.2 step 2(a), apply here.

   (c) If the new IMSI field in the home network database entry for this subscriber is non-null then a new IMSI has already been assigned, and it is not necessary to choose another new value. Otherwise a new IMSI is selected from the pool of unused IMSIs; the status of this IMSI is changed from *free* to *in transit* and the new IMSI field in the database is given the chosen value. We assume that the MCC and MNC of the IMSI are known to the USIM (since they are fixed for this network operator) and hence only the 9- or 10-digit MSIN needs to be sent embedded in $RAND$. The MSIN is encoded as a 36- or 40-bit value using binary coded decimal, the 'standard' way of encoding IMSIs, and the result is padded to 48 bits by an agreed padding scheme.

   (d) The 48-bit MSIN block is XORed with the encryption key EK, and we refer to the result as the concealed MSIN.

   (e) The AuC generates a 16-bit random number R using the same process as normally used to generate 128-bit $RAND$ values.

(f) The AuC sets $RAND$ to be the concatenation of the concealed MSIN, R and $SMAC$.

If an IMSI transfer is not required, the AuC generates $RAND$ in the normal way.
3. The AuC follows the standard steps to generate the authentication vector from the $RAND$ value, and sends it (including $RAND$) to the serving network.

On receipt of an authentication request, the USIM proceeds using the standard AKA procedure. After successful completion of the AKA protocol, the USIM checks whether the challenge value contains an embedded IMSI in the following way (as shown in Fig. 4(b)).

1. The USIM uses the received SQN and its stored long term key K to regenerate $SMAC$.
2. It compares the computed $SMAC$ with the appropriate part of $RAND$.
3. If they do not agree then the USIM terminates the checking process. However, if they agree then the USIM performs the following steps.
   (a) The USIM retrieves the concealed MSIN from $RAND$.
   (b) The USIM regenerates the encryption key EK using $f5$ with the value of SQN retrieved during the AKA processing and its long-term stored key K as inputs.
   (c) The EK is XORed with the concealed MSIN to recover the cleartext encoded MSIN.
   (d) The USIM generates the new IMSI by prefixing the decoded MSIN with the MCC and MNC.
   (e) The USIM checks whether the new IMSI is the same as the value it is using already; this is essential since it may receive the change instruction more than once. If they are the same it does nothing. If they are different it keeps a record of the new IMSI and later updates its IMSI using the procedure described in Sect. 5.

To reduce signalling costs, it appears to be standard practice for the AuC to generate a small set of authentication vectors for provision to a serving network. If the procedure specified above is followed to generate this set of vectors, and an IMSI change is scheduled for the subscriber, then all the $RAND$ values in the set will contain an embedded concealed MSIN. Whilst this will cause minimal additional overhead for the USIM, since $RAND$ values are always checked for an embedded $SMAC$ value, it will have the benefit of maximising the chance that the IMSI change will be performed by the USIM.

As discussed in Sect. 6.2, there may be a significant delay in the IMSI change signal embedded in $RAND$ reaching the USIM. However, this will not affect IMSI synchronisation between the home network and the USIM since the home network will not update the current IMSI entry in the subscriber database until it receives a request for authentication vectors from a visited network using this new IMSI. As discussed above, once a new IMSI has been assigned to a subscriber (with the *in transit* status), every $RAND$ generated for that USIM will contain the embedded IMSI value until the success of the change has been observed.

Finally, as in Sect. 6.2, there is the chance that a randomly chosen *RAND* could contain a 'correct' *SMAC*, triggering an unauthorised IMSI change. However, the probability of such an event is vanishingly small, and certainly orders of magnitude smaller than the probability of a USIM failure. We therefore do not consider it further here.

8   Experimental Validation

Testing the schemes is challenging due to the unavailability of a test UMTS network. However, the availability of the *SIMtrace* [1] hardware and the software implementations of the *Osmocom project* [30] support testing of the necessary modifications to the USIM. We used SIMtrace to trace USIM-ME communications, together with *SysmoUSIM-SJS1* [2], a standards-compliant test UMTS UICC card.

To validate the proposed modification to the USIM, we first ran an experiment to update the IMSI value in the test USIM. Using a standard contact smart card reader, smart card scripting tool, and a custom script we were able to modify the IMSI value. The process is similar to updating any file in the USIM file structure given the appropriate access grant. We subsequently developed a *SIM toolkit applet* using the Java card framework and the packages included with the 3GPP technical specification covering USIM API for Java card [5]. The SIM toolkit applet was designed to use the *REFRESH* proactive command to arrange for the ME to fetch the new IMSI. We chose to use the *REFRESH* command, as the command is understood by all MEs which support proactive commands. We tested the full range of modes of the *REFRESH* command [15], i.e. USIM initialization and file change notification, file change notification, and UICC reset.

We loaded the applet in the USIM to carry out further tests. We connected the USIM and the ME to the SIMtrace device, which was connected to a laptop to record the APDU commands exchanged between the USIM and the ME. We observed that, when a *REFRESH* is executed in the USIM initialization and file change notification mode, a series of read commands are issued by the ME. Although the record of commands exchanged showed that the IMSI file is read, because of our lack of a test UMTS environment we were unable to confirm that the read operation actually updated the IMSI value stored in the ME. When the mode is changed to UICC reset, the ME simply restarted its session, as expected. These observations confirmed that, if the *REFRESH* command is used in the UICC reset mode, the ME is made aware of the new IMSI. As a result, all future authentication procedures performed by the UE will use the new IMSI, which will have the effect of notifying the home network of use of the new IMSI.

During the experiments, we tested a range of standard MEs, all of which support the proactive command. As mentioned earlier, due to the unavailability of a test UMTS network we were unable to implement the modified home network. However the changes required at the home network are purely minor software changes to the operation of the AuC database, which should not require significant additional computing resource.

9   Analysis

The above proposals raise privacy and availability issues, which we now discuss.

9.1   User Privacy

The use of multiple IMSIs does not provide a complete solution to user identity confidentiality. While in use, the IMSI still functions as a pseudonym, potentially enabling the interactions of a single phone to be tracked for a period; of course, this is always true for any mobile network when a subscriber resides in a single location area, even where only a privacy-preserving TMSI is used. Of course, the more frequently IMSIs are changed the less the impact of possible tracking, but frequent IMSI changes have an overhead in terms of database management. The use of a predefined set of IMSIs further restricts the degree of user identity confidentiality protection. In this case, over a period of time it might be possible for an eavesdropper to link at least some of the fixed IMSIs.

The design of the schemes ensures an eavesdropper is unable to infer any confidential information from the value of $RAND$. As discussed in Sects. 6.2 and 7, in schemes where the $RAND$ is used to signal to the USIM, the $RAND$ is constructed so that it is indistinguishable from a truly random value; this is based on the assumption that a MAC generated using $f1$ and a data string encrypted using the output from $f5$ are indistinguishable from random data. Moreover, in the scheme described in Sect. 7 where the IMSI is sent embedded in $RAND$, the IMSI (actually the MSIN) is encrypted to prevent an eavesdropper observing it.

Overall the IMSI-changing proposal can be seen as allowing a trade-off between user privacy and the cost of implementing frequent IMSI changes.

9.2   IMSI Synchronisation

If, in the modifiable multiple IMSIs case, an active adversary is able to persuade the USIM to change its IMSI to an unauthorised value, then the USIM (and the UE) will cease to be able to access the network. It is therefore essential that robust cryptographic (and other) means are used to guarantee the correctness and timeliness of the new IMSI.

In the predefined IMSIs schemes described in Sect. 6, loss of synchronisation cannot arise, as even if the USIM is persuaded to make an unauthorised change the new IMSI will be known to the home network. In any event, as argued in Sect. 6.2, the probability of such an event is vanishingly small.

Loss of synchronisation appears to be a more significant threat in the case where new IMSIs are sent embedded in the $RAND$ value, as in the scheme described in Sect. 7. However, as discussed there, for similar arguments to those given in Sect. 6.2, the probability of a random $RAND$ giving a correct $SMAC$ is negligible. Also, malicious changes to a valid $RAND$, e.g. involving changing the encrypted MSIN whilst leaving the $SMAC$ unchanged, will be detected by the AKA network authentication process.

10   Related Work

To the best of our knowledge, no user privacy enhancing scheme for mobile telephony has previously been proposed that does not require changes to the existing network infrastructure, i.e. the serving network and/or the mobile equipment. While other authors observe that significant changes to widely deployed infrastructure are unlikely to be feasible [8,22], realistic and practical proposals have not been made. Choudhury et al. [8] proposed a scheme to improve user identity confidentiality in the LTE network. Their scheme involves significant changes to the air interface protocol. They propose the use of a frequently changing dynamic mobile subscriber identity (DMSI) instead of the IMSI across the air interface. The DMSI is constructed by concatenation of the MCC, MNC, a random number chosen by the mobile operator, and a 128-bit encrypted version of the chosen random number. As a result, the structure of the DMSI differs significantly from the structure of the IMSI. The DMSI is updated on every run of the authentication. The DMSIs are managed by the home network and the USIM. However, the use of the DMSI imposes changes in the protocol messages, mobile equipment, and the serving network. Table 1 summarises the change impact of this scheme, where a 'Yes' indicates that changes are required to the indicated part of the system.

Køien [22] has recently proposed a privacy enhanced mutual authentication scheme for LTE. Although the author claims to use existing signalling mechanisms, the author introduces identity-based encryption to encrypt the IMSI when sent across the air interface. The scheme does not introduce any new system entities, but does make significant changes in system operation. The scheme suggests that the ME generates a public key ID computed from the home network ID, serving network ID, and an agreed expiry value, where the home network computes the corresponding private key. The serving network broadcasts the key components of the public ID to its subscribers. Moreover, a mobile device needs to generate a random number to be used in encryption. All these functional components require major modification to the mobile device. The home network shares the private key with the serving network as the key is used by the serving network to decrypt the encrypted IMSI. A serving network has to maintain individual private keys for each home network. In the event of updating the public key ID for any serving network, all partner home networks need to recompute the private key and ensure a secure exchange of keys. This will introduce new signalling messages into the core network. As a result it appears that the scheme is not easily deployable as it is not transparent to the mobile equipment or the serving network. Thus the privacy-protecting enhancements to the system come at the cost of significant modifications to all the major elements of the system; it is therefore more appropriate to consider it as a proposal for a future network. Table 1 summarises the change impact of the Køien scheme.

Sung et al. [26] have proposed a scheme to provide location privacy which uses multiple IMSIs for a single (U)SIM in some ways similar to our proposal. However, it involves an additional party in its operation, needs support by the ME, and requires wireless data connectivity for sending and receiving calls. The

Table 1. Comparison of change impact of the proposed scheme with other proposals

Domain/Proposal	Choudhury et al. [8]	Køien [22]	Our proposal
USIM	Yes	Yes	Yes
Mobile equipment	Yes	Yes	No
Signaling message	Yes	Yes	No
Serving network	Yes	Yes	No
Home network	Yes	Yes	Yes

threat model is also very different, in that the home network is a potential adversary. The scheme employs phones without a local SIM; instead the phone's software retrieves a virtual SIM offered by an Internet-accessible third party, which is used for a limited period and paid for using an anonymous Internet payment system where messages are sent via Tor.

Tagg and Campbell [27] describe a scheme to use multiple IMSIs for multiple networks with a single USIM. Their scheme involves the use of an update server to provide a suitable IMSI as and when it is required. The objective is to avoid roaming charges by dynamically switching network provider. Marsden and Marshall [24] propose a similar approach. The focus of their work is thus very different to the schemes described above; they also do not provide a means of transferring new IMSIs transparently to a USIM.

Dupré [10] presents a process to control a subscriber identity module (SIM) for mobile phone systems. Generic guidance regarding the transmission of control information from the network to the SIM is provided. The schemes described in this paper extend Dupré's idea in a more concrete way.

11   Conclusions

In this paper we propose two general approaches to using multiple IMSIs for a mobile telephony subscriber. The goal of these proposals is to improve user privacy by reducing the impact of IMSI disclosure on the air interface. The approaches do not require any changes to the existing deployed network infrastructures, i.e. to the serving network, air interface protocols or mobile devices. The overhead introduced is modest and should be feasible to manage in real-world networks. One major advantage is that the proposed schemes could be deployed immediately since they are completely transparent to the existing mobile telephony infrastructure.

The proposed schemes provide a form of pseudonymity on the air interface, even when it is necessary to send the IMSI in cleartext. The schemes reduce the impact of user privacy threats arising from IMSI capture.

Future work could include setting reliable rules for triggering an IMSI change. A formal security analysis of the complete proposal could provide additional confidence in its robustness.

References

1. Osmocom SIMtrace. http://bb.http://osmocom.org/trac/wiki/SIMtrace. Accessed 20 May 2015
2. SysmoUSIM-SJS1 SIM + USIM. http://www.sysmocom.de/products/sysmousim-sjs1-sim-usim. Accessed 20 May 2015
3. 3rd Generation Partnership Project: 3GPP TS 23.003 Version 3.14.0 (2003–12): 3rd Generation Partnership Project; Technical Specification Group Core Network; (Numbering, addressing and identification), December 2003
4. 3rd Generation Partnership Project: 3GPP TR 21.905 Version 10.3.0; 3rd Generation Partnership Project; Technical Specification Group Services and System Aspects; Vocabulary for 3GPP Specifications (2011)
5. 3rd Generation Partnership Project: 3GPP TS 31.130 Version 10.0.0; Technical Specification Group Core Network and Terminals; (U)SIM Application Programming Interface (API); (U)SIM API for Java Card (Release 10) (2011)
6. Arapinis, M., Mancini, L., Ritter, E., Ryan, M., Golde, N., Redon, K., Borgaonkar, R.: New privacy issues in mobile telephony: fix and verification. In: Yu, T., Danezis, G., Gligor, V.D. (eds.) ACM Conference on Computer and Communications Security, CCS '12, Raleigh, NC, USA, 16–18 October 2012, pp. 205–216. ACM (2012)
7. Arapinis, M., Mancini, L.I., Ritter, E., Ryan, M.: Privacy through pseudonymity in mobile telephony systems. In: 21st Annual Network and Distributed System Security Symposium, NDSS 2014, San Diego, California, USA, 23–26 February 2014 (2014). http://www.internetsociety.org/doc/privacy-through-pseudonymity-mobile-telephony-systems
8. Choudhury, H., Roychoudhury, B., Saikia, D.K.: Enhancing user identity privacy in LTE. In: IEEE 11th International Conference on Trust, Security and Privacy in Computing and Communications (TrustCom), 2012, pp. 949–957. IEEE (2012)
9. Deng, Y., Fu, H., Xie, X., Zhou, J., Zhang, Y., Shi, J.: A novel 3GPP/SAE authentication and key agreement protocol. In: IEEE International Conference on Network Infrastructure and Digital Content, 2009 (IC-NIDC 2009), pp. 557–561. IEEE (2009)
10. Dupré, M.: Process to control a Subscriber Identity Module (SIM) in mobile phone system, US Patent 6,690,930 (2004)
11. European Telecommunications Standards Institute (ETSI): ETSI TS 121 133 Version 4.1.0 (2001–12): Universal Mobile Telecommunications System (UMTS); 3G Security; Security threats and requirements, December 2001
12. European Telecommunications Standards Institute (ETSI): ETSI TS 131.102 Version 4.15.0 Release 4; Universal Mobile Telecommunications System (UMTS); Characteristics of the USIM application (2005)
13. European Telecommunications Standards Institute (ETSI): ETSI TS 121 111 Version 8.0.1 (2008–01): Universal Mobile Telecommunications System (UMTS), USIM and IC card requirements, January 2008
14. European Telecommunications Standards Institute (ETSI): ETSI TS 102 221 Version 8.2.0; Smart Cards; UICC–Terminal Interface; Physical and logical characteristics (2009)
15. European Telecommunications Standards Institute (ETSI): ETSI TS 131 111 Version 7.15.0: Digital cellular telecommunications system (Phase 2+); Universal Mobile Telecommunications System (UMTS); LTE; Universal Subscriber Identity Module (USIM) Application Toolkit (USAT) (2010)

16. European Telecommunications Standards Institute (ETSI): ETSI TS 102 223 Version 11.1.0; Smart Cards; Card Application Toolkit (CAT) (2012)
17. European Telecommunications Standards Institute (ETSI): ETSI TS 133 102 Version 11.5.1 (2013–07): Digital cellular telecommunications system (Phase 2+); Universal Mobile Telecommunications System (UMTS); 3G Security; Security architecture, July 2013
18. Forsberg, D., Horn, G., Moeller, W.D., Niemi, V.: LTE Security. Wiley, Chichester (2010)
19. International Organization for Standardization: ISO/IEC 7816–3; Identification cards – Integrated circuit cards; Part 3: Cards with contacts – Electrical interface and transmission protocols, November 2006
20. Juang, W.S., Wu, J.L.: Efficient 3GPP authentication and key agreement with robust user privacy protection. In: Wireless Communications and Networking Conference, WCNC 2007, pp. 2720–2725. IEEE (2007)
21. Khan, M.S.A., Mitchell, C.J.: Another look at privacy threats in 3G mobile telephony. In: Susilo, W., Mu, Y. (eds.) ACISP 2014. LNCS, vol. 8544, pp. 386–396. Springer, Heidelberg (2014)
22. Køien, G.M.: Privacy enhanced mutual authentication in LTE. In: 2013 IEEE 9th International Conference on Wireless and Mobile Computing, Networking and Communications (WiMob), pp. 614–621. IEEE (2013)
23. Kóien, G.M., Oleshchuk, V.A.: Aspects of Personal Privacy in Communications: Problems, Technology and Solutions. River Publishers, Denmark (2013)
24. Marsden, I., Marshall, P.: Multi IMSI system and method, US Patent App. 13/966,350, 20 February 2014. http://www.google.com/patents/US20140051423
25. Samfat, D., Molva, R., Asokan, N.: Untraceability in mobile networks. In: Proceedings of the 1st Annual International Conference on Mobile Computing and Networking, MobiCom '95, pp. 26–36. ACM, New York, NY, USA (1995). http://doi.acm.org/10.1145/215530.215548
26. Sung, K., Levine, B.N., Liberatore, M.: Location privacy without carrier cooperation. In: IEEE Workshop on Mobile Security Technologies, MOST 2014, San Jose, CA, USA, 17 May 2014
27. Tagg, J., Campbell, A.: Identity management for mobile devices, US Patent App. 13/151,942, 6 December 2012. http://www.google.com/patents/US20120309374
28. Telecommunication Standardization Sector of ITU: ITU-T E.212: International operation Maritime mobile service and public land mobile service (The international identification plan for public networks and subscriptions), May 2008
29. Valtteri, N., Nyberg, K.: UMTS Security. Willey, Chichester (2003)
30. Various Contributors: Osmocom Project. http://osmocom.org. Accessed 20 May 2015
31. Vintila, C.E., Patriciu, V.V., Bica, I.: Security analysis of LTE access network. In: The 10th International Conference on Networks ICN 2011, pp. 29–34 (2011)
32. Xiehua, L., Yongjun, W.: Security enhanced authentication and key agreement protocol for LTE/SAE network. In: 7th International Conference on Wireless Communications, Networking and Mobile Computing (WiCOM), pp. 1–4. IEEE (2011)

Generating Unlinkable IPv6 Addresses

Mwawi Nyirenda Kayuni, Mohammed Shafiul Alam Khan, Wanpeng Li,
Chris J. Mitchell$^{(\boxtimes)}$, and Po-Wah Yau

Information Security Group, Royal Holloway, University of London, Egham, UK
{Mwawi.NyirendaKayuni.2011,Wanpeng.Li.2013}@live.rhul.ac.uk,
shafiulalam@gmail.com, {C.Mitchell,P.Yau}@rhul.ac.uk

Abstract. A number of approaches to the automatic generation of IPv6
addresses have been proposed with the goal of preserving the privacy of
IPv6 hosts. However, existing schemes for address autoconfiguration do
not adequately consider the full context in which they might be implemented, in particular the impact of low quality random number generation. This can have a fundamental impact on the privacy property of
unlinkability, one of the design goals of a number of IPv6 address autoconfiguration schemes. In this paper, the potential shortcomings of previously proposed approaches to address autoconfiguration are analysed
in detail, focussing on what happens when the assumption of strong randomness does not hold. Practical improvements are introduced, designed
to address the identified issues by making the random generation requirements more explicit, and by incorporating measures into the schemes
designed to ensure adequate randomness is used.

1 Introduction

The move from IPv4 to IPv6 brings with it a range of challenging security and
privacy issues. Of course, the vastly larger address space of IPv6 is a huge advantage, allowing the use of globally unique identifiers for all Internet-connected
devices. However, this very advantage brings with it possible user privacy problems [1].

That is, if each device has a long-term and globally unique identifier, then use
of this identifier enables devices to be tracked. As stated in RFC 4941 [1], if part
of the IPv6 address remains fixed then privacy problems arise for mobile devices,
since the fixed part of the address can be used to track use of a particular device
across networks.

This privacy threat has become increasingly serious with the proliferation of
network-enabled personal devices, including phones and tablets. That is, tracking
of IP addresses on such devices could enable the movements and activities of a
single user to be recorded. This threat will become even more apparent as an
increasing variety of devices become IP-enabled, particularly with the emergence
of the Internet of Things (IoT).

As a result, a method is needed to enable devices to generate new unique IPv6
addresses on a regular basis with the property that pairs of addresses generated

L. Chen and S. Matsuo (Eds.): SSR 2015, LNCS 9497, pp. 185–199, 2015.
DOI: 10.1007/978-3-319-27152-1_10

by the same device are *unlinkable*. That is, given two IPv6 addresses generated by the method, it should not be possible for a third party to learn anything from the addresses themselves regarding whether or not they belong to the same or distinct devices.

Of course, there is already a substantial body of work addressing this problem, including RFC 4941 [1], discussed further in Sect. 2 below. However, as we discuss in this paper, there are serious practical problems with all the existing approaches. In essence, the existing solutions all depend on the availability of either high quality random bit streams or long-term state (or both) within the device generating its own IPv6 addresses. Meeting these requirements could be very challenging in certain classes of device, particularly those small portable platforms for which the privacy threat may well be greatest. As a result, new solutions are required which can work on a wide variety of platforms while still providing acceptable levels of address privacy.

In this paper, as well as pointing out the scale and scope of the 'randomness' problems with the prior art, we make a detailed proposal for the use of randomness in the existing address generation schemes. The solutions proposed are designed to be readily implemented on current platforms, and should enable significant improvements in the level of privacy offered by the various approaches to dynamic IP address generation.

The remainder of this paper is structured as follows. Section 2 describes previous work on IPv6 autoconfiguration, focussing on proposals for addressing the privacy issue. The limitations of previous approaches are considered in Sect. 3, which leads to Sect. 4 in which new approaches to IPv6 address autoconfiguration are explored. Section 5 summarises the main findings and recommendations, and notes possible directions for future work.

2   Background

2.1   Stateless Address Autoconfiguration (SLAAC)

An IPv6 address is a 128-bit identifier [2] for a specific network interface within a device (referred to as a *host* throughout). That is, a network interface cannot communicate in an IPv6 network unless it has one or more suitably configured IPv6 addresses. Hosts may need to automatically generate (*autoconfigure*) their own IPv6 addresses. This need is addressed by the *IPv6 Stateless Address Autoconfiguration* protocol, or *SLAAC*, specified in RFC 4862 [3]. SLAAC involves a host first generating a *global address* via stateless address autoconfiguration, and then using the *Duplicate Address Detection (DAD)* procedure to verify the local uniqueness of the global address. The mechanism 'allows a host to generate its own [global] addresses using a combination of locally available information and information advertised by routers'.

SLAAC operates in the following general way. A router advertises a 64-bit prefix that identifies the subnet to which the host is attached. The host then generates a 64-bit *interface identifier*, uniquely identifying the host on the subnet. The 128-bit IPv6 address is simply the concatenation of these two values.

The source of the interface identifier, which must be in *modified EUI-64 format* [2], will depend on the underlying link-layer protocol. In many cases, an interface's identifier will be derived directly from that interface's link-layer address [2]. For example, it may be derived from an IEEE 802 48-bit MAC layer address.

This approach is appropriate 'when a site is not particularly concerned with the exact addresses hosts use, so long as they are unique and properly routable'. SLAAC is an alternative to the *Dynamic Host Configuration Protocol for IPv6 (DHCPv6)* [4], appropriate when a site requires tighter control over exact address assignments.

2.2   Privacy Extensions to SLAAC

We first observe that, although the first 64 bits of a SLAAC-generated address will change when a host switches subnets, the last 64 bits will stay constant, since they are generated from a fixed interface identifier. This issue has motivated the development of RFC 4941 [1]. As stated in Sect. 2.3 of this RFC, problems arise if the interface identifier contained within the IPv6 address remains fixed and, in such a case, 'the interface identifier can be used to track the movement and usage of a particular machine' (this threat is, of course, particularly relevant to mobile devices). More detailed discussions of the privacy issues arising from the use of SLAAC are provided in Sect. 1 of Gont [5] and in Cooper, Gont and Thaler [6].

The goal of RFC 4941 is to describe methods for a host to automatically generate IPv6 addresses that change over time and which cannot be linked to each other, thereby giving a level of pseudonymity to a host. The focus of RFC 4941 is on the case where the interface identifier used in SLAAC is generated from a fixed IEEE MAC layer address. RFC 4941 seeks to propose new methods for address generation that minimise the changes to SLAAC, and that enable a sequence of apparently random addresses to be generated. RFC 4941 addresses are expected to be used for a 'short period of time (hours to days)' ([1], Sect. 3).

The main change to SLAAC is to replace the fixed interface identifier with a randomised value. Two approaches are described for generating such a randomised identifier.

1. Method 1 (*When stable storage is present*). As the title suggests, this approach assumes that the software responsible for generating the randomised interface identifiers has access to a means of storing changeable data long-term. More specifically, the scheme requires the storage of a 64-bit *history value H*. The scheme also requires the software to have access to a 64-bit random value which is used to initialise the history value. It is further assumed that the 64-bit fixed interface identifier I is available, e.g. as derived from the MAC layer address.

   Whenever a new randomised interface identifier is required, the following steps are performed.

(a) Compute $V = h(H||I)$, where h is the MD5 cryptographic hash function [7], and here, as throughout, $||$ denotes concatenation of bit-strings. Hence V is a 128-bit value.

(b) Set the new history value H to be the rightmost 64 bits of V, and store this value.

(c) Let J be the leftmost 64 bits of V, after setting the 7th bit (counting from the left) to zero to indicate an address of local significance only.

(d) Compare J against a list of reserved interface identifiers and those already assigned to an address on the host (on a different network interface). If J matches a forbidden address then restart the process; otherwise use J as the randomised interface identifier.

The use of MD5 is not mandatory; that is, h could be instantiated as any other suitable cryptographic hash function with an output length of at least 128 bits (longer output lengths can be truncated).

2. Method 2 (*In the absence of stable storage*). In this case it is proposed that the interface identifier can simply be generated 'at random'. No method is specified for random generation, although it is suggested that host-specific configuration information (such as a user identity, security keys, and/or serial numbers) could be concatenated with random data and input to MD5 to generate the interface identifier.

2.3   The Gont Approach

Gont [5] notes that temporary addresses, as proposed in RFC 4941, bring difficulties. From a network management perspective, 'they tend to increase the complexity of event logging, trouble-shooting, enforcement of access controls and quality of service, etc. As a result, some organizations disable the use of temporary addresses even at the expense of reduced privacy [8]. Temporary addresses may also result in increased implementation complexity, which might not be possible or desirable in some implementations (e.g., some embedded devices)'.

As a result, Gont [5] proposes another approach to generating user-privacy-protecting interface identifiers. This scheme generates interface identifiers that are stable within a subnet, but which vary between subnets. That is, when a host migrates from one subnet to another, both the first and second 64-bit components of the IPv6 address change, preventing tracking of hosts as they migrate. As with the RFC 4941 scheme, it is intended that a generated interface identifier cannot be linked to a long-term host identifier (such as the SLAAC interface identifier).

Use of the scheme requires choice of a pseudorandom function f giving a 64-bit output that must be difficult to invert. The choice for f is not mandated, but it is suggested that it could be computed by taking the 64 least significant bits of the output of SHA-1 or SHA-256 [9]. The scheme also requires the address-generating software to have access to a host-unique secret key K (of length at least 128 bits), which is chosen at random at system installation time. It is further assumed that, as in RFC 4941, a fixed network interface identifier I is available, e.g. as derived from the MAC layer address. The scheme then operates as follows.

1. Compute $J = f(P||I||N||D||K)$, where f, I and K are as above, P is the 64-bit SLAAC prefix, e.g. as obtained from a router advertisement message, N is an identifier for the network interface for the generated identifier, and D is a counter used to resolve DAD conflicts (initialised to zero every time this process is run). Hence J is a 64-bit value.
2. Compare J against a list of reserved interface identifiers and those already assigned to an address on the host (on a different network interface). Also perform DAD. If J matches a forbidden address or DAD fails then increment D and restart the process; otherwise use J as the subnet-specific interface identifier.

Including P in the computation ensures that J is subnet-specific; similarly, including N ensures, with high probability, that different network interfaces on the same host have different values of J.

2.4   The Rafiee-Meinel Scheme

In a recent paper, Rafiee and Meinel [10] propose yet another approach to randomised interface identifier generation. They reject the Gont approach (see Sect. 2.3) on the basis that fixing the interface identifier for a given subnet is potentially privacy-compromising, since all accesses to this subnet will be trackable. They also criticise method 1 of RFC 4941 [1] on the basis that stable storage may not be available.

The Rafiee-Meinel scheme can be regarded as a specific instantiation of method 2 of RFC 4941, i.e. it is a specific method of generating randomised interface identifiers that does not make use of stable storage. It assumes that the system generating the identifier has access to the current system time T in the form of a 64-bit integer denoting the number of milliseconds since the beginning of 1970. The scheme operates as follows.

1. Generate a 128-bit random value R.
2. Compute $V = h(R||T||P)$, where T is a timestamp (as above), P is the 64-bit subnet prefix, e.g. as obtained from a router advertisement message, and h is SHA-256 [9]. V is thus a 256-bit value.
3. Let J be the leftmost 64 bits of V.
4. Perform DAD. If DAD fails then increment R and restart the process; otherwise use J as the subnet-specific interface identifier.

2.5   Other Schemes

Before proceeding we also briefly mention two other papers which describe IPv6 address generation schemes which are apparently relevant. Al'Sadeh, Rafiee and Meinel [11] and Rafiee and Meinel [12] describe modified versions of *Cryptographically Generated Addresses (CGA)* [13] designed to address the privacy problem discussed above. CGA is a method of generating 64-bit IPv6 interface identifiers designed to enable a host that owns an identifier to prove its owner-ship. To use CGA, a host must generate an asymmetric signature key pair and then calculate the interface identifier as a SHA-1 hash of the public key and certain other parameters. If a third party challenges the host to prove ownership of the identifier, the host can release both the public key and a signature on a third-party-provided challenge created using the signature key.

Clearly CGA-generated interface identifiers are, by definition, random in appearance, and hence appear to address the privacy issue. Thus regular use of CGA would provide 'unlinkable' short-term IPv6 addresses. However, the gen-eration of a key pair is a non-trivial operation, and it would seem that the only reason to adopt such an approach is if the security provided by CGA is required. Of course, improvements in the efficiency of CGA (as claimed in the two papers referred to above) are welcome, but do not change this conclusion. Thus, since the resource requirements of implementing CGA limit its applicability as a gen-eral solution, we do not consider CGA, and variants thereof, further here.

2.6 A Summary

If we ignore the CGA variants, three basic approaches have been proposed to generate privacy-protecting interface identifiers:

- RFC 4941 [1] method 1, which enables the generation of a sequence of ran-domised interface identifiers based on an initial random value;
- RFC 4941 method 2, including a specific instance due to Rafiee and Meinel [10], which enables the generation of a sequence of random identifiers based on 'one off' random values;
- the approach due to Gont [5] which involves the generation of fixed, but unlinkable, subnet-specific interface identifiers.

3   Practical Limitations to Privacy

In practice the schemes we have described all have potential shortcomings arising from poor use of randomness. Before analysing the individual schemes we first consider the use and abuse of random values.

3.1   Use of Randomness

Perhaps the first question that springs to mind when considering the prior art is 'Why not just generate interface identifiers at random?' Indeed, the tech-niques we have described all, to some extent at least, require the generation of random numbers. This issue is addressed in 3.2.1 of RFC 4941 [1], where it is stated that 'In practice, however, generating truly random numbers can be tricky. Use of a history value [as in method 1] is intended to avoid the particular scenario where two nodes generate the same randomized interface identifier, both detect the situation via DAD, but then proceed to generate identical randomized interface identifiers via the same (flawed) random number generation algorithm. The above algorithm avoids this problem by having the interface identifier (which will often be globally unique) used in the calculation that generates subsequent randomized interface identifiers'.

That is, the authors of the RFC were very well aware of the difficulties of generating random values, and the possibility that, in practice, a flawed random number generator might be used. Examining the various proposals in more detail, it is clear that in no case are precise instructions provided covering how to generate the necessary random values.

- The specifications of the two methods in RFC 4941 simply contain pointers to RFC 4086 [14] for guidance on how to generate random values. RFC 4086 certainly contains much excellent advice, but does not contain a specific proposal for a random number generator.
- Exactly the same situation holds for Gont [5], who simply refers to RFC 4086 for advice on generating random values.
- Rafiee and Meinel [10] do not address the issue of randomness generation at all.

In the absence of very clear and specific instructions on how random numbers must be generated, or at least a reference to such instructions, there is a great danger that implementers will choose simple, but ineffective, methods for 'random' number generation. Certainly, past experience suggests that implementers cannot be relied upon to make good security decisions, particularly when called upon to generate random values. Examples demonstrating this include the following.

- After conducting a large scale survey of RSA public keys, Lenstra et al. [15] showed that a small but significant proportion offered no security whatever; specifically, 12720 of 4.7 million sampled RSA moduli had a single large prime factor in common. Moreover, 'of 6.4 million distinct RSA moduli, 71052 (1.1 %) occur more than once, some of them thousands of times'. This could only occur because the RSA key generation software used by significant numbers of users makes very poor use of 'randomness'.
- Bond et al. [16] have shown that many EMV (chip and PIN) terminals have a very worrying defect. The EMV protocol requires the terminal to send an 'unpredictable number' to a payment card, which is then used to compute a response to the terminal; the terminal uses this response to authenticate the card, with the unpredictable number being a guarantee of the response's freshness. However, in practice, many terminals (including those from highly reputable manufacturers) generate this unpredictable number in a very predictable way, i.e. very little genuine randomness is involved, meaning that security vulnerabilities result.
- In fact, even when security specifications are apparently precise, implementers cannot be relied upon to implement security correctly. As part of research into the security of IPsec, Degabriele and Paterson [17] looked at six open source IPsec implementations, including those for Linux, FreeBSD and OpenSolaris.

Their surprising, and very worrying, finding was that not one of them correctly implemented a security-critical part of the protocol. Further evidence of poor use of security specifications has been provided by two separate recent studies [18,19], which have shown that a wide range of serious vulnerabilities can be found in SSL implementations.

This experience suggests that specifications of security protocols need to be absolutely explicit about measures to be taken by implementers. Providing pointers to good advice is not enough.

As a result of the lack of clear specifications of randomness generation in all the schemes we have examined, there is a danger that the unlinkability property of addresses generated by these schemes will be compromised. We examine each of the schemes in greater detail below, following the ordering given in Sect. 2.6 above.

3.2   Privacy Goals

Before analysing the effectiveness of the various schemes, it is important to understand their privacy goals. The two methods proposed in RFC 4941 and the Rafiee-Meinel scheme all aim to provide a degree of privacy protection both within a subnet and between subnets. That is, they provide a means of generating pseudonymous addresses for devices so that no two addresses can be linked either when they are used on the same subnet or when used on different subnets. Of course the degree of privacy obtained from these approaches will depend on a range of other factors, including how long an address is used, but these are outside the scope of the discussion here — that is we focus here purely on the linkability of addresses.

The privacy goal of the other scheme we examine, namely the Gont scheme, is rather different. It proposes use of a fixed address on each subnet, and the only privacy goal is unlinkability of addresses used on different subnets.

In the remainder of this section we consider for each scheme the degree to which its privacy goals are met, and in the next section we consider how the various schemes can be improved to try to more effectively meet their goals.

3.3   RFC 4941 Method 1

The provision of privacy of this scheme clearly relies on the initial assignment of a random value to H. In RFC 4941 it is simply stated that the the initial history value should be hard to guess, and a reference to RFC 4086 is given. All the randomised interface identifiers J are derived as a function of H and I (the fixed interface identifier, e.g. derived from the MAC address).

If the initial value of H has full 64-bit entropy, i.e. it is a 64-bit truly random value, and we assume that h is one-way (and, despite its shortcomings with respect to collisions, MD5 is not known to be not one-way), then the scheme appears secure, assuming that a search of size 2^{63} is infeasible.

However, if H has much less entropy and the method of generation is known to an attacker, then the privacy properties of the scheme are at grave risk. To see why, suppose that the initial value of H has k bits of entropy ($k << 64$) and that an attacker knows how to search through the possible initial values of H in 2^k steps. Now suppose also that such an attacker is temporarily on the same subnet as the target host, and is thus able to observe both the current temporary interface identifier J and also the host's MAC address (and hence can compute I).

If we assume that the host changes addresses once a day, and that the device was initialised less than a year ago, the attacker can perform a simple search through all possible values of H, in each case generating all 365 possible temporary addresses and comparing the generated values with J. Such a search has complexity 365×2^k hash operations (and comparisons). If, for example, we suppose that $k = 32$, this means that an exhaustive search for the initial value of H can be completed in a little over 2^{40} operations. Once the initial value of H is known then all future interface identifiers for this host are simple to compute, i.e. the scheme has been broken.

This analysis makes clear that the address unlinkability property provided by of the scheme is at significant risk if anything other than a very robust method for initialising H is used. Unfortunately, as previous experience shows, this seems to be a very strong and risky assumption.

3.4   RFC 4941 Method 2 and the Rafiee-Meinel Scheme

There is not much one can say about method 2 as described in RFC 4941, except to reiterate the difficulties of generating random values. We instead turn our attention to the Rafiee-Meinel scheme as an example of an attempt to provide a specific implementation of method 2.

This approach requires the host to generate a 128-bit random value R. The correct operation of the scheme depends to a considerable extent on the quality of this value, but no guidance is provided. One is tempted to suspect that in practice this value may be taken from a pseudorandom number function provided by the development environment, which could mean that R has very little entropy. That is, if two devices both attempt to generate a temporary address on the same subnet at the same instant, then they may very well generate the same value J. If replicated across large numbers of devices this could cause significant duplicate address problems, which is precisely why RFC 4941 method 1 was proposed. Whilst this address-collision issue is not privacy-threatening, anything that threatens network connectivity is a major problem, which raises significant doubts about this approach.

3.5   The Gont Scheme

This scheme, like method 1 of RFC 4941, requires the generation of an initial random secret key K, but does not use randomness thereafter. Assuming the robustness of the function f, the security of the scheme rests completely on the entropy in K. Gont [5] simply states that K shall not be known by the attacker, and points to RFC 4086 [14] for advice on generating random values.

If K has close to 128 bits of entropy, then the scheme appears to be secure. However, if K has much less entropy and the method of generation is known to an attacker (including knowledge of N, the identifier for the network interface used in this particular implementation), then the privacy properties of the scheme are at serious risk. Demonstrating why is rather similar to the attack on RFC 4941 method 1 given above. Suppose the value of K has k bits of entropy ($k << 128$) and that an attacker knows how to search through the possible values of K in 2^k steps. Now suppose also that such an attacker is temporarily on the same subnet as the target host, and is thus able to observe both the current temporary interface identifier J and also the host's MAC address (and hence can compute I).

Then, for each candidate value K^* for K (from a set of size 2^k) the attacker computes $V^* = f(P||I||N||0||K^*)$, which is possible since we assume that the attacker knows P, I and N. The attacker then simply compares V^* against J; if they agree then there is a high probability that $K = K^*$, i.e. the attacker has found K.

Thus the privacy property of this scheme, like RFC 4941 method 1, is at significant risk if anything other than a very robust method for initialising K is used. As discussed above, this appears to be a very risky assumption.

4   Practical Measures to Improve Randomness Generation

Our objective here is to consider ways in which the operational privacy of previous proposals could be improved, even when the host device has very limited capabilities for generating random values. We start by considering the randomness generation problem and the nature of randomness sources that might be available to an implementer. We then consider ways in which the privacy properties of RFC 4941 method 1 and the Gont scheme might be improved. We do not consider the Rafiee-Meinel scheme further here because of the issues with regard to recurring address collisions.

4.1   Generating Randomness

We start by observing that internationally standardised means of generating random bits are given in ISO/IEC 18031 [20]. The models introduced there for *random bit generators (RBGs)* are particularly relevant. The means used in RFC 4941 method 1 to generate the sequence of history values H falls into the class of *Pure Deterministic RBGs (Pure DRBGs)*. The scheme is a *pure* DRBG since entropy is only used once, to generate the initial 'seed value' H, and the method to generate subsequent values of H is purely deterministic. This contrasts with what ISO/IEC 18031 calls a *Hybrid* DRBG, in which a source of entropy is also used as part of the state update function. ISO/IEC 18031 ([20], 7.3) discusses the security advantages of such hybrid DRBGs.

Any DRBG, whether pure or hybrid, relies on a source of randomness to initialise it, and possibly to provide further input during use. We therefore briefly consider possible sources of randomness that are likely to be available to almost any platform. It is important to note that combining a number of sources of randomness, each yielding a modest number of bits of entropy, is just as effective as using a single source of larger quantities of randomness.

- We start by considering the use of timestamps from a system clock, as incorporated into the scheme of Rafiee and Meinel discussed in Sect. 2.4. Such an approach has the great advantage that almost any device will incorporate a system clock, and hence this approach is universally applicable. Moreover, if the clock has a resolution to the millisecond level, then, assuming that the precise time of sampling is not available to an attacker, use of a clock would appear to be able to yield between 10 and 20 bits of entropy.
  However, there are issues with the use of a clock value as a source of entropy. RFC 4086 [14] observes that 'One version of an operating system running on one set of hardware may actually provide, say, microsecond resolution in a clock, while a different configuration of the same system may always provide the same lower bits and only count in the upper bits at much lower resolution. This means that successive reads of the clock may produce identical values even if enough time has passed that the value should change based on the nominal clock resolution'.
  Note that this issue raises further doubts about the operation of the Rafiee-Meinel scheme, i.e. in certain implementations address collisions may be more likely than one might expect. Nonetheless, and despite the words of caution in RFC 4086, a millisecond-accurate clock would appear to be a very valuable and almost ubiquitous source of a modest number of bits of randomness (entropy).
- Memory state information, in particular the number of free (or used) bytes in long-term storage or in RAM, would appear to be a possible source for a few bits of randomness. Again, whilst the number of bits available from each sampling may be modest, this would appear to be a reliable and ubiquitous source of randomness.
- Timings and values of external events make up another source of randomness that is discussed in RFC 4086. One example might be the timings of packet arrivals. In circumstances where an 'entropy-harvesting' process is running continuously in the background, e.g. as part of a hybrid DRBG, such an approach could again be a valuable contributor of modest numbers of bits of entropy.
- Modern mobile devices are equipped with a range of sensors, any of which could be used as a source of randomness. Microphones and cameras will generate large volumes of data likely to be highly unpredictable. A GPS receiver will similarly generate hard to predict data. Even a simple motion sensor, e.g. as used to determine the screen orientation for a portable device, could generate useful material. Of course, some sensors are highly privacy-compromising and hence may not be usable by the address generation software; however others, such as motion sensors, are far less sensitive, and could be readily available.

– Of course, hardware-based non-deterministic sources of randomness, such as those built into implementations of the Trusted Platform Module (TPM) incorporated into large numbers of notebook and desktop PCs (see, for example, Gallery [21]), would be ideal, and should clearly be employed where available. However, not all devices performing address autoconfiguration will have access to such random sources, and the main purpose of this paper is to make provision for devices without a good single source of randomness.

4.2   A Simple Improvement to RFC 4941 Method 1

As we have discussed above, problems potentially arise with RFC 4941 method 1 if the 64-bit 'random value' used to initialise the history value H contains insufficient entropy. Because a pure DRBG is used, if any instance of the history value ever becomes known, then all future outputs can be determined. This is clearly undesirable.

It should be clear that adoption of a hybrid DRBG, incorporating new randomness whenever a new address is generated, would address this problem. Over the long term entropy will 'accumulate', making future address prediction impossible unless almost every address is tracked.

Such an approach is also simple to achieve. Whatever source of entropy is available to generate the initial value of H can be re-used to provide new entropy for each subsequent history value update. We therefore propose the following very simple change to the generation of the value V in step 1, namely to put:

$$V = h(H||I||R)$$

where R is a 'random' value containing new entropy. This should not significantly increase the complexity of using this method, but will ensure a sufficient level of entropy is used to generate each new address, irrespective of the randomness properties of H.

We further propose that R should be mandated to be constructed as the concatenation of:

– a timestamp accurate to the nearest millisecond (guaranteeing 10–15 bits of entropy)[1];
– (optionally, but highly recommended) the number of bytes free in short-term and/or long-term memory;
– (optionally) any other values which contain unpredictable information, notably including the outputs of any device sensors available to the DRBG.

Further items could be added to the list if they are deemed to be likely to be readily available.

---

[1] One possible issue with using this as a source of randomness in this context is that address updates may occur at fixed times, e.g. at the same time every day. If this is the case then the number of bits of randomness obtained is likely to be significantly reduced.

4.3   Making the Gont Scheme More Robust

A second challenge is to find ways of making the Gont scheme more robust against attacks arising from poor sources of randomness. This is more problematic, since one goal of the scheme is that the same 'randomised' interface identifier will be generated whenever the device is attached to the same subnet. This makes it highly problematic to introduce new randomness during the lifetime of the system.

The only practical solution would therefore appear to be to require the gathering of entropy over a period before generating the key value K. This would involve building an 'entropy-harvesting' hybrid DRBG, with a state of at least 128 bits. The initial state would be set using whatever sources of randomness are available. The system would then be required to be cycled through a number of iterations over a period of hours. On each iteration, additional randomness should be included as part of the state update function. At the completion of such a process, the state of the DRBG should contain a large number of bits of entropy, preventing key-guessing attacks of the type discussed in Sect. 3.5.

A question that naturally arises is in this context is 'How many iterations would be required in practice'? Of course this depends on the number of bits of entropy introduced in each iteration. As a result, one way of deciding on the number of iterations would be to require the implementer to make an estimate for the number of bits of entropy, b say, that are harvested during any one iteration. To try to ensure that the DRBG 128-bit state is 'fully randomised', a minimum of $\lceil 128/b \rceil$ entropy-harvesting iterations will be required.

During the initial period while the key K is being generated, the fixed IPv6 address provided by SLAAC could be used by the device. After all, the main privacy threat arises from use of a single address over a long period of time and across multiple networks. As a result, use of the fixed address for a day or two is unlikely to pose a significant threat.

5   Summary and Conclusions

We have examined three proposed methods for 'randomised' IPv6 address auto-configuration. Significant shortcomings have been identified in all three of these methods. Two of them do not adequately protect user privacy if only weak sources of randomness are available. The other approach appears likely to give problems with address collisions, at least in some operational environments.

Modifications to two of the three methods have been proposed which are designed to mitigate the threats arising from implementations of systems on devices without hardware RBGs. These modifications have been deliberately designed to involve only minor changes, and should not significantly increase implementation complexity. It would therefore appear reasonable to explore ways of modifying RFC 4941 and the Gont internet draft to incorporate the simple modifications proposed.

Possible future work would include looking at real-life implementations of the schemes we have examined in this paper. It would be particularly interesting to test the degree of entropy actually being deployed in a range of devices implementing RFC 4941. In some cases, e.g. on smart phones or PCs, implementers may choose to use the random number generation facilities provided by the operating system, in which case the robustness of the solution will very much depend on the quality of the provided random numbers. However, the situation may be very different for small, low-power devices. Finally, it would also appear to be worth building prototype implementations of the proposed modified schemes, to test their randomness properties in practice.

References

1. Narten, T., Draves, R., Krishnan, S.: Privacy extensions for stateless address autoconfiguration in IPv6. RFC 4941, Internet Engineering Task Force (2007)
2. Hinden, R., Deering, S.: IP version 6 addressing architecture. RFC 4291, Internet Engineering Task Force (2006)
3. Thomson, S., Narten, T., Jinmei, T.: IPv6 stateless address autoconfiguration. RFC 4862, Internet Engineering Task Force (2007)
4. Droms, R., Bound, J., Volz, B., Lemon, T., Perkins, C., Carney, M.: Dynamic host configuration protocol for IPv6 (DHCPv6). RFC 3315, Internet Engineering Task Force (2003)
5. Gont, F.: A method for generating semantically opaque interface identifiers with IPv6 Stateless address autoconfiguration (SLAAC). Internet Engineering Task Force, Internet draft-ietf-6man-stable-privacy-addresses-17 (2014)
6. Cooper, A., Gont, F., Thaler, D.: Privacy considerations for IPv6 address generation mechanisms. Internet Engineering Task Force, Internet draft-ietf-6man-ipv6-address-generation-privacy-01 (2014)
7. Rivest, R.L.: The MD5 message-digest algorithm. RFC 1321, Internet Engineering Task Force (1992)
8. Broersma, R.: IPv6 everywhere: living with a fully IPv6-enabled environment. Presentation at the Australian IPv6 Summit 2010, Melbourne, Australia (2010)
9. International Organization for Standardization Genève, Switzerland: ISO/IEC 10118–3, Information technology – Security techniques – Hash-functions – Part 3: Dedicated hash-functions. 3rd edn. (2004)
10. Rafiee, H., Meinel, C.: Privacy and security in IPv6 networks: challenges and possible solutions. In: Elci, A., Gaur, M.S., Orgun, M.A., Makarevich, O.B. (eds.) The 6th International Conference on Security of Information and Networks, SIN 2013, 26–28 November 2013, Aksaray, Turkey, pp. 218–224. ACM (2013)
11. AlSa'deh, A., Rafiee, H., Meinel, C.: IPv6 stateless address autoconfiguration: balancing between security, privacy and usability. In: Garcia-Alfaro, J., Cuppens, F., Cuppens-Boulahia, N., Miri, A., Tawbi, N. (eds.) FPS 2012. LNCS, vol. 7743, pp. 149–161. Springer, Heidelberg (2013)
12. Rafiee, H., Meinel, C.: SSAS: a simple secure addressing scheme for IPv6 autoconfiguration. In: Castella-Roca, J., Domingo-Ferrer, J., Garcia-Alfaro, J., Ghorbani, A.A., Jensen, C.D., Manjon, J.A., Onut, I.V., Stakhanova, N., Torra, V., Zhang, J. (eds.) Eleventh Annual International Conference on Privacy, Security and Trust, PST 2013, 10–12 July 2013, Tarragona, Catalonia, Spain, pp. 275–282. IEEE (2013)
13. Aura, T.: Cryptographically generated addresses (CGA). RFC 3972, Internet Engineering Task Force (2005)

14. Eastlake, D., Schiller, J., Crocker, S.: Randomness requirements for security. RFC 4086, Internet Engineering Task Force (2005)
15. Lenstra, A.K., Hughes, J.P., Augier, M., Bos, J.W., Kleinjung, T., Wachter, C.: Ron was wrong, Whit is right. Cryptology ePrint Archive: Report 2012/62 (2012)
16. Bond, M., Choudary, O., Murdoch, S.J., Skorobogatov, S., Anderson, R.: Chip and Skim: cloning EMV cards with the pre-play attack (2012). arXiv:1209.2531 [cs.CY]
17. Degabriele, J.P., Paterson, K.G.: Attacking the IPsec standards in encryption-only configurations. In: Proceedings of the 2007 IEEE Symposium on Security and Privacy (S&P 2007), 20–23 May 2007, Oakland, California, USA, pp. 335–349. IEEE Computer Society Press, Los Alamitos (2007)
18. Fahl, S., Harbach, M., Muders, T., Smith, M., Baumgärtner, L., Freisleben, B.: Why Eve and Mallory love Android: an analysis of Android SSL (in)security. In: Yu, T., Danezis, G., Gligor, V.D., (eds.) ACM Conference on Computer and Communications Security, CCS 2012, 16–18 October 2012, Raleigh, NC, USA, pp. 50–61. ACM (2012)
19. Georgiev, M., Iyengar, S., Jana, S., Anubhai, R., Boneh, D., Shmatikov, V.: The most dangerous code in the world: validating SSL certificates in non-browser software. In: Yu, T., Danezis, G., Gligor, V.D. (eds.) ACM Conference on Computer and Communications Security, CCS 2012, 16–18 October 2012, Raleigh, NC, USA, pp. 38–49. ACM (2012)
20. International Organization for Standardization Genève, Switzerland: ISO/IEC 18031:2011, Information technology – Security techniques – Encryption algorithms – Random bit generation. 2nd edn. (2011)
21. Gallery, E.: An overview of trusted computing technology. In: Mitchell, C.J. (ed.) Trusted Computing, pp. 29–114. IEE Press, London (2005)

Trust and Formal Analysis

A Practical Trust Framework: Assurance Levels Repackaged Through Analysis of Business Scenarios and Related Risks

Masatoshi Hokino[1], Yuri Fujiki[1], Sakura Onda[1], Takeaki Kaneko[1],
Natsuhiko Sakimura[2], and Hiroyuki Sato[3]([⊠])

[1] JIPDEC, Tokyo, Japan
{hokino-masatoshi,fujiki-yuri,onda-sakura,kaneko-takeaki}@jipdec.or.jp
[2] Nomura Research Institute, Tokyo, Japan
n-sakimura@nri.co.jp
[3] The University of Tokyo, Tokyo, Japan
schuko@satolab.itc.u-tokyo.ac.jp

Abstract. In cyberspace, standards for the expression of the trustworthiness of identities have been developed by various parties. This trustworthiness is often referred to as entity authentication assurance, and its degree is often called LoA (levels of assurance, or assurance levels). There are two prominent LoA standards: NIST SP800-63-2 and ISO/IEC 29115:2013. LoAs are designed to express different levels of assurance. Multiple viewpoints are set in assessment, and related assessment criteria for each viewpoint are packaged into one LoA. For deployment of LoAs in enterprise business scenarios, the choice of assessment criteria in a given LoA must match the specific business requirements. We perform a field survey on business scenarios in which trust in identities is a major problem. In the survey, we focus on two key factors of assessment: identity proofing and authentication process. In addition, we observe the overall fit and gap in business scenarios. Results indicate that raising the assurance of the authentication process is effective for raising the overall assurance level. Based on the investigations performed, we repackage light weight identity proofing and LoA 2 equivalent credential management and usage into a new assurance level, LoA 1+, for the "right" cost benefit balance.

1 Introduction

The importance of trusted identities in cyberspace has become widely recognized in recent years. Standards for the expression of the trustworthiness of identities have been developed by various parties. Many government led activities exist, e.g., FICAM TFPAP [6] and NSTIC [14] of the U.S.A., GOV.UK Verify of U.K. [7], etc. While some of these activities are for government use, others do target both efficient e-governments and more efficient commercial sector activities promoting the formation of new industries.

In the case of FICAM TFPAP [6], major objectives include the establishment of well-defined identity and credential management at the identity provider (IdP)

© Springer International Publishing Switzerland 2015
L. Chen and S. Matsuo (Eds.): SSR 2015, LNCS 9497, pp. 203–217, 2015.
DOI: 10.1007/978-3-319-27152-1_11

which is proportionate to the risk of compromise of the business that consumes the resulting identities, referred to as the relying party (RP). The identity that is conveyed from an IdP to an RP is called "federated identity" and a group of such IdPs and RPs are called a "federation." To establish mutual trust among the participants, technical and operational standard should be followed. The trust establishment also requires enforcement functions for violators. The combination of these "tools and rules" is referred to as a "trust framework."

A trust framework enables its stakeholders to trust claims made from the other stakeholders under condition of information asymmetries. For example, when an IdP provides a set of attributes related to an entity to an RP, in general, it has no means of evaluating the trustworthiness of the information that it receives. Under such circumstances, the transaction will typically not occur and the market breaks down, as shown in the Market of Lemons introduced by [1].

Establishing a measurement unit for the quality of identities along with a kind of operational framework that assures the truthfulness of the providers would be a solution. The trust framework can be applied for this purpose.

In order for a trust framework to be practical, it must be applicable to a variety of business scenarios. To express the required assurance level that is proportionate to the risk of the RP's business, the concept of LoAs (level of assurance, or assurance level) have been introduced. An RP requires a level of assurance as the minimum requirement under which it can accept identities from an IdP. Furthermore, an IdP will provide such identities if it can. Standardizing this expression in a small number of variance enables the trust framework to scale up the participation, which is an important requirement for cyberspace applications.

There are many standards that define the aspects of LoAs. The combination of OMB M-04-04 [16] and NIST SP800-63-2 [3] is a prime example. It defines four levels of risks and corresponding LoAs. However, the adoption of these standards in the private sector is not widespread, possibly because the requirements are focused towards the U.S. government entities and its direct adoption is difficult for private sector entities especially those outside U.S.

ISO/IEC 29115 [9] generalizes the older version of NIST SP800-63-2 that originally targets the US government usage. It reflects the demand of extending trust frameworks to business scenarios. Unlike NIST SP800-63-2, it does not mandate the use of government issued photo IDs nor trusts such IDs. It is more risk based and process oriented. Only photo IDs produced in a documented process deemed to produce a sufficient confidence in the document are trusted. While this certainly expands the scope of applicability, the adoption is still not widespread in the private sector. One of the reasons appears to be that for most business, LoA 2 and above are not cost effective.

In this paper, we begin our examination from the fact that both standards set multiple viewpoints in assessment, and package assessment criteria for each viewpoint into one LoA. The choice of assessment criteria in a given LoA must be examined to discover the match of business requirements.

First, we examine business use cases to determine possible reasons for low adoption of the standards. Next, we undertake a field survey on business scenarios in which trust in identities represents a major problem. The survey focuses on two key factors of assessment, identity proofing and authentication process, and observe the fit and gap for those business scenarios.

From the investigation, we repackage light weight identity proofing and LoA 2 equivalent credential management and usage into a new assurance level, LoA 1+ for the "right" cost and benefit balance.

As the result, we show that the process of field survey, investigation and repackaging is a subject of engineering.

The rest of this paper is organized as follows: Sect. 2 surveys related work and standardization. In Sect. 3, we analyze the assessment criteria of existing standards. In Sect. 4, we explain our field survey of business scenarios. In Sect. 5, we discuss LoA 1+, and the repackaging process based on the result of Sect. 4. Section 6 concludes this paper.

2   Related Work on Trust Framework

The majority of identity trust frameworks have two facets: technical requirements that define LoAs and operational rules that ensures the adherence to the technical requirements. This combination is often referred to as "tools and rules."

Technical requirements are further decomposed into a credential issuance process that includes identity proofing and an authentication process. Identity proofing and related issues (especially privacy) is discussed in [19]. In [13], the assurance of authentication is described.

Until now, there have been a number of technical proposals related to trust frameworks and assurance levels. In [29], assurance of attributes has been proposed in addition to assurance of authentication. [30] gives a discussion of digital identities in general. Today, assurance levels are considered a topic of engineering which includes trust elevation [15,18] that aims at collecting evidence of low assurance in order to give higher assurance. Furthermore, [17] proposes a fine tuning of assurance levels. However, such proposals need to be applied to enterprise business scenarios to obtain feedback for standardization. In this regard, standards for operations of practical trust frameworks are of high significance.

The U.S. has a long history of defining LoAs. The combination of OMB M04-04 [16] and NIST SP800-63-2 [3] sets risk and control criteria for building a trust of governmental agencies. From their inception, multiple levels are incorporated to cover a wide spectrum of trust ranging from id/password authentication to PKI.

Japan has also created the guidelines for risk analysis, digital signing and authentication for on-line applications and processing [4].

ISO/IEC 29115 [9] and its ITU-T version ITU-T X.1254 [10] are a framework for managing assurance levels of entity authentication. As in NIST SP800-63-2, four assurance levels and criteria are defined. In all of these, a final level of assurance is defined as the lowest of the process.

[2] discusses identity assurance in another scheme that includes the audit process.

Deployment of this kind scheme for the healthcare sector is discussed in [5].

3   Assessment Criteria of Assurance Levels

There are two significant standards for the assessment of assurance levels: NIST SP800-63-2 and ISO/IEC 29115.

This paper focuses on identity proofing, credential issuance, and the authentication process in the set of assessment criteria.

3.1   Credential Issuance and Identity Proofing Process Requirements

In both standards [3,9], the identity proofing process is defined as a prerequisites for the issuance of credentials. Here, concrete threats are analyzed, and controls corresponding to each threat are considered.

At the credential management phase, there is some difference in the levels of protection as shown in Table 1.

Table 1. Controls on credential management

Level	Control at issuance	Secure storage
1	–	access control
2	mechanism to protect the credential from being handed to a wrong person	not to be stored in clear text
3	stricter mechanism to protect the credential from being handed to a wrong person	mechanism to protect the credential

As most modern systems provide a secure mechanism to protect credentials, there is little difference in the evaluation at this point. In practice, the most difference is derived from the identity proofing process. NIST SP800-63-2 and ISO/IEC 29115 define similar criteria. Table 2 shows ISO/IEC 29115 identity proofing objectives and controls.

Nonetheless, there is only a slight difference between NIST SP800-63-2 and ISO/IEC 29115. As NIST SP800-63-2 is created to meet the requirements of the U.S. government, where with most government related uses, there is a requirement to map the identity at the front door of the service to the identity stored in the backend database, using such identifying attributes such as name, date of birth, gender, and address as the keys. Because mis-matching of the keys would cause risks, in higher levels, showing "government issued" identity documents is demanded, They are more likely than others to include those "keys" that correctly map to the governmental backend database. Furthermore, the U.S.

Table 2. Requirements of identity proofing in ISO/IEC 29115 (Summary)

Level	Objectives	Controls
1	Self-claimed or self-asserted	Self-claimed or self-asserted
2	Identity is unique within context and the entity to which the identity pertains exists objectively	Proof of identity through use of identity information from an authoritative source
3	Identity is unique within context, entity to which the identity pertains exists objectively, identity is verified, and identity is used in other contexts	Proof of identity through use of identity information from an authoritative source + identity information verification

government typically knows the quality of the government issued identity documents. In Table 2, the "authoritative source" is replaced with "government" in NIST SP800-63-2.

On the other hand, ISO/IEC 29115 aims at being useful to the private sector internationally. In private sector use cases, it does not often matter whether the business exactly knows the customer's real name and date of birth or other attributes. Instead, it is more important to know whether that person has actually completed payment, which is the decisive factor for the entitlement of that specific service. As a result, ISO/IEC 29115 introduces the concept of "policy compliant identity document." The attributes to be proofed and verified depend on the business context. The business should document them according to the policy.

Another aspect of ISO/IEC 29115 is its tendency to be more process oriented. In an international context, there are government issued identity documents that are sometimes produced by low-quality processes. Such identity documents are not trustworthy even if they have been issued by an agency of government. This is another reason why ISO/IEC 29115 is asking for "policy compliant" identity documents that have adequately addressed the threats.

3.2 Authentication Process Requirements

In the standards [3,9], there are specified requirements for the authentication process. There are a number of proposed and deployed authentication mechanisms and processes which include passwords, one-time passwords, biometrics, and public key authentication. Their risks have been analyzed, and as a result, the strength of each authentication method can be objectively discussed.

NIST SP800-63-2 not only sets the threat that each LoAs should mitigate, but it maps specific type of credentials to be used for each LoAs.

ISO/IEC 29115 takes a slightly different approach. Instead of assigning specific types of credentials to each level, it only presents the technical requirements. ISO/IEC 29115 does not specify the credential type to accommodate the combination of various techniques. Furthermore, there is a specified control selection in the credential usage. Reflecting that any number of combinations of those controls can be used to mitigate the specific risk, it does not specify what has to be done at each level, but leaves those decisions for implementation.

3.3   Requirements for Certification

To establish a trust framework, only defining the levels of assurance is not sufficient. A trust framework must define a mechanism that provides a sufficient level of confidence of the members' adherence to the rules.

The combination of FICAM TFPAP [6] and certified trust frameworks such as Kantara Initiative Identity Assurance Program [12] are prime examples of such works. This combination lays out the audit requirements for each level of assurance. A third party audit is required for any levels including and above LoA 2.

Similarly, InCommon Federation [8] provides a program for certifying levels, while federations in Europe and Japan provide a limited program of certification.

4   Analysis of Business Scenarios in Terms of Assurance Levels

As NIST SP800-63-2 is designed for use by federal agencies, its applicability to the private sectors is limited in nature. The design of ISO/IEC 29115 is more generalized. However, its usefulness to the private sector has yet to be thoroughly investigated. Especially because they have a structure of packaging requirements of different viewpoints, this structure should be examined to determine whether the combination of requirements in the standards is appropriate, or covers a wide range of businesses. To identify the fit and gap, we have conducted a field survey in Japan to identity applicable business scenarios for trust frameworks.

4.1   Design Objectives of Field Survey

The objectives of this field survey is to identify the structure of assurance levels in terms of cost and effectiveness. In previous sections, we have presented that the structure of assurance levels is determined by the identity proofing and authentication processes together with the objectivity of the assessment.

Therefore, in order to have an appropriate coverage of business sectors. the survey has collected a wide range of business scenarios that are consumer oriented and where identity proofing is either legally required or required through industry self-regulation.

The data collected for each business type is listed below:

Market size this reflects the influence that the business sector has on the economy. This data is basically from [26].

Business practice (on-line/off-line/both) off-line business practice is also included because it is a future on-line business candidate, and the importance is not affected by the business practices.

In terms of assurance levels, we have collected the information below:

Authentication method/process criteria defined in [6] are used. In the criteria, the method of identity proofing is classified as non-technical, and shows the most conspicuous difference between levels. The criteria of identity proofing for levels 1 to 3 listed in Table 2 are used.

Regulations some processes are enforced by law. For example, in Japan, when opening a bank account, the identity proofing of level 2 is required by law. However, even if there are no regulations, some industry associations define self-regulation of identity proofing to achieve safer transactions and to protect the reputation of the business. in this survey, both processes enforced by law and processes self-regulated by industry associations have been collected.

As regulations are closely related to the quality (objectivity) control of the assessment, we also discuss this problem in Sect. 4.3.

4.2   Classification of Business Scenarios

Combining both evaluation criteria in terms of business type and assurance levels, the classification of services surveyed is presented in Table 3.

In Table 3, the first column expresses the levels of identity proofing. Services classified as 1-{1, 2, 3, 4} require level 1 identity proofing. Similarly, services classified as 2, and 3 require level 2, and level 3, respectively.

For services that require level 1 identity proofing, we have found that there are different regulation stipulations which are given a separate class listing from 1-1 through 1-4.

In examining the table, some significant categories begin to emerge:

1. The first category is the case where identity proofing is entirely self asserted. In this category, customers are requested to fill their information by themselves. There is no stipulation on identity proofing. This category is marked as 1-1. In the case of a hotel stay, the customer is required by the Inns and Hotels Act to inform the hotel one's true identity. Failure to do so may result in detention of less than 30 days or a fine of less than JPY 10,000.
   Note that this category contains scenarios whose business size is very large.
2. In the second category, some kind of identity proofing is required either by law or self-regulation, it is not strictly enforced in practice. Here, a wide range of identity proofing processes are adopted. Examples include checking photo ID in any form. This may include the IDs that are not issued by the government (e.g. student ID issued by a university), and inspecting the validity of a credit card. For business processes that only require age verification, an identity document is requested only in cases where the age is under suspicion. In these cases, the identity proofing method is often specified by the self-regulatory bodies, not by law. Penalties for being non compliant seems to be relatively minor. (marked from 1-2 through 1-4).

Example 1 (horse racing†). In Japan, by law, minors under the age of 20 are not allowed to bid on horse races. However, the method of identity proofing is not stipulated. The promoters perform the identity proofing by inspecting the

Table 3. Classification of businesses by types of identity proofing regulation

Class	Services	on-line/off-line	Size (M JPY)	Regulation (Publication of self-regulation on identity proofing by industrial association ✓)	
1-1	Hotel booking	both	4,045,618	Customers are required to fill out the name and address form (self-assert)	
	On-line shopping	on-line	[24] 12,800,000	practice not stipulation	✓
	On-line games	on-line	[23] 577,100		✓
1-2	Gov.controlled gambling†	on-line	1,834,110		✓
	Sport based lottery	on-line	[a] 110,797		✓
	Shopping (tobacco, liquor)	on-line	1,741,853	minors are not allowed	
	Late show (cinema, karaoke)	offline	[b] 319,329	(practice not stipulated)	
	adult (cinemas, magazines)	on-line	N/A		
1-3	Rental (video, autos, etc.)‡	both	1,867,196		
	Certification	offline	N/A	identity proofing required	
	Shopping (tobacco)	offline	150,539	(practice by self-regulation)	
	Marriage matching	on-line	18,167		✓
1-4	On-line dating	on-line	N/A	identity and age proofing required	
	Pawnshop	offline	N/A	(practice stipulated)	
2	Cell phones§	both	[c] 6,775,517		
	Bank account		[11] 15,881,400		
	Life insurance		[21] 41,981,800		
	Non-life insurance		[20] 9,667,900	identity proofing required	
	Credit card	offline	[22] 57,069,076	(enforced by law)	
	Real estate brokerage		9,824,601	(practice stipulated)	
	Precious metal trading		444,552		
	Secondhand articles dealer	on-line	303,844		
	Private office		N/A		
	Hotel booking (for foreigners)		[d] 4,045,618		
3	Digital certificate issuance	on-line	227,993		

Unless specified, from [26].

[a] Official publication of the Sports Promotion Lottery "Toto."

[b] total market size (not restricted to late show)

[c] Calculated by reference to [25]

[d] total market size (not restricted to foreigners)

credit card presented. A significant penalty exists for mis-identification. Promoters knowingly selling the race tickets to a minor will be subject to a fine of less than JPY 500,000 (Horse Racing Law[1], Article 34).

Example 2 (DVD rental‡). In Japan, there is no legal requirement for identity proofing in the DVD rental business. However, some business sectors voluntarily define regulations in which identity proofing should usually be performed by using photo IDs issued by public sectors.

---

[1] Horse Racing Law (in Japanese) http://law.e-gov.go.jp/htmldata/S23/S23HO158.html.

Table 4. Mapping identity proofing requirements of types of businesses to FICAM and ISO LoAs

Class	Evidences/procedures used by the business	FICAM LoA	ISO LoA
1-1	Self-Claimed	1	1
1-2	Documented procedure on Authoritative Sources[a]		2
1-3	Inspection of Photo ID (non-government ID allowed)		
1-4	Inspection of publicly issued documents		
2	Government Issued[b]Photo ID inspection	2	
3	Government Issued[a]Photo ID validation	3	3

[a] In case of age confirmation, IDs may not be required where determination is obvious,

[b] Depending on the business, some government issued photo IDs are not accepted.

3. Additional categories require the identity proofing methods corresponding to levels 2 and level 3 (marked as 2 and 3, respectively in Table 3). Most of these identity proofing processes are required by Japanese law.

Example 3 (cell phones§). To purchase mobile phones in Japan, a customer is required to show a government issued photo ID for identity proofing, or at least two pieces of evidence from public services for the proofing of one's name and address. Furthermore, the customer's address is verified by sending something to that address using the postal service. This is a typical process of identity proofing for assurance level 2.

By analyzing identity proofing processes of typical business scenarios, we can conclude that the criteria defined for government use [6] or the ISO standard [9] could also be used in many business scenarios.

Table 4 maps the identity proofing level of each category to the FICAM TFPAP and ISO/IEC 29115 LoA.

Note that the classification in Table 4 is based on the assurance levels of FICAM, which is represented by the matching of the major number of class and the FICAM LoAs.

However, these findings highlight differences to ISO LoAs in level 1 and 2. Actually, ISO/IEC 29115 expands the types of evidence of identity to accommodate private sector reality. On the other hand, however, it does not necessarily accept government issued identity documents. The difference lies in the importance of the process adopted during the creation of the document. This explains the fact that some businesses in Japan do not accept certain kinds of identity documents issued by some government agencies because they consider the possibilities of fraudulent issuance of those identity documents unacceptably high. This would not happen for government agencies, because any document produced by another governmental body is deemed accurate under Japanese law.

What is important, however, is that the identity proofing methods adopted by each class are covered by the ones specified by FICAM or ISO/IEC, even if we find some differences between the two.

In the remainder of this paper, we present our proposal to adopt the identity proofing methods of ISO/IEC 29115 to design a new class of assurance levels.

4.3 Self-Regulation and Objectivity

Table 3 includes the survey of the self-regulations. The objectivity of a claim is a common issue of self-regulations. Using independent audit or assessment is counted as a solution, which has still a problem on cost.

In Table 3, we see that some of them make effort in defining and publishing their own regulation as a form of their industrial associations (marked with ✓), which raises assurance of adherence to the regulation.

4.4 Effectiveness of High Level Authentication Processes

From the survey results of Sect. 4.1, use cases fall under ISO/IEC 29115 LoA2 and above are clustered as an important economic sector. However, we should not dismiss the cluster of businesses in Class 1-1 which correspond to ISO/IEC 29115 LoA 1. The mere fact that the market size of on-line shopping entities of Class 1-1 far exceeds that of shopping at the LoA 2 and 3 indicates its importance.

On-line games are an example of one such service and their prevalence merits analysis. The second survey is a case study on the effectiveness of raising assurance of an authentication processes.

On-line games are usually considered privacy sensitive. Our survey has found that the identity used there are usually self asserted. However, attacks that use previously collected username and password pairs from elsewhere, referred to as a list-based attack, saw a sharp increase in 2011. Statistics showed that some content providers in Japan received over 200,000 attacks per month.

A press release from the National Police Agency (NPA) on March 4, 2010 states:

Notes on the prevention of illegitimate access. Access controllers should improve their security (e.g. improving user authentication through introduction of One-Time-Password)

Actions to be taken by the NPA. NPA should influence the businesses (e.g., On-line game providers and Internet banks) by requesting them to improve the user authentication method.

Responding to the request by the NPA, the Japan On-line Games Association (JOGA) started operating a shared identity platform and helped each on-line game provider to adopt a two factor authentication process. This platform helped greatly to reduce the introduction and operation costs for each provider. In addition, if the provider uses the platform, any economic damage incurred from account compromise will be covered. Today, this platform is widely used by many game providers in Japan.

From a game user's standpoint, using the same authenticator across providers has reduced the difficulty of provider-wise authentication. Moreover, the fact that there is a reimbursement for account compromise seems to have made the acceptance rate high, which offsets any additional actions required by the platform.

The "On-line Game Security Guideline" published by JOGA on August 15, 2012 [28] contains the following points:

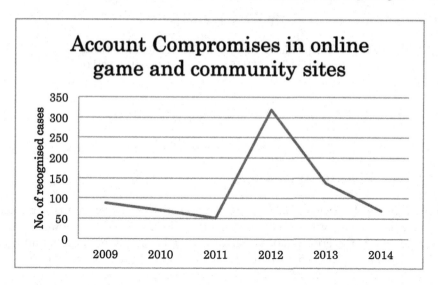

Fig. 1. Account compromises in on-line game and community sites

- Information Sharing guidelines when a security incident occurs,
- Guidelines for dealing with list-based account cracking,
- Security Solution Guidelines for One-Time Password etc., and
- Guidelines on coordination with security vendors and related associations.

As shown in Fig. 1, the introduction of the guidelines, in conjunction with the higher assurance authentication measures drastically led to a decrease in the number of incidents and countered the increasing trend of account compromise, this showing marked improvement. The data of Fig. 1 is taken from the series of annual reports published by the National Police Agency from 2010 to 2015 [27]. The reports published in March each year cover incidents for the previous year. In 2013, a 56 % decrease compared to 2012 is observed, and in 2014, a 78 % decrease is observed.

On-line games are classified as Class 1-1. This indicates the significance and usefulness of combining higher level authentication measures with a lower identity proofing level. This practice is distinguished from the business scenarios that use lower security authentication measures with the same level of identity proofing. Thus, even when low level identity proofing is adopted, high assurance authentication measures can be used effectively to raise security. In addition these results prove that there are some authentication measures which support high assurance levels and are also cost-effective.

However, in neither SP800-63 [3] nor ISO/IEC 29115 [9], the difference between "self-claimed identity proofing + low security authenticator" and "self-claimed identity proofing + higher security authenticator," cannot be distinguished despite the latter being useful for risk management.

Thus, we have identified a new class of assurance levels: low assurance in identity proofing and high assurance in authentication process. We conclude that the results from these surveys call for a new category of level of assurance: LoA 1+.

Table 5. LoA table

LoA	Level of identity proofing (as of ISO/IEC 29115)	Level of authentication (as of ISO/IEC 29115)
1	level 1	level 1
1+	level 1	level 2
2	level 2	level 2
3	level 3	level 3

5 Level of Assurance 1+

As discussed in the previous section, the results of our survey suggest the creation of a new assurance level is useful for important business sectors. However, as a natural request, the criteria for assessment must be simply organized, and that a rise in security level should be easily understood. Therefore, we propose the creation of a new assurance level by re-packaging components of assessment criteria. To further justify this, we consider identity proofing, authentication method/process, and objectivity assurance.

In the previous section, we have shown that in level 1 (self-assertion) in identity proofing, a higher security authentication method can be a factor for raising security. Here, our proposal departs form the idea of taking the "minimum level" of the processes as the resulting LoA, and we define LoA 1+ as the one requiring higher authentication process, while keeping the level of identity proofing the same as LoA 1. The result of this re-packaging is shown in Table 5.

Note that there is no combination of level 2 and above for identity proofing with a low level authentication process (specifically, level 1). There is no point of using those combinations because if the level of authentication is low, the resulting identity may be compromised even if higher level identity proofing is adopted.

The following examples examine the kind of business scenarios that can be entitled to LoA 1+ in order to achieve their advanced security.

Statistics from JOGA [28] justify the effectiveness of adopting LoA 1+. Thus, we can conclude that LoA 1+ meets the requirements of JOGA, and verifies its effectiveness in preventing fraudulent use in on-line shopping and on-line game sites. Moreover, these businesses can claim higher security by acquiring a new LoA certification which is stronger than LoA 1.

Table 3 shows that on-line shopping and games are classified as 1-1 (lowest), yet have a significantly large market size. By acquiring LoA 1+, these businesses can claim that their security is higher than 1-1.

Certification Requirement. For operating LoAs in a given trust framework, it is important that some assurance is given related to the policy compliance. As seen in Sect. 4.3, service providers pay attention to acquiring trust in their business. Obtaining public certification and announcements of their policies and procedures are usually used for this purpose.

In trust frameworks obeying FICAM TFPAP, such assurance is given by attestation through an independent audit performed by designated assessors. The separation is drawn between LoA 1 and LoA 2, in that, an independent audit is required by FICAM TFPAP to obtain LoA 2 certification. While an independent audit gives objectivity to one's claim, however, it also incurs a higher operation cost.

However with LoA 1, because the identity proofing is based on self-assertion, there is not much to audit and an objective external/independent audit is not required. However, there is inherent limitation in objectivity.

In the case of LoA 1+, the identity proofing situation is the same as in LoA 1, but relying parties may want to have some assurance as to the trustworthiness of the authentication measure used there.

The control that we propose in this paper is to adopt a reputation model to build trust. More specifically, we propose to have the operating body publicly announce the operating policies, procedures and its adherence to them. The procedures themselves will be the target of public evaluation. If an operating body does not operate as declared, it will face a reputation risk as well as other legal consequences. Furthermore, there is made some effort such as endorsement by related industrial associations that can be considered as a kind of cost effective social system for assurance raising.

As discussed in the last section of the survey, this reputation model is widely adopted and has proved effective to some extent. We conclude that this model is enough for an operating body to keep itself compliant with the policies and procedures declared in advance. Table 6 summarizes this discussion.

Table 6. Certification requirement for each LoA

LoA	Requirement	Grounds
1	No requirement	–
1+	No requirement	reputation risk
≥2	External/independent audit	objectivity

6   Concluding Remarks

In this paper, we have first analyzed the assessment criteria of existing standards of FICAM and ISO/IEC 29115, examining the key factors of identity proofing and authentication process. Next, conducting a fit and gap survey of various factors in business scenarios, we have listed typical instances with their characteristics and market size together with the adopted identity proofing and authentication process. In the analysis, we have discussed and proposed the creation of a new class of assurance level, which we refer to as LoA 1+ formed by re-packaging components of existing standards to serve an important business type.

Results from our survey have shown that conventional LoAs are useful in many situations. However, our data indicates that conventional LoAs have overlooked an important category of use case: the combination of self-claimed identity and high level authenticator. While the market sizes of the sectors that rely on self-claimed identity are huge, they face significant risk as shown in the on-line games case. Data from that example has proved that it is possible to significantly reduce the risk by just upgrading the authentication process without upgrading the identity proofing process. Observing these results, we have proposed a repackaging of existing LoA framework and created LoA 1+ that combines self-claimed identity and high level authentication.

It is also important to address the information asymmetry. While an identity provider may claim that it has used a high level authenticator to authenticate the user, it may well not be the case. A third party audit would be certainly effective to prove the claim, but this is a heavy weight process. Instead, we have proposed to rely on transparency and reputation risk for compliance of LoA 1+.

This level of assurance seems to fulfill trust and security requirements both from the service providers and from the users as evident in the on-line games use cases. Other industries are likely to benefit from following a similar strategy in adopting the LoA 1+ entity authentication assurance level.

References

1. Akerlof, G.A.: The market for "lemons": quality uncertainty and the market mechanism. Q. J. Econ. **84**(3), 488–500 (1970)
2. Baldwin, A., Mont, M.C., Beres, Y., Shiu, S.: On Identity assurance in the presence of federated identity management systems. In: Proceedings of the International ACM Workshop on Digital Identity Management 2007, pp. 27–35 (2007)
3. Burr, W.E., Dodson, D.F., Newton, E.M., Perlner, R.A., Polk, W.T., Gupta, S., Nabbus, E.A.: Electronic Authentication Guidance. NIST SP 800–63-2 (2013)
4. Cabinet of Japan: Guideline for Risk Analysis, Digital Signing, and Authentication for On-line Applications and Processing (2010) (in Japanese). http://www.kantei. go.jp/jp/singi/it2/guide/guide_line/guideline100831.pdf
5. Coats, B., Acharya, S.: The forecast for electronic health record access: partly cloudy. In: Proceedings of the IEEE/ACM International Conference on Advances in Social Networks Analysis and Mining, pp. 937–942 (2013)
6. Federal Identity, Credential, and Access Management Trust Framework Solutions: Trust Framework Provider Adoption Process (TFPAP) For All Levels of Assurance (2014). http://www.idmanagement.gov/sites/default/files/documents/ FICAM_TFS_TFPAP_0.pdf
7. GOV.UK: Introducing GOV.UK Verify (2015). https://www.gov.uk/government/ publications/introducing-govuk-verify/introducing-govuk-verify
8. InCommon: The inCommon Assurance Program. http://www.incommon.org/ assurance/
9. ISO: ISO/IEC 29115:2013, Entity authentication assurance framework (2013)
10. ITU-T: Recommendation X.1254, Entity authentication assurance framework (2012)
11. Japanese Bankers Association: FY2013 Financial Statements of All Banks (2014)

12. Kantara: Identity Assurance. https://kantarainitiative.org/idassurance/
13. Noor, A.: Identity protection factor (IPF). In: Proceedings of the IDtrust 2008, pp. 8–18 (2008)
14. NSTIC: National Strategy for Trusted Identities in Cyberspace. http://www.nist.gov/nstic/
15. OASIS: Electronic Identity Credential Trust Elevation Framework V 1.0 (2014). http://docs.oasis-open.org/trust-el/trust-el-framework/v1.0/trust-el-framework-v1.0.pdf
16. Office of Management and Budget: M-04-04: E-Authentication Guidance for Federal Agencies (2003)
17. Sato, H.: $N\pm\epsilon$: reflecting local risk assessment in LoA. In: Meersman, R., Dillon, T., Herrero, P. (eds.) OTM 2009, Part II. LNCS, vol. 5871, pp. 833–847. Springer, Heidelberg (2009)
18. Sato, H.: A formal model of LoA elevation in online trust. ASE Sci. J. $\mathbf{1}$(4), 166–178 (2012)
19. Slomovic, A.: Privacy issues in identity verification. IEEE Secur. Priv. $\mathbf{12}$, 71–73 (2014)
20. The General Insurance Association of Japan: Income Statement (2015) (in Japanese)
21. The Life Insurance Association of Japan: Life Insurance Fact Book 2014 (2014) (in Japanese)
22. The Ministry of Economy, Trade and Industry: 2013 Survey of Selected Service Industries (2014) (in Japanese)
23. The Ministry of Economy, Trade and Industry: Digital Content White Paper 2014 (2014) (in Japanese)
24. The Ministry of Economy, Trade and Industry: Market Research on Electronic Commerce 2015 (2015) (in Japanese). http://www.meti.go.jp/press/2015/05/20150529001/20150529001-3.pdf
25. The Ministry of Internal Affairs and Communications: White Paper on Information and Communications in Japan (2014) (in Japanese)
26. The Ministry of Internal Affairs and Communications and the Ministry of Economy, Trade and Industry: 2012 Economic Census for Business Activity (2012) (in Japanese)
27. The National Police Agency (2010–2015) (in Japanese). https://www.npa.go.jp/cyber/statics/h2{2-6},/pdf041.pdf
28. Third Networks Co.: JOGA Security System for On-line Games and Smartphone Games (2011) (in Japanese). http://www.jssec.org/dl/111117_4_amemiya.pdf
29. Thomas, I., Meinel, C.: An attribute assurance framework to define and match trust in identity attributes. In: Proceedings of the 2011 IEEE International Conference on Web Services, pp. 580–587 (2011)
30. Yong, J., Bertino, E.: Digital identity enrolment and assurance support for VeryIDX. In: Proceedings of the 14th International Conference on Computer Supported Cooperative Work in Design, pp. 734–739 (2010)

First Results of a Formal Analysis of the Network Time Security Specification

Kristof Teichel[1]([✉]), Dieter Sibold[1], and Stefan Milius[2]

[1] Physikalisch-Technische Bundesanstalt, Bundesallee 100,
38116 Braunschweig, Germany
{kristof.teichel,dieter.sibold}@ptb.de
[2] Chair for Theoretical Computer Science, Friedrich-Alexander Universität
Erlangen-Nürnberg, Martensstr. 3, 91058 Erlangen, Germany
stefan.milius@fau.de

Abstract. This paper presents a first formal analysis of parts of a draft version of the Network Time Security specification. It presents the protocol model on which we based our analysis, discusses the decision for using the model checker ProVerif and describes how it is applied to analyze the protocol model. The analysis uncovers two possible attacks on the protocol. We present those attacks and show measures that can be taken in order to mitigate them and that have meanwhile been incorporated in the current draft specification.

Keywords: Time synchronization · Security protocols · Formal verification · Model checking · ProVerif

1 Introduction

In networked infrastructures, time synchronization protocols are often used to synchronize clocks. Many technical infrastructures require reliable time synchronization. Therefore, it is essential to be able to secure time synchronization from malicious attacks. One approach in this area is the Network Time Security (NTS) specification [29], which aims to provide sufficiently generic mechanisms to secure two of the most important time synchronization protocols: the Network Time Protocol (NTP) [21] and the Precision Time Protocol (PTP) [14]. Since the NTS specification is still under development, it is beneficial to apply some form of formal protocol analysis. This might serve to find weaknesses which still exist, as well as to establish which of its components can already be considered secure.

It has been asserted that the difficulty of the formal analysis of a security protocol increases with its dependence on timing [3,4,12]. This is even more the case for NTS, which not only depends on timing, but it additionally distinguishes itself from other timing-dependent security protocols because the underlying secured time synchronization protocol actively influences the clocks of its participants. A thorough analysis has to take into account the impact of timing on the security properties of the specification.

© Springer International Publishing Switzerland 2015
L. Chen and S. Matsuo (Eds.): SSR 2015, LNCS 9497, pp. 218–245, 2015.
DOI: 10.1007/978-3-319-27152-1_12

In this paper, we limit the scope and present the first steps in the analysis of NTS, namely an evaluation of parts of the specification with timing aspects abstracted away. For the analysis, we apply the model checker ProVerif [6] to analyze version 03 of the NTS specification. This unveils two possible attacks. The corresponding vulnerabilities have been removed in the next version of the NTS specification.

The paper is organized as follows. Section 2 provides an overview of the role of timing in the context of secure time synchronization. Furthermore it discusses criteria for different stages of analysis of secure time synchronization specifications, explaining our chosen scope, and it illuminates the choice of the model checker ProVerif as the formal analysis tool for the particular context of this paper. In Sect. 3, we introduce the notation and modeling assumptions that are required for the rest of the paper. Section 4 provides some background on the NTS specification, and presents the protocol steps that are modeled for our analysis. Section 5 describes the resulting model for ProVerif. The obtained results (attacks and countermeasures) are discussed in Sect. 6, and Sect. 7 concludes the paper.

2   Security for Packet-Based Time Synchronization

2.1   Time Synchronization Methods

Figure 1 displays a two-way time transfer message exchange that a client (Alice) and a server (Bob) can use to achieve clock synchronization. At time t_1, Alice initiates a time request message (a) to Bob, where it arrives, after a delay of $\xi\delta$, at time t_2. At time t_3, Bob sends his time response (b) back to Alice, where it arrives after a delay of $(1-\xi)\delta$, at time t_4. The timestamps $T_1, \ldots, T_4$ are added to the exchanged messages by Alice and Bob respectively.[1] Upon arrival of the time response, Alice can calculate her time offset Δ with respect to Bob and the network round-trip delay δ to

$$\Delta = \frac{(T_1 + T_4) - (T_2 + T_3)}{2} + \left(\xi - \frac{1}{2}\right)\delta, \quad 0 < \xi < 1, \tag{1}$$

$$\delta = (T_4 - T_1) - (T_3 - T_2), \tag{2}$$

where δ denotes the round-trip travel time and the asymmetry parameter ξ quantifies the degree of asymmetry in the travel time between the time request and time response messages [16]. Without specific knowledge about the network infrastructure components on the path between Alice and Bob the asymmetry parameter is *a priori* unknown. Hence, in most cases symmetric message delays are assumed, i.e. $\xi = 0.5$. In this case, the second term in the right-hand side of (1) vanishes and Alice can calculate the time offset Δ. Since the ratio between the message delays is important for the time offset calculation, the cryptographic means employed by any security specification must not unduly influence the asymmetry parameter ξ [22].

---

[1] Timestamp T_i corresponds to the reading of the respective system clock at time t_i.

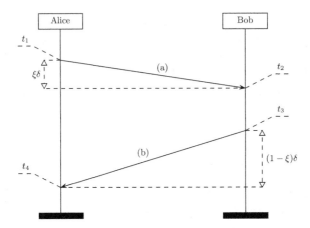

Fig. 1. This is a schematic depiction of a two-way time transfer message exchange that is used for unicast time synchronization between two participants. In the depiction, δ denotes the round-trip time of the complete message exchange and ξ, with $0 < \xi < 1$, quantifies its asymmetry.

There is also a necessity of considering attacks on the timeliness of the exchanged messages, most importantly delay attacks (also called "pulse delay attacks") [11,22]. This kind of attack is characterized by the fact that an adversary can delay the delivery of a message, by first preventing it from being delivered and later replaying it. Such an attack degrades the performance of the time synchronization, it is very simple to perform, and it is not preventable by purely cryptographic means since the content of the exchanged messages does not have to be modified by the adversary [22].

2.2   Criteria for Different Stages of Analysis

We first discuss the criteria underlying the choice of methods and tools for our analysis. It is highly desirable that a full evaluation of a secure time synchronization protocol such as NTS-secured NTP take into account clocks and time as well as properties of cryptographic security protocols. Not only does the information transmitted by the time synchronization protocol concern time data, but the protocol's purpose is to achieve time synchronization. Consequently, the amount of time that it takes for messages to be transmitted is important. Furthermore, time and clock information is relevant for the TESLA protocol [25], which is employed in the NTS specification for broadcast/multicast messages to ensure integrity protection. As a consequence, there is complex interaction between the synchronization operations that the time synchronization protocol performs and the security measures that it uses. All this indicates that the methods and tools used for the analysis of a specification like NTS should ideally be able to consider the notion of time and clocks. Also, the model for time and clocks should be as detailed as possible, without compromising the automated analysis of the model.

However, for parts of the NTS specification, a clean distinction between timing and security-related properties is possible. Specifically, the majority of goals for the unicast mode of NTS, including server authentication, secure key exchange, and integrity protection of time synchronization messages, can be evaluated without considering time or clocks as a part of the model. An analysis using this self-contained scope can already be regarded as a substantial step in the analysis of the whole protocol.

The analysis in this paper was performed at an early stage of the specification process of NTS [29, version 03]. At such a stage, an important criterion is that the methods and tools used for the analysis can find existing weaknesses quickly and point them out clearly. In this case, the specification efforts can benefit from the results of a formal analysis, if the analysis unveils security vulnerabilities in the specification which subsequently can be communicated and considered for the next redrafting of the specification.

These considerations lead us to present an analysis of the unicast mode of NTS, with all of its timing and clock aspects abstracted away, matching the above-mentioned self-contained scope. Since this scope does not require consideration of timing or clocks, the ability to do so can be neglected as a requirement in the choice of methods and tools. We are aware that with this scope, the detection of delay attacks is not possible. The more intricate interactions between cryptographic security and timing would have to be included for a full evaluation of NTS. For such a full evaluation, one could also include an analysis of the broadcast mode of NTS. Further research in that direction is in progress and will be published at a later time [19].

2.3   Choice of Tool for the Analysis

Several approaches are available to perform a formal analysis of a security protocol, the most established being theorem proving and model checking [1]. We first applied Paulson's Inductive Approach [24] for our analysis of NTS. This approach is based on theorem proving and works with the theorem proving tool Isabelle [23]. Using it was motivated, inter alia, by developments which permit physical properties to be considered, especially timing and clocks [4]. Being based on theorem proving, the approach has the drawback that failure to construct a security proof is inconclusive. The reason might either be that the proof is too difficult to find, although the investigated specification is secure, or it might be that the specification is simply not secure. Although counterexample finders exist for Isabelle [7], we were unable to obtain any attack traces for our analysis with the help of such tools. This potential ambiguity is not suitable if the analysis shall support the ongoing drafting process of the considered security specification. Moreover, we found that the pace at which this approach delivers results is slow, at least for non-experts in theorem proving.

The model checking approach offers faster and easier ways to identify vulnerabilities in protocol specifications. With regard to the scope discussed in Subsect. 2.2, we base our formal analysis in this paper on model checking by applying the designated protocol verifer ProVerif [6], which applies a variant of the applied π-calculus with support of types [2,27] as its input language. Note that ProVerif does not support consideration of time and clocks for modeling: although there are time and clock related extensions of the π-calculus [28], none of them are applicable within ProVerif. Note also that it is impossible to simply reproduce the timing-related extension of the Inductive Approach for use with ProVerif. Its lack of support for time and clock consideration would constitute a serious disadvantage of using ProVerif for a full evaluation of NTS. However, for the scope as described in Sect. 2.2 this is negligible, especially since ProVerif fulfills the other criterion very well: it is tailored to be a verification tool for security protocols and offers help in the detection of attack scenarios. Since our results could readily be obtained with ProVerif, we did not investigate the use of other protocol verifiers such as Scyther [8] and Tamarin [18].

3   Basic Assumptions and Protocol Notation

In this section we establish assumptions to simplify the considered model and introduce the essential notation.

Assumption 1. There is exactly one trusted entity, called Trent. Every agent knows Trent's identity, as well as his public key K_T. Every honest agent trusts Trent completely. Certificates (see the list of cryptographic operators below) signed by Trent are assumed to authenticate agents beyond any doubt. In short form, Trent is denoted as T.

Remark 2. For the model, it is a helpful simplification to leave out any intermediate certificates. Therefore, certificate chains that certify an agent X and have Trent as their root authority are seen as equivalent to a certificate for X signed directly by Trent.

Assumption 3. There is exactly one attacker, called Mallory.[2] She complies with the Dolev-Yao attacker model [9], which means that she "controls the network." Explicitly, this means that the attacker has the following capabilities:

– She overhears and intercepts any message sent on the network. In particular, this also means that she can choose to prevent any message from being delivered in its original form.
– She can send messages to any agent on the network, claiming to possess any identity she chooses.
– She can synthesize messages by:
  • inventing new values (it is, however, out of her power to guess secret values like keys or nonces),
  • assembling multiple values known to her into a tuple value,

---

[2] The assumption that there is only one attacker is made for simplification. The assumed situation is equivalent to a situation where several attackers are cooperating, or to a situation where one attacker is being helped by one or more dishonest agents, see Reference [32].

- disassembling any tuple value that she knows into its single component values,
- applying any operator to any value known to her, possibly using any keys, as long as they are also in her knowledge.

In short form, Mallory is denoted as M.

Remark 4. Note that the Dolev-Yao model assumes cryptographic operations to be unbreakable. Thus, although the attacker can claim any chosen identity on a network level, it is still possible to verify authorship of a message by cryptographic means, through appropriate use of secrets.

Notation 5. For convenience and readability, we often use the following "box notation". Concatenation in box notation is displayed by writing the concatenated messages below each other in one box, separated by a dashed line; for example, the concatenation of values x_1 and x_2 is expressed as $\boxed{\begin{array}{c} x_1 \\ \hline x_2 \end{array}}$. The box notation displays use of cryptographic operators as follows. The expression $\mathrm{Op}\left\langle \boxed{x}\right.$ evaluates to $\mathrm{Op}(x)$, where Op can be any one of the cryptographic operators presented below, and where x is usually a concatenated term.

Notation 6. We now provide some notation on cryptographic operators as used in the further presentation of the protocol. Note that a term inside square brackets [ ] generally denotes a key.

- The expression $\mathrm{Enc}[K](m)$ stands for the ciphertext that results from using asymmetric cryptography to encrypt the message m with the key K.
- Following the notion of asymmetric cryptography, it is assumed that for every key K, there exists an inverse key K^{-1}, such that both of the equations $\mathrm{Enc}[K^{-1}]\left(\mathrm{Enc}[K](m)\right) = m$ and $\mathrm{Enc}[K]\left(\mathrm{Enc}[K^{-1}](m)\right) = m$ are satisfied.
- The expression $\mathrm{Sign}[K](m)$ evaluates to $\mathrm{Sign}[K](m) = \mathrm{Enc}[K^{-1}](m)$, which describes the signature of the message m created in such a way that it can be validated with the key K.
- The expression Cert_X evaluates to $\mathrm{Cert}_X = \mathrm{Sign}[K_T]\left\langle \boxed{\begin{array}{c} X \\ \hline K_X \end{array}}\right.$, which represents the certificate for X's public key K_X, issued by Trent.
- It is assumed that all participants have agreed to use a fixed cryptographic hash function h.[3] For a given key value K and message m, the expression $\mathrm{HMAC}[K](m)$ stands for the keyed-hash message authentication code computed over m, with K as key and h as the cryptographic hash function, as defined in RFC 2104 [15].

---

[3] Note that the NTS specification includes negotiation of hash functions as part of the protocol. However, as stated under Assumption 3, all cryptography is assumed to be unbreakable. Therefore, algorithm negotiation has been ignored for our analysis. It might be interesting to include these steps in future analysis, to look for downgrade attacks.

- The expression Seed_X represents an agent X's seed, a random secret value that is known only to X.
- The expression $\text{Cook}_X(v)$ evaluates to $\text{Cook}_X(v) = \text{HMAC}[\text{Seed}_X](v)$, which represents the cookie that agent X generates, using its own secret seed as well as the given input value v.

4   The Protocol Steps Under Analysis

4.1   The Network Time Security Project

The Network Time Security (NTS) specification aims at secure time synchronization over networks like the Internet. It is motivated by the fact that neither of the predominant time synchronization protocols, in particular the Network Time Protocol (NTP) [21] and the Precision Time Protocol (PTP) [14] currently provide adequate security mechanisms (see, e.g., Reference [26] for an analysis of the NTP's most current security measure, the Autokey protocol [20]). It has the goal of being usable to secure at least those two protocols [29]. Currently the specification is in the standardization process in the Internet Engineering Task Force (IETF).

Time synchronization in NTP unicast mode is secured via a secret value (called "cookie") which is unique for each client-server association. The nature of this cookie is such that the server can always deterministically re-generate it from an input value given by the client, while the client simply memorizes it. It is used as a key for generating a keyed-hash message authentication code (MAC) for each time synchronization response message going from the server to the client. Under the requirement that the cookie has been exchanged securely (its secrecy and authenticity need to be guaranteed), this procedure shall give authenticity and integrity guarantees for the time synchronization response messages. Having cookie and MAC generation based on keyed-hash mechanisms shall keep these steps fast enough to not significantly influence the quality of time synchronization [29].

4.2   Overview of the Protocol Sequence

We now describe the protocol messages and the protocol sequence. We present the protocol steps as specified in the NTS draft version 03, which constitutes the basis for the verification described in this paper. We only present those messages that are relevant for the scope of our analysis here. Furthermore, as discussed in Sect. 2.2, we abstract away any timing and clock aspects of the message exchanges that are presented.

The **certification** message exchange serves to supply the client Alice with the server Bob's certificate. The message exchange proceeds according to the protocol steps depicted in Table 1.

After the message exchange, Alice performs two checks. First, she verifies the validity of Cert_B. Second, she verifies the validity of the included signature. If all of this is successful, Alice trusts any subsequent message if it contains a signature that she is able to verify as originating from Bob.

Table 1. (Certification) These are the two message formats used for transmitting the certificate for server Bob (B) to client Alice (A) according to protocol version 0.3.0.

Name	Direction	Contents
client_cert	$A \rightarrow B$:	A
server_cert	$B \rightarrow A$:	$\text{Sign}[K_B](A)$ Cert_B

The **cookie establishment** message exchange establishes the shared secret called the *cookie*, which is calculated by the server (deterministically and based on data supplied by the client) and then securely transmitted back to the client. The message exchange is performed according to the steps visible in Table 2.

After the message exchange, Alice decrypts the received response, and then performs two checks on the encrypted data. First, she validates that the nonce included in the response is the same one that she included in her request. Second, she confirms the validity of the signature included in the response. If all of this is successful, Alice trusts any subsequent message if it contains a MAC calculated with the cookie she received via this message exchange.

The purpose of the **unicast time synchronization** message exchange is to perform the actual time synchronization. Security is based on the secret cookie, which must be exchanged prior to this message exchange (via a cookie establishment exchange). The time synchronization message exchange follows the steps listed in Table 3, where S_A and S_B denote the time synchronization data transmitted by the client and the server, respectively. Note that with respect to the unicast time synchronization procedure described in Sect. 2 (see Fig. 1), S_A includes a timestamp of t_1, whereas S_B includes timestamps of t_1, t_2 and t_3.

After the message exchange, Alice performs two checks on the response. First, she validates that the nonce included in the response is the same one that she included in her request. Second, she confirms the validity of the MAC included in the response. If all of this is sucessful, Alice uses the received time data S_B for time synchronization.

5   Performing the Analysis

In order to use ProVerif for analysis of the NTS protocol, we model its participant roles (client and server) as processes in ProVerif's input language. The resulting specification is fed into ProVerif, which is able to analyze the protocol and prove reachability properties, correspondence assertions, as well as observational equivalence [6]. Note that having ProVerif check any single goal and, if applicable, generate an attack trace took merely a few seconds on a standard laptop computer.[4]

---

[4] The computer was running a 64 bit version of Windows 7 on an Intel i5 dual core at 2.6 GHz, with 8 GB of RAM.

Table 2. (Cookie Exchange) These are the two message formats used for the cookie exchange between client Alice (A) and server Bob (B) according to protocol version 0.3.0.

Name	Direction	Contents
client_cook	$A \to B$:	$\boxed{\begin{array}{c} N \\ \hline K_A \end{array}}$
server_cook	$B \to A$:	$\mathrm{Enc}[K_A]\left\langle \boxed{\begin{array}{c} N \\ \hline \mathrm{Cook}_B(h(K_A)) \\ \hline \mathrm{Sign}[K_B]\left\langle \boxed{\begin{array}{c} N \\ \hline \mathrm{Cook}_B(h(K_A)) \end{array}}\right. \end{array}}\right.$

Table 3. (Unicast Time Synchronization) These are the two message formats used for synchronizing the clock of client Alice (A) with that of server Bob (B) according to protocol version 0.3.0.

Name	Direction	Contents
time_request	$A \to B$:	$\boxed{\begin{array}{c} S_A \\ \hline N \\ \hline h(K_A) \end{array}}$
time_response	$B \to A$:	$\boxed{\begin{array}{c} S_B \\ \hline N \\ \hline \mathrm{HMAC}[\mathrm{Cook}_B(h(K_A))]\left\langle \boxed{\begin{array}{c} S_B \\ \hline N \end{array}}\right. \end{array}}$

ProVerif allows the user to specify so-called "queries", which can be used to specify the required protocol goals (Reference [6, Sect. 2] provides a short introduction to how ProVerif interprets queries). When a goal has been specified in this way, ProVerif can look for a protocol state which violates the condition corresponding to that goal. It can then return one of three possible results:

– There is no state which violates the goal condition.
– A state could be constructed which violates the given goal condition. In this case, ProVerif shows a trace of events leading to that state. The violation of the given condition might represent a viable attack on the protocol but it might also uncover an error in the model underlying the ProVerif code.
– ProVerif cannot prove that the goal is correct but it cannot construct a state violating the goal condition either.

The last case can occur because ProVerif works on an over-approximation of the state space of the input specification. Thus, ProVerif may find a state that violates the goal condition but that is not part of the state space of the input specification. Over-approximation is unavoidable since the decision problem underlying ProVerif is known to be undecidable [10] (however, it is of course semi-decidable). See Reference [6, Sect. 3.3.1] for some more details on the different possible results and how to interpret them, in particular for the third of the above cases.

We now provide an overview of how client and server are modeled in ProVerif. The ProVerif code relevant to the model of the originally analyzed version can be seen in Appendix A). Alternatively, the full code for use with ProVerif can be downloaded under https://www8.cs.fau.de/staff/milius/ProVerif-NTS.rar. The client is modeled via two ProVerif processes:

– The **inner client process** performs, exactly once, the message exchange of the time synchronization phase.
– The **outer client process** runs through the appropriate message exchanges for certificate exchange and cookie establishment exchange in chronological order. When this is done, it ultimately starts a bulk of "infinitely many" iterations[5] of the inner client process running in parallel.

Accurately modeling the server is more complex than modeling the client. This is a consequence of the fact that the server is required to be stateless (except for the global server seed which is client independent). Nevertheless, the server must be able to reply at any time to any of the specified protocol messages, for an arbitrary number of clients. In the model, this is achieved by having one process type for each possible message exchange and running all of these process types in parallel, each with "infinitely many" iterations.

Consequently, the server is modeled via the following ProVerif Processes:

– There are three **inner server processes**, which perform the respective message exchanges for certification, cookie exchange, and time synchronization.
– The **outer server process** just starts "infinitely many" iterations of each of the inner processes, all running in parallel.

6   Results of the First Analysis

In the course of the analysis, two attacks were discovered by checking the protocol goals, listed below as Goals 7–9, with ProVerif. The first discovery is an attack detected by ProVerif mostly due to imprecision in the model underlying the code version c030ut, which was used at the time. This version does not support enough types to distinguish between a single hostname, on the one hand, and a complex structure of arbitrarily many arbitrary data types, on the other hand. The second (and more practically relevant) discovery is a Man-in-the-Middle attack. It exploits the fact that in NTS, as specified in reference [29, version 03],

---

[5] Having "infinitely many" iterations is what is modeled for the analysis. In practice, this will obviously be a finite number.

a client Alice is given no indication as to whether a cookie has been generated based on her own public key or someone else's.

The goals for this analysis were derived during development of the specification of the protocol, an unusual situation which is due to the fact that the analysis was performed by people who were also authors of the specification. The analysis with ProVerif also involved checks for protocol sanity goals, e. g. that it is possible for the protocol to finish successfully, i. e. with the client accepting some time synchronization data. It additionally involved a weak authenticity goal for the cookie exchange (a weaker version of Goal 9 listed below), but this was only helpful to the extent that it pointed toward the improtance of Goal 8. The ProVerif source code for the queries associated with all of the mentioned goals can be inspected in Appendix A.8. We do not discuss the sanity goals or the weak authenticity goal in more detail, as they do not contribute essential results for the analysis. Instead, we now present the specific security goals directly relevant for this analysis. The first of these goals is the one that represents the full extent of positive security affirmation that our analysis can provide. Both of the attacks that were found lead to a violation of this goal.

Goal 7 (Time Synchronization Authenticity). If Alice accepts the data from a *time_response* message as authentic from Bob, then Bob has indeed sent a *time_response* message with the same time data and the same nonce, secured with the correct cookie for the association between Alice and Bob.

We now present two intermediary security goals which aid with getting more specific information about the nature of the attacks. Both of these goals are necessary preconditions for Goal 7, since the integrity of time synchronization messages depends on the cookie being shared securely between Alice and Bob.

Goal 8 (Cookie Secrecy). If Alice accepts a cookie C from a *server_cook* message as being legitimate, then C is unknown to Mallory.

Goal 9 (Cookie Authenticity). If Alice accepts a cookie C from a *server_cook* message as being legitimate for her association with Bob, then Bob has indeed sent a *server_cook* message in which he has transmitted C and which he has intended for Alice.

We now describe the first of the two attacks that were discovered. This attack violates even the very specific Goal 9, which gives a hint about the underlying weakness: this is not an attack where a legitimate cookie is spied upon, but where a completely new cookie is invented and Alice is deceived into accepting it.

Attack 10. ProVerif discovered an event trace which violated Goal 9 as well as Goal 8. From that trace, the following attack scenario could be derived:

- Mallory intercepts a *client_cook* message CC_1 containing a nonce N.
- She memorizes both values and prevents the message from being delivered in its original form.
- She invents some cookie value X and concatenates it with N.

- She sends a *client_cert* message to the server Bob, which instead of her name M (see the appropriate message diagram for *client_cert* in Sect. 4) contains the concatenated value $\dfrac{N}{X}$.
- The server sends the appropriate *server_cert* message back to Mallory. It contains two values: Bob's certificate, which is irrelevant, and additionally a signature $\mathrm{Sign}[K_B]\left\langle\begin{array}{c}N\\X\end{array}\right.$, which Mallory memorizes.
- Mallory then employs concatenation of values known to her as well as encryption under Alice's public key to create a *server_cook* message SC_3 which reads

$$\mathrm{Enc}[K_A]\left\langle\begin{array}{c}N\\X\\\mathrm{Sign}[K_B]\left\langle\begin{array}{c}N\\X\end{array}\right.\end{array}\right. .$$

- She sends SC_3 back to Alice, who decrypts it, validates the nonce and signature, and finally accepts the cookie X as the legitimate cookie sent to her from Bob.

As mentioned before, this attack works only if Bob does not recognize when a *client_cert* message contains the concatenation of a nonce and a cookie-length value instead of a hostname. Under this assumption, anyone can get a signature from an arbitrary server for an arbitrary message by abusing the certification message exchange. However, this is unlikely in any practical implementation.

Countermeasure 11. The introduction of proper identifiers for the different message components completely defeats this attack. Successful ProVerif verification, performed on a more strongly typed model, supports this, as at least Goal 9 has been shown to hold there.[6]

In the following we describe the second of the two revealed attacks.

Attack 12. ProVerif discovered an event trace in which Goal 8 (Cookie Secrecy) was violated. From this trace, the following attack scenario could be deduced:

- Mallory intercepts a *client_cook* message CC_1 containing a nonce N and Alice's public key K_A.
- She memorizes both values and prevents the message from being delivered in its original form.
- She creates a new *client_cook* message CC_2 by concatenating the received nonce N with her own public key K_M.
- She sends this newly created *client_cook* message CC_2 to the server Bob.

---

[6] The reason why Goals 7 and 8 do not hold for this model is that Countermeasure 11 only defends against Attack 10, not against Attack 12.

- Bob sends back the *server_cook* message SC_1, which has the format

$$\text{Enc}[K_M]\left\langle \begin{array}{l} N \\ \text{Cook}_B(h(K_M)) \\ \text{Sign}[K_B]\left\langle \begin{array}{l} N \\ \text{Cook}_B(h(K_M)) \end{array} \right. \end{array} \right. ,$$

which can hence be decrypted by Mallory and which contains a cookie and a valid signature confirming that Bob has created and sent this cookie as an answer to the request with nonce N.

- Mallory intercepts SC_1 and may prevent it from being delivered in its original form. She decrypts it and then instantly re-encrypts it with Alice's public key K_A, resulting in a new *server_cook* message SC_2.
- She sends SC_2 to Alice, who decrypts it, validates the nonce and signature, and finally accepts the cookie $\text{Cook}_B(K_M)$ as the legitimate cookie sent to her from Bob.

As mentioned above, the given attack is only viable due to the following two reasons:

1. On receipt of a *server_cook* message, it is impossible for anyone to know which public key was used as input value for the generation of the given cookie (because of the one-way property of HMAC and the secrecy of the server seed).
2. On receipt of such a message, it is also impossible for anyone to know for whom the message was originally encrypted (since decrypting and then re-encrypting a message leaves no trace).

The reasons above are interdependent, because an honest server like Bob will always encrypt a *server_cook* message with exactly the same public key upon which the generation of the cookie included in that message is based. We provide two possible countermeasures against Attack 12, both supported by ProVerif validations.

Countermeasure 13. The simple addition of the input value used for the cookie generation into the set of values covered by the signature prevents the described attack. This measure was taken for model version 0.3.1 of the protocol. The resulting message format for *server_cook* can be seen in Table 4. With the ProVerif model updated accordingly, Goal 8 and even Goal 7 can be verified to hold.

Countermeasure 14. An alternate countermeasure is to switch the order of encrypting and signing for the *server_cook* message to encrypting first, then signing. This measure was taken for model version 0.3.2 of the protocol. The resulting message format for *server_cook* can be seen in Table 5. Goal 8 and also Goal 7 can then be verified to hold for an accordingly updated model of the protocol.

Table 4. These are the two updated message formats used for the cookie exchange between client Alice (A) and server Bob (B) according to protocol version 0.3.1.

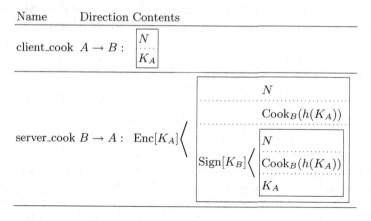

Table 5. These are the two message formats used for the cookie exchange between client Alice (A) and server Bob (B) according to protocol version 0.3.2.

Countermeasure 14 has some small advantages over Countermeasure 13: it is simpler to include in a specification, and it also has the benefit that an invalid signature can be detected at a slightly lower computational cost, because no decryption operation is necessary in order to obtain the signature value.

7   Conclusion

We have used ProVerif for the analysis of an early version of the NTS specification. We were able to discover considerable weaknesses in draft version 03 of the specification. Those discoveries were considered in the further development NTS and by version 04, the specification was updated [29, Version 04, Sect. 6.3.2] with a measure conforming to Countermeasure 13, which mitigates Attack 12.

In version 05, the usage of the Cryptographic Message Syntax (CMS) [13] was introduced [29, Version 05, Sect. 6] and a new separate document was created for details on how to use this [30, Version 00]. The usage of the CMS conforms to Countermeasure 11 and provides sufficient data type information to help prevent Attack 10. Moreover, the specification has adapted changes in the cookie exchange that match what is described under Countermeasure 14 [29, Version 05, Sect. 6.2.2], [30, Version 00, Sects. 2 and 4.2.2.2]. Overall, these changes prevent both of the attacks presented in Sect. 6, and therefore represent important improvements.

The discoveries are quite common (and, for experts, perhaps obvious) attacks. An important lesson learned in the course of this analysis is: when designing a new security protocol, it is sensible to check its basic structure with a formal tool, even if this tool may not be sophisticated enough to analyze the finer details. A rough formal analysis may already yield valuable results, as even very basic weaknesses may be overlooked when just considering a protocol on paper. Therefore, it can be useful for the analysis to use abstraction in order to dismiss a complex aspect (like timing in our case), as long as the remaining aspects enable a self-contained examination. Furthermore, a question that one might deduce from the analysis is: to what extent could attacks like Attack 10, which are preventable through the use of well defined data types, be relevant in practice? Do protocol developers need to sufficiently consider the question of data types, or is this a matter that should be left to the implementers? It is also noteworthy that we skipped over a deeper evaluation of different model checkers in the course of this analysis, which leaves room for further research comparing the relative merits of different protocol verifiers for the analysis of protocols and draft standards.

For a complete analysis of the full NTS specification, timing and clocks need to be considered (see Sect. 2). Therefore, further work is necessary to properly consider the timing aspect in interaction with security. To this end, ProVerif is not the suitable tool, because it is unable to consider timing and clocks. Model-checkers other than ProVerif (specifically UPPAAL [5] and TAuth [17]) are being considered, but it is yet unclear if they indeed support the considerations of all necessary aspects for a thorough analysis of secure time synchronization techniques. In particular, finding the right approach and tool is important to analyze the security aspects of broadcast time synchronization with NTS, where TESLA is used for authenticity and integrity protection. Using TESLA to secure time synchronization is especially tricky, since TESLA requires rough time synchronization not only initially, but continuously. Hence, successfully disturbing time synchronization just by a small amount might endanger the security of the whole protocol.

We have started a detailed analysis [19] focusing on an intricate attack using delay techniques to compromise time synchronization when it is secured with TESLA-like mechanisms. A draft version of this work is available under https://www8.cs.fau.de/staff/milius/AttackPossibilityTimeSyncTESLA.pdf.

An automated (or semi-automated) formal analysis of the full NTS specification would be beneficial because it could be re-used for other time synchronization protocols which employ TESLA-like mechanisms, for example the TinySeRSync protocol [31] for wireless sensor networks.

A   ProVerif Source Code

In this appendix, we present some of the relevant ProVerif source code used for the preparation of this paper. There have been different code versions involved in the analysis:

- **Code version c030ut:** It models protocol version 0.3.0 but has no dedicated type for hostnames.
- **Code version c030:** It models protocol version 0.3.0 and does have a dedicated hostname type.
- **Code version c031:** It models protocol version 0.3.1.
- **Code version c032:** It models protocol version 0.3.2.

All of the code below is taken from code version c030, which formed the basis of the analysis. Presenting the other versions in full would take up a lot of space, whereas presenting only the differences is difficult. This is because although the changes between the different code versions are minor, they still include numerous lines that are spread far apart. Some comment lines and structuring have been left out for the presentation.

For any reader who is interested in access to the complete code (all code versions, ready for use with ProVerif), it is available under https://www8.cs.fau.de/staff/milius/ProVerif-NTS.rar.

A.1   Cryptographic Primitives

The first lines of the ProVerif source code form the cryptographic primitives that are needed.

```
(* Basics: Keys and Hostnames *)
   type key.
   type hostname.

(* Symmetric Encryption *)
   fun senc(bitstring, key): bitstring.
   reduc forall m: bitstring, k: key;
        sdec(senc(m,k),k) = m.

(* Asymmetric Encryption *)
   type skey.
   type pkey.
   fun sk_of(hostname): skey [private].
   fun pk(skey): pkey.
   letfun pk_of(X: hostname) = pk(sk_of(X)).
```

```
fun aenc(bitstring, pkey): bitstring.
reduc forall m: bitstring, k: skey;
      adec(aenc(m, pk(k)), k) = m.
```

```
(* Asymmetric Signatures *)
  type sskey.
  type spkey.
  fun ssk_of(hostname): sskey [private].
  fun spk(sskey): spkey.
  letfun spk_of(X: hostname) = spk(ssk_of(X)).

  fun sign(bitstring, sskey): bitstring.
  reduc forall m: bitstring, k: sskey;
        getmess(sign(m,k)) = m.
  reduc forall m: bitstring, k: sskey;
        checksign(sign(m,k), spk(k)) = m.
  letfun signed_message(m: bitstring, k: sskey)
        = (m, sign(m, k)).
```

```
(* Hash and HMAC Functions *)
  fun hash(bitstring): bitstring.
  fun keyhash(pkey): bitstring.

  fun seed_of(hostname): bitstring [private].

  fun cookie_gen(bitstring, bitstring): key.
  fun hmac(key, bitstring): bitstring.
```

A.2   Global Variables and Constants

Next we have the declarations of channels and other global variables and constants.

```
(* The Channel for All"Network" Protocol Communications *)
  free c: channel.
```

```
(* The Channel and Hostname for Communication with the TA *)
  free ta: channel.
  free TA: hostname.
```

```
(* Two Possible Version Identifiers *)
  free version_1: bitstring.
  free version_2: bitstring.
```

A.3   Events

The declarations of ProVerif "events" follow that are used for better traceability of what is happening in what order.

```
(* Server Side Events *)
   event serverError().

   event serverHasOwnCert(hostname, pkey, spkey).
   event serverRespondsTime(bitstring, bitstring,
                            bitstring).

   event serverSaysCookie(key, pkey, hostname).
   event serverGeneratesCookie(key, hostname).

   event serverAcceptsCert(hostname).
   event serverRejectsCert().

(* Client Side Events *)
   event clientError().
   event clientHasOwnCert(hostname, pkey, spkey).

   event clientAcceptsCookie(key, hostname, hostname).
   event clientAcceptsSomeCookie().

   event clientRejectsCookie(key, hostname, hostname).
   event clientRejectsSomeCookie().

   event cookieCompromised(key).

   event clientAcceptsCert(hostname, spkey).
   event clientRejectsCert().

   event client_timereq(bitstring).
   event clientAcceptsTime(bitstring, bitstring,
                           bitstring).
   event clientDiscards(bitstring).

(* Authority Side Event(s) *)
   event authorityGivesCert(hostname, pkey, spkey).
```

A.4   The Trusted Authority Process

The code then moves on to the processes by which the participants are modeled. We first present the process that models the trusted authority. Its only purpose is to generate and distribute certificates for server and client type participants.

```
(* [Process] ::: TRUSTED AUTHORITY ::: [NTS v0.3.0]
   |: Issues certificates on request. *)
   let authority() =
   let skTA = ssk_of(TA) in

     (* The authority receives a certificate request. *)
       in(ta, H: hostname);
```

```
(* The authority determines the correct public keys
   for the requester. *)
let pkH = pk(sk_of(H)) in
let spkH = spk(ssk_of(H)) in

(* From that information, the authority assembles
   and signs the appropriate certificate. *)
let certificate = (H, pkH, spkH) in
let signature = sign(certificate, skTA) in
let certificate_sig = (certificate, signature) in

(* The authority sends the response back to the
   requesting participant. *)
event authorityGivesCert(H, pkH, spkH);
out(ta, certificate_sig).
```

A.5   The Server Side Processes

Inner Server Processes. Next are those processes that make up the model of the server. The first process represents the server module which deals with the certification message exchange.

```
(* [Process] ::: SERVER CERTIFIER MODULE ::: [NTS v0.3.0]
   |: Replies to a client_cert message with a server_cert
   |: message as specified. *)
let server_certifier(B: hostname, pkB: spkey,
                     skB: sskey) =

  (* The server acquires the TA's public key. *)
    let pkTA = spk(ssk_of(TA)) in

  (* The server receives a client's certification request. *)
    in(c, X_client_cert: bitstring);

  (* The server extracts the necessary information. *)
    let (version_x: bitstring, A_x: hostname)
        = X_client_cert in

  (* The server requests its certificate chain from the
     trusted authority and performs a validity check on
     the response that it receives. *)
    out(ta, B);
    in(ta, Z_certificate: bitstring);
    let (=B, some_key: pkey, =pkB,
         Z_cert_signature: bitstring)
        = Z_certificate in
    if (B, some_key, pkB)
        <> checksign(Z_cert_signature, pkTA)
    then event serverError()
```

```
         else event serverHasOwnCert(B, some_key, pkB);

      (* The server creates a server_cert response
         as specified. *)
         let msg_server_cert = (A_x, Z_certificate) in
         let msg_server_cert_sign
            = (msg_server_cert,
                sign(msg_server_cert, skB))
            in

      (* The server sends the composed response to the
         requesting client. *)
         out(c, msg_server_cert_sign).
```

The next process involves the server module whose purpose it is to execute the cookie message exchange as well as the required calculations.

```
(* [Process] ::: SERVER COOKIE MODULE ::: [NTS v0.3.0]
   |: Takes a client_cook request, generates the appropriate cookie
   |: and replies with a server_cook message as specified. *)
   let server_cookie(B: hostname, pkB: spkey, skB: sskey,
                     seed: bitstring) =

   (* The server acquires the TA's public key. *)
      let pkTA = spk(ssk_of(TA)) in

   (* The server receives a client's cookie request. *)
      in(c, X_cook: bitstring);

   (* The server matches it to the specified message
      pattern and extracts the necessary information. *)
      let (n_x: bitstring, pkA_x: pkey) = X_cook in

   (* The server builds the cookie for the received client
      (identified via its public [encryption] key pkA. *)
      let cookie = cookie_gen(keyhash(pkA_x), seed) in

   (* It builds the appropriate response *)
      let response = (cookie, n_x) in

   (* It constructs its signature and attaches it to the response. *)
      let signature = sign(response, skB) in
      let response_sig = (response, signature) in

   (* It encrypts it. *)
      let response_sig_enc = aenc(response_sig, pkA_x) in

   (* It sends it back to the client. *)
      event serverGeneratesCookie(cookie, B);
      event serverSaysCookie(cookie, pkA_x, B);
```

```
out(c, response_sig_enc).
```

Then the server module follows that takes care of the time synchronization message exchange.

```
(* [Process] ::: SERVER TIMESYNC MODULE ::: [NTS v0.3]
  |: Replies to a time_request message with a time_response message as
  |: specified. *)
  let server_time_response(B: hostname, pkB: spkey,
                           skB: sskey, seed: bitstring) =

    (* The server receives a time_request message from a client. *)
    in(c, Y: bitstring);

    (* It extracts the necessary data. *)
    let (t1_y: bitstring, n_y: bitstring,
         pkA_hash_y: bitstring) = Y in

    (* It creates the appropriate time sync data for its response. *)
    new t2: bitstring;

    (* It re-computes the cookie. *)
    let cookie = cookie_gen(pkA_hash_y, seed) in

    (* It composes its response. *)
    let response = (n_y, t1_y, t2,
                    hmac(cookie, (n_y, t1_y, t2)))
            in

    (* It sends its response back to the requesting client. *)
    event serverRespondsTime(n_y, t1_y, t2);
    out(c, response).
```

Outer Server Process. We then see the "outer" server process whose purpose is simply to execute iterations of all the "inner" processes (the modules listed above in Appendix A.5).

```
(* [Process] ::: SERVER GLOBAL PROCESS ::: [NTS v0.3.0]
  |: Executes all server modules at once, running arbitrarily many
  |: instantiations of each of them in parallel. *)
  let server(B: hostname) =

    (* Before running any modules, the server generates an
       unpredictable seed value and remembers its own key pair. *)
    let seed = seed_of(B) in
    let skB = ssk_of(B) in
    let pkB = spk(skB) in

    (* The server then runs all modules. *)
```

```
      !server_certifier(B, pkB, skB)
    | !server_cookie(B, pkB, skB, seed)
    | !server_time_response(B, pkB, skB, seed).
```

A.6   The Client Side Processes

Inner Client Process. Moving on to the client side processes, there is first the "inner" process which takes care of the time synchronization message exchange, including the necessary checks on the MAC.

```
(* [Process] ::: CLIENT TIMESYNC MODULE ::: [NTS v0.3.0]
  |: Generates time_request messages as specified and sends them to a
  |: time server. It then awaits a time_response message on which it
  |: performs the necessary checks as specified. *)
  let client_time_request(A: hostname, pkA: pkey,
    B: hostname, cookie: key) =

    (* The client generates time data and a nonce. *)
      new t1: bitstring;
      new n1: bitstring;

      event client_timereq(t1);

    (* The client constructs its time_request message and sends it. *)
      let request = (t1, n1, keyhash(pkA)) in
      out(c, request);

    (* It receives a time_response message and extracts the necessary
       information. *)
      in(c, X: bitstring);
      let (=n1, =t1, t2x: bitstring, hmacx: bitstring)
          = X in

    (* Depending on the result of validity checks, it either accepts
       the response as authentic or discards it. *)
      if hmacx = hmac(cookie, (n1, t1, t2x))
          then event clientAcceptsTime(n1, t1, t2x)
          else event clientDiscards(X).
```

Outer Client Process. The "outer" client side process follows, which performs the initial message exchanges (server certification and cookie exchange) and then executes instantiations of the inner process.

```
(* [Process] ::: CLIENT GLOBAL PROCESS ::: [NTS v0.3.0]
  |: Executes the steps for association, certification and cookie
  |: exchange, one of each, sequentially. Then it runs arbitrarily
  |: many instances of the client timesync module in parallel. *)
  let client(A: hostname, B: hostname) =
```

```
let skA = sk_of(A) in
let pkA = pk(skA) in

let pkTA = spk(ssk_of(TA)) in

(* CERTIFICATE PHASE ---------------------------*)

    (* The client sends a client_cert message as specified *)
        let msg_client_cert = (version_1, A) in
        out(c, msg_client_cert);

    (* The client receives a response of type server_cert. *)
        in(c, X_server_cert: bitstring);

    (* The client extracts data from the response. *)
        let (=A, certificate_x: bitstring,
             signature_x: bitstring)
            = X_server_cert in

    (* The client reads the certificate. *)
        let (=B, other_key: pkey, spkB_x: spkey,
             cert_signature: bitstring)
            = certificate_x in

    (* The client performs the necessary test. On failure, it
       exits with an error. On success, the client accepts the
       key given in the certificate as B's public key. *)
        event check();
        if ((B, other_key, spkB_x)
            <> checksign(cert_signature, pkTA))
           || ((A, certificate_x)
               <> checksign(signature_x, spkB_x))
        then event clientRejectsCert()

        else event clientAcceptsCert(B, spkB_x);

            (* COOKIE PHASE ----------------------*)
                let pkB = spkB_x in


            (* The client sends a client_cook message as specified. *)
                new n_cook: bitstring;
                let msg_client_cook = (n_cook, pkA)
                    in
                out(c, msg_client_cook);

            (* The client receives a response of type server_cook. *)
                in(c, X_server_cook: bitstring);

            (* It decodes the response and extracts the data from it. *)
```

```
let X_dec = adec(X_server_cook, skA)
    in
let ((cookie_x: key, =n_cook),
     signature_x: bitstring) = X_dec
    in
```

```
(* It performs the necessary checks as specified. On success,
   it starts sending time_request messages as specified.*)
if (cookie_x, n_cook)
    = checksign(signature_x, pkB)
then event
    clientAcceptsCookie(cookie_x, A, B)
  (* TIMESYNC PHASE ------------------*)
   | !client_time_request(A, pkA,
                          B, cookie_x)
   | ( in(c, =cookie_x);
       event cookieCompromised(cookie_x)
```

```
else event
    clientRejectsCookie(cookie_x, A, B)).
```

Note that the first **else**-branch includes all the code below it. Note also the dedicated listener process given by

```
   | ( in(c, =cookie_x);
       event cookieCompromised(cookie_x)
```

which is started when the client accepts a cookie and does not really represent client behavior according to the protocol, but only listens for the cookie on an open channel. This enables us to check for the loss of a cookie by querying whether the event `cookieCompromised()` is ever executed at all (see Appendix A.8).

A.7   The Environment Process

Here, we present the global ProVerif process which takes care of instantiating all participants.

```
(* [Process] MAIN OVERALL PROCESS ::: [NTS v0.3.0]
   |: Runs everything that needs to be run. *)
process

   (* More strongly typed version with hostnames,
      would otherwise be"bitstring" type variables *)
   new B: hostname;
   new A: hostname;

   (* There are arbitrarily many clients running, but only one server,
      for simplification in the earlier phase of this analysis. *)
      !server(B) | !client(A, B) | !authority()
   | out(c, A)  | out(c, B) )
```

A.8   ProVerif Queries

Now we present the ProVerif queries. We first consider those queries that concern the cookie exchange.

Sanity – Cookie Exchange. This query makes sure that there is some cookie x which is accepted by the honest client A as coming from the honest server B, i.e. the cookie exchange can be completed successfully.

```
query x: key;
      event(clientAcceptsCookie(x, new A, new B)).
```

This query holds for all four code versions.

Weak Authenticity – Cookie. This query ensures that if the honest client A accepts a cookie x for communication with the honest server B, then B has in fact generated x and released it into the network (note that this gives no guarantee that B intended x for communication with A in particular).

```
query x: key;
      event(clientAcceptsCookie(x, new A, new B))
  ==> event(serverGeneratesCookie(x, new B)).
```

Applying this query to the different ProVerif code versions yields the following results:

- It does **not** hold for code version c030ut. The attack that ProVerif discovers is the first one described in Sect. 6.
- It holds for the code versions c030, c031, and c032.

Authenticity – Cookie. This query strengthens the guarantee acquired with the previous query: it ensures that if the honest client A accepts a cookie x for communication with the honest server B, then B has in fact issued x based on A's public key, and has also encrypted the appropriate message with said public key. This is the query that corresponds to Goal 9.

```
query x: key;
      event(clientAcceptsCookie(x, new A, new B))
  ==> event(serverSaysCookie(x, pk(sk_of(new A)), new B)).
```

The results for this query are as follows:

- It does **not** hold for code version c030ut. Authenticity for the cookie would require weak authenticity for it, which is not given (see above). Also, the Man-in-the-Middle attack described in Sect. 6 works on this version.
- It does **not** hold for code version c030. Again, the corresponding attack is the Man-in-the-Middle attack described in Sect. 6.
- It holds for the code versions c031, and c032

Secrecy – Cookie. This query asserts that if the honest client A accepts a cookie x, then the attacker does not know x. This is realized via the event `cookieCompromised()` from the dedicated listener process as described in Appendix A.6. This query corresponds to Goal 8.

```
query x: key;
    event(cookieCompromised(x)).
```

- This query does **not** hold for code version c030ut and **neither** does it hold for c030: Both of the possible attacks enable Mallory to make Alice accept a cookie that Mallory knows. In the case of the Blind-signature attack she manufactures said cookie herself; in the case of the Man-in-the-Middle attack she maliciously re-distributes a valid key, signed by Bob.
- This query holds for the code versions c031 and c032.

Next, we take a look at some queries that concern the time synchronization message exchange.

Sanity – Time Synchronization. This query checks whether it is possible for the protocol to be run such that the honest client A successfully accepts a timesync response as valid and authentic from the honest server B.

```
query nonce: bitstring, x: bitstring, y: bitstring;
    event(clientAcceptsTime(new A, new B, nonce, x, y)).
```

This query holds for all four code versions.

Authenticity – Time Synchronization. This query ensures that if a time-sync message t is accepted by the honest client A as authentic from an honest server B, then B has really issued a message with the exact time data as in t and secured it with the cookie which is generated based on A's public key. This query corresponds to Goal 7.

```
query nonce: bitstring, x: bitstring, y: bitstring;
    event(clientAcceptsTime(new A, new B, nonce, x, y))
 ==> event(serverRespondsTime(keyhash(pk(sk_of(new A))),
                              new B, nonce, x, y)).
```

- As might be expected due to the lack of cookie secrecy, this query also does **not** hold for code version c030ut and **neither** does it hold for c030. Since for these code versions Mallory can gain access to cookies that Alice accepts as valid, she can use those cookies to generate time synchronization packets with maliciously manufactured synchronization information that Alice will accept.
- This query holds for the code versions c031 and c032.

References

1. Abadi, M.: Security protocols: principles and calculi. In: Aldini, A., Gorrieri, R. (eds.) FOSAD 2007. LNCS, vol. 4677, pp. 1–23. Springer, Heidelberg (2007)
2. Abadi, M., Fournet, C.: Mobile values, new names, and secure communication. In: Proceedings of the 28th ACM SIGPLAN-SIGACT Symposium on Principles of Programming Languages, POPL 2001, pp. 104–115. ACM, New York (2001). http://doi.acm.org/10.1145/360204.360213
3. Archer, M.: Proving correctness of the basic TESLA multicast stream authentication protocol with TAME. In: Workshop on Issues in the Theory of Security, pp. 14–15 (2002)
4. Basin, D., Capkun, S., Schaller, P., Schmidt, B.: Formal Reasoning About Physical Properties of Security Protocols. ACM Trans. Inf. Syst. Secur. **14**(2), 16:1–16:28 (2011)
5. Behrmann, G., David, A., Larsen, K.G.: A tutorial on UPPAAL. In: Bernardo, M., Corradini, F. (eds.) SFM-RT 2004. LNCS, vol. 3185, pp. 200–236. Springer, Heidelberg (2004). http://dx.doi.org/10.1007/978-3-540-30080-9_7
6. Blanchet, B., Smyth, B., Cheval, V.: ProVerif 1.88: automatic cryptographic protocol verifier, user manual and tutorial. Technical report, INRIA Paris-Rocquencourt, 08 2013
7. Blanchette, J.C., Nipkow, T.: Nitpick: a counterexample generator for higher-order logic based on a relational model finder. In: Kaufmann, M., Paulson, L.C. (eds.) ITP 2010. LNCS, vol. 6172, pp. 131–146. Springer, Heidelberg (2010). http://dblp.uni-trier.de/db/conf/itp/itp2010.html#BlanchetteN10
8. Cremers, C.J.F.: The scyther tool: verification, falsification, and analysis of security protocols. In: Gupta, A., Malik, S. (eds.) CAV 2008. LNCS, vol. 5123, pp. 414–418. Springer, Heidelberg (2008). http://dx.doi.org/10.1007/978-3-540-70545-1_38
9. Dolev, D., Yao, A.: On the security of public key protocols. IEEE Trans. Inf. Theory **29**(2), 198–208 (1983)
10. Durgin, N.A., Lincoln, P., Mitchell, J.C.: Multiset rewriting and the complexity of bounded security protocols. J. Comput. Secur. **12**(2), 247–311 (2004). http://content.iospress.com/articles/journal-of-computer-security/jcs215
11. Ganeriwal, S., Pöpper, C., Capkun, S., Srivastava, M.B.: Secure time synchronization in sensor networks (E-SPS). In: Proceedings of 2005 ACM Workshop on Wireless Security (WiSe 2005), pp. 97–106. ACM, Sept 2005
12. Hopcroft, P., Lowe, G.: Analysing a stream authentication protocol using model checking. Int. J. Inf. Secur. **3**(1), 2–13 (2004)
13. Housley, R.: Cryptographic Message Syntax (CMS). RFC 5652, RFC Editor, September 2009. http://www.rfc-editor.org/rfc/rfc5652.txt
14. IEEE: IEEE Standard for a Precision Clock Synchronization Protocol for Networked Measurement and Control Systems (2008). http://ieeexplore.ieee.org/servlet/opac?punumber=4579757
15. Krawczyk, H., Bellare, M., Canetti, R.: HMAC: Keyed-Hashing for Message Authentication. RFC 2104, RFC Editor, 02 1997. http://www.rfc-editor.org/rfc/rfc2104.txt
16. Levine, J.: A Review of Time and Frequency Transfer Methods. Metrologia **45**(6), 162–174 (2008)
17. Li, L., Sun, J., Liu, Y., Dong, J.S.: TAuth: verifying timed security protocols. In: Merz, S., Pang, J. (eds.) ICFEM 2014. LNCS, vol. 8829, pp. 300–315. Springer, Heidelberg (2014). http://dx.doi.org/10.1007/978-3-319-11737-9_20

18. Meier, S., Schmidt, B., Cremers, C., Basin, D.: The TAMARIN prover for the symbolic analysis of security protocols. In: Sharygina, N., Veith, H. (eds.) CAV 2013. LNCS, vol. 8044, pp. 696–701. Springer, Heidelberg (2013)
19. Milius, S., Sibold, D., Teichel, K.: An Attack Possibility on Time Synchronization Protocols Secured with TESLA-Like Mechanisms, Draft version: https://www8.cs.fau.de/staff/milius/AttackPossibilityTimeSyncTESLA.pdf
20. Mills, D., Haberman, B.: Network Time Protocol Version 4: Autokey Specification. RFC 5906, RFC Editor, 06 2010. http://www.rfc-editor.org/rfc/rfc5906.txt
21. Mills, D., Martin, J., Burbank, J., Kasch, W.: Network Time Protocol Version 4: Protocol and Algorithms Specification. RFC 5905, RFC Editor, 06 2010. http://www.rfc-editor.org/rfc/rfc5905.txt
22. Mizrahi, T.: Security Requirements of Time Protocols in Packet Switched Networks. RFC 7384, RFC Editor, 10 2014. http://www.rfc-editor.org/rfc/rfc7384.txt
23. Nipkow, T., Wenzel, M., Paulson, L.C.: Isabelle/HOL: A Proof Assistant for Higher-order Logic. Springer, Heidelberg (2002)
24. Paulson, L.C.: The inductive approach to verifying cryptographic protocols. J. Comput. Secur. 6(1–2), 85–128 (1998)
25. Perrig, A., Song, D., Canetti, R., Tygar, J.D., Briscoe, B.: Timed Efficient Stream Loss-Tolerant Authentication (TESLA): Multicast Source Authentication Transform Introduction. RFC 4082, RFC Editor, 06 2005. http://www.rfc-editor.org/rfc/rfc4082.txt
26. Röttger, S.: Analysis of the NTP Autokey Procedures, 2 2012. http://zero-entropy.de/autokey_analysis.pdf
27. Ryan, M.D., Smyth, B.: Applied pi calculus. In: Cortier, V., Kremer, S. (eds.) Formal Models and Techniques for Analyzing Security Protocols, Chap. 6. IOS Press (2011). http://www.bensmyth.com/files/Smyth10-applied-pi-calculus.pdf
28. Saeedloei, N., Gupta, G.: Timed π-calculus. In: Abadi, M., Lluch Lafuente, A. (eds.) TGC 2013. LNCS, vol. 8358, pp. 119–135. Springer, Heidelberg (2014)
29. Sibold, D., Teichel, K., Röttger, S.: Network time security. Technical report, IETF Secretariat, 07 2013. https://datatracker.ietf.org/doc/draft-ietf-ntp-network-time-security/history/
30. Sibold, D., Teichel, K., Röttger, S., Housley, R.: Protecting network time security messages with the cryptographic message syntax (CMS). Technical report, IETF Secretariat, 10 2014. https://datatracker.ietf.org/doc/draft-ietf-ntp-cms-for-nts-message/history/
31. Sun, K., Ning, P., Wang, C.: TinySeRSync: secure and resilient time synchronization in wireless sensor networks. In: Proceedings of the 13th ACM Conference on Computer and Communications Security, CCS 2006, pp. 264–277. ACM, New York (2006)
32. Syverson, P., Meadows, C., Cervesato, I.: Dolev-yao is no better than machiavelli. In: Degano, P. (ed.) First Workshop on Issues in the Theory of Security – WITS 2000, pp. 87–92, Jul 2000. http://theory.stanford.edu/~iliano/papers/wits00.ps.gz

Formal Support for Standardizing Protocols with State

Joshua D. Guttman, Moses D. Liskov$^{(\boxtimes)}$, John D. Ramsdell,
and Paul D. Rowe

The MITRE Corporation, Bedford, MA, USA
mliskov@mitre.org

Abstract. Many cryptographic protocols are designed to achieve their goals using only messages passed over an open network. Numerous tools, based on well-understood foundations, exist for the design and analysis of protocols that rely purely on message passing. However, these tools encounter difficulties when faced with protocols that rely on non-local, mutable state to coordinate several local sessions.

We adapt one of these tools, CPSA, to provide automated support for reasoning about state. We use Ryan's Envelope Protocol as an example to demonstrate how the message-passing reasoning can be integrated with state reasoning to yield interesting and powerful results.

Keywords: Protocol analysis tools · Stateful protocols · TPM · PKCS #11

1 Introduction

Many protocols involve only message transmission and reception, controlled by rules that are purely local to a session of the protocol. Typical protocols for authentication and key establishment are of this kind; each participant maintains only the state required to remember what messages must still be transmitted, and what values are expected in messages to be received from the peer.

Other protocols interact with long-term state, meaning state that persists across different sessions and may control behavior in other sessions. A bank account is a kind of long-term state, and it helps to control the outcome of protocol sessions in the ATM network. Specifically, the session fails when we try to withdraw money from an empty account. Of course, one session has an effect on others through the state: When we withdraw money today, there will be less remaining to withdraw tomorrow.

Hardware devices frequently participate in protocols, and maintain state that helps control those protocols. For example, PKCS#11 devices store and use keys, and are constrained by key attributes that control e.g. which keys may be used to wrap and export other keys. Trusted Platform Modules (TPMs) maintain Platform Configuration Registers (PCRs) some of which are modified only by certain special instructions. Thus, digitally signing the values in these

© Springer International Publishing Switzerland 2015
L. Chen and S. Matsuo (Eds.): SSR 2015, LNCS 9497, pp. 246–265, 2015.
DOI: 10.1007/978-3-319-27152-1_13

registers attests to the history of the platform. Some protocols involve multiple state histories; for instance, an online bank transfer manipulates the state of the destination account as well as the state of the source account.

State-based protocols are more challenging to analyze than protocols in which all state is session-local. Among the executions that are possible given the message flow patterns, one must identify those for which a compatible sequence of states exists. Thus, to justify standardizing protocols involving PKCS#11 devices or TPMs, one must do a deeper analysis than for stateless protocols. Indeed, since these devices are themselves standardized, it is natural to want to define and justify protocols that depend only on their required properties, rather than any implementation specific peculiarities.

The goal of this paper is to explain formal ideas that can automate this analysis, and to describe a support tool that assists with it.

Contributions of this paper. We make four main contributions:

- We identify two central axioms of state that formalize the semantics of state-respecting behaviors (Defnition 3). Each time a state is produced,
  1. it can be consumed by at most one subsequent transition.
  2. it cannot be observed after a subsequent transition consumes it.

  The first axiom is the essence of how the state-respecting analysis differs from standard message-based analysis. By contrast, once a message has been transmitted, it can be delivered (or otherwise consumed) repeatedly in the future.

  The second axiom, like the reader/writer principle in concurrency, allows observations to occur without any intrinsic order among them, so long as they all occur while that state is still available. It preserves the advantages of a partial order model, as enriched with state.
- We provide an alternative model of execution that maintains state in a family of traditional state machines, whose transitions are triggered by synchronization events in a state-respecting manner (see the extended version [17] for definitions and proofs). The justification for our two axioms is that they match this alternative, explicit-state-machine model exactly.
- We incorporated these two axioms into the tool CPSA [24], obtaining a tool that can perform state-respecting enrich-by-need protocol analysis.
- We applied the resulting version of CPSA to an interesting TPM-based protocol, the Envelope Protocol [2], verifying that it meets its security goal. We have also analyzed some incorrect variants, obtaining attacks.

Roadmap. After giving some background, we describe the Envelope Protocol and the TPM behaviors it relies on (Sect. 2). We introduce our protocol model (Sect. 3) in both its plain form, and the form enriched by the axioms in Contribution 1. Section 4 describes the CPSA analysis in the original model where state propagation is not distinguished from message-passing, and in the enriched model. We turn to related work in Sect. 5. Section 6 addresses a logical interpretation of enrich-by-need analysis and observes that this framework may be used, unmodified, for stateful protocols as we model them. We end with a brief comment on conclusions and future work.

Background: Strand spaces. We work within the strand space framework. A *strand* is a (usually short) finite sequence of events, where the events are

message transmission nodes;
message reception nodes; and
state synchronization nodes.

Each message transmission and reception node is associated with a message that is sent or received. State synchronization nodes were introduced into strand spaces recently [15]. Including them des not alter key definition such as bundles (Definition 1), and they allow us to flag events that, though the protocol principals perform them, are not message events. State synchronization nodes will be related to states via two different models in Sect. 3.

The behavior of a principal in a single, local run of one role of a protocol forms a strand. We call these *regular strands*. We also represent basic actions of an adversary as strands, which we call *adversary strands*. Adversary strands never need state synchronization nodes, since our model of the adversary allows it to use the network as a form of storage that never forgets old messages.

A protocol Π is represented by a finite set of strands, called the *roles* of the protocol, together with some auxiliary information about freshness and non-compromise assumptions about the roles. We write $\rho \in \Pi$ to mean that ρ is one of the roles of the protocol Π. The *regular strands of* Π are then all strands that result from any roles $\rho \in \Pi$ by applying a substitution that plugs in values in place of the parameters occurring in ρ.

For more information on strand spaces, see e.g. [14,27]. For the version containing state synchronization events as well as transmissions and receptions, see [15,23].

Background: Enrich-by-need analysis. In our form of protocol analysis, the input is a fragment of protocol behavior.

The output gives zero or more executions that contain this fragment. We call this approach "enrich-by-need" analysis (borrowed from our [16]), because it is a search process that gradually adds information as needed to explain the events that are already under consideration.

An analysis begins with an execution fragment $\mathbb{A}$, which may, for instance, reflect the assumption that one participant has engaged in a completed local session (a strand); that certain nonces were freshly chosen; and that certain keys were uncompromised. The result of the analysis is a set S of executions enriching the starting fragment $\mathbb{A}$. An algorithm implementing this approach is sound if, for every possible execution $\mathbb{C}$ that enriches $\mathbb{A}$, there is a member $\mathbb{B} \in S$ such that $\mathbb{C}$ enriches $\mathbb{B}$.

We do not require S to contain all possible executions because there are infinitely many of them if any. For instance, executions may always be extended by including additional sessions by other protocol participants. Thus, we want the set S to contain representatives that cover all of the *essentially different* possibilities. We call these representatives S the *shapes* for $\mathbb{A}$.

In practice, the set S of shapes for $\mathbb{A}$ is frequently finite and small.

When we start with a fragment $\mathbb{A}$ and find that it has the empty set $S = \emptyset$ of shapes, that means that no execution contains all of the structure in $\mathbb{A}$. To use this technique to show confidentiality assertions, we include a disclosure event in $\mathbb{A}$. If $\mathbb{A}$ extends to no possible executions at all, we can conclude that this secret cannot be revealed. If S is non-empty, the shapes are attacks that show how the confidentiality claim could fail.

The set S of shapes, when finite, also allows us to ascertain whether authentication properties are satisfied. If each shape $\mathbb{B} \in S$ satisfies an authentication property, then every possible execution $\mathbb{C}$ enriching $\mathbb{A}$ must satisfy the property too: They all contain at least the behavior exhibited in some shape, which already contained the events that the authentication property required.

This style of analysis is particularly useful in a partially ordered execution model, such as the one provided by strand spaces. In partially ordered models, when events e_1, e_2 are causally unrelated, neither precedes the other. In linearly ordered execution models, both interleavings $e_1 \prec e_2$ and $e_2 \prec e_1$ are possible, and must be considered. When there are many such pairs, this leads to exponentially many interleavings. None of the differences between them are significant.

2   The Envelope Protocol

We use Mark Ryan's Envelope Protocol [3] as a concrete example throughout the paper. The protocol leverages cryptographic mechanisms supported by a TPM to allow one party to package a secret such that another party can either reveal the secret or prove the secret never was and never will be revealed, but not both.

It is a particularly useful example to consider because it is carefully designed to use state in an essential way. In particular, it creates the opportunity to take either of two branches in a state sequence, but not both. In taking one branch, one loses the option to take the other. In this sense, it utilizes the non-monotonic nature of state that distinguishes it from the monotonic nature of messages. Additionally, although the Envelope Protocol is not standardized, it demonstrates advanced and useful ways to use the TPM. Standardization of such protocols is under the purview of the Trusted Computing Group (TCG). It will be very useful to understand the fundamental nature of state and to provide methods and tools to support the future standardization of protocols involving devices such as the TPM.

Protocol motivation. The plight of a teenager motivates the protocol. The teenager is going out for the night, and her parents want to know her destination in case of emergency. Chafing at the loss of privacy, she agrees to the following protocol. Before leaving for the night, she writes her destination on a piece of paper and seals the note in an envelope. Upon her return, the parents can prove the secret was never revealed by returning the envelope unopened. Alternatively, they can open the envelope to learn her destination.

The parents would like to learn their daughter's destination while still pretending that they have respected her privacy. The parents are thus the adversary. The goal of the protocol is to prevent this deception.

Necessity of long-term state. The long-term state is the envelope. Once the envelope is torn, the adversary no longer has access to a state in which the envelope is intact. A protocol based only on message passing is insufficient, because the ability of the adversary monotonically increases. Initially, the adversary has the ability to either return the envelope or tear it. In a purely message-based protocol the adversary will never lose these abilities.

Cryptographic version. The cryptographic version of this protocol uses a TPM to achieve the security goal. Here we restrict our attention to a subset of the TPM's functionality. In particular we model the TPM as having a state consisting of a single PCR and only responding to five commands.

A `boot` command (re)sets the PCR to a known value. The `extend` command takes a piece of data, d, and replaces the current value s of the PCR state with the hash of d and s, denoted $\#(d, s)$. In fact, the form of `extend` that we model, which is an `extend` within an encrypted session, also protects against replay. These are the only commands that alter the value in a PCR.

The TPM provides other services that do not alter the PCR. The `quote` command reports the value contained in the PCR and is signed in a way as to ensure its authenticity. The `create key` command causes the TPM to create an asymmetric key pair where the private part remains shielded within the TPM. However, it can only be used for decryption when the PCR has a specific value. The `decrypt` command causes the TPM to decrypt a message using this shielded private key, but only if the value in the PCR matches the constraint of the decryption key.

In what follows, Alice plays the role of the teenaged daughter packaging the secret. Alice calls the `extend` command with a fresh nonce n in an encrypted session. She uses the `create key` command constraining a new key k' to be used only when a specific value is present in the PCR. In particular, the constraining value cv she chooses is the following:

$$cv = \#(\mathsf{obt}, \#(n, s))$$

where obt is a string constant and s represents an arbitrary PCR value prior the extend command. She then encrypts her secret v with k', denoted $\{\!|v|\!\}_{k'}$.

Using typical message passing notation, Alice's part of the protocol might be represented as follows (where we temporarily ignore the replay protection for the `extend` command):

$$
\begin{array}{rcl}
A & \to \text{TPM} & : \ \{\!|\mathsf{ext}, n|\!\}_k \\
A & \to \text{TPM} & : \ \mathsf{create}, \#(\mathsf{obt}, \#(n, s)) \\
\text{TPM} \to & A & : \ k' \\
A & \to \text{Parent} & : \ \{\!|v|\!\}_{k'}
\end{array}
$$

The parent acts as the adversary in this protocol. We assume he can perform all the normal Dolev-Yao operations such as encrypting and decrypting messages when he has the relevant key, and interacting with honest protocol participants. Most importantly, the parent can use the TPM commands available in any order with any inputs he likes. Thus he can extend the PCR with the string `obtain` and use the key to decrypt the secret. Alternatively, he can refuse to learn the secret and extend the PCR with the string `ref` and then generate a TPM quote as evidence the secret will never be exposed. The goal of the Envelope Protocol is to ensure that once Alice has prepared the TPM and encrypted her secret, the parent should not be able to both decrypt the secret and also generate a refusal quote, $\{\!|\,\mathsf{quote}, \#(\mathsf{ref}, \#(n, s)), \{\!|v|\!\}_{k'}\,|\!\}_{aik}$.

A crucial fact about the PCR state in this protocol is the collision-free nature of hashing, ensuring that for every x

$$\#(\mathsf{obt}, \#(n, s)) \quad \neq \quad \#(\mathsf{ref}, x) \tag{1}$$

Formal protocol model. We formalize the TPM-based version of the Envelope Protocol using strand spaces [14]. Messages and states are represented as elements of a crypto term algebra, which is an order-sorted quotient term algebra. Sort T is the top sort of messages. Messages of sort A (asymmetric keys), sort S (symmetric keys), and sort D (data) are called *atoms*. Messages are atoms, tag constants, or constructed using encryption $\{\!|\,\cdot\,|\!\}_{(\cdot)}$, hashing $\#(\cdot)$, and pairing $(\cdot, \cdot)$, where the comma operation is right associative and parentheses are omitted when the context permits.

We represent each TPM command with a separate role that receives a request, consults and/or changes the state and optionally provides a response. As shown in Fig. 1, we use $m{\rightarrow}\bullet$ and $\bullet{\rightarrow}m$ to represent the reception and transmission of message m respectively. Similarly, we use $s \rightsquigarrow\!\circ$ and $\circ \rightsquigarrow s$ to represent

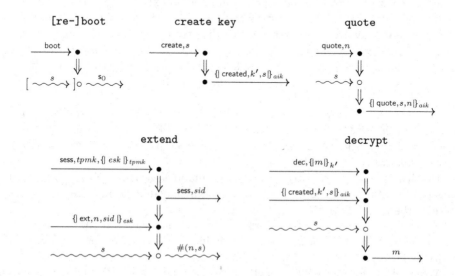

Fig. 1. TPM roles

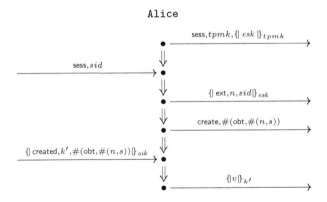

Fig. 2. Alice's role

the actions of reading and writing the value s to the state. We write $m \Rightarrow n$ to indicate that m precedes n immediately on the same strand.

As noted above, the **boot** role and the **extend** role are the only two roles that alter the state. This is depicted with the single event $\rightsquigarrow\circ\rightsquigarrow$ that atomically reads and then alters the state. The **boot** role receives the command and resets any current state s to the known value s_0. An alternate version of **boot** is needed to ensure that our sequences of state are well-founded. This version has a single state write event $\circ\rightsquigarrow s_0$.

The **extend** role first creates an encrypted channel by receiving an encrypted session key esk which is itself encrypted by some other secured TPM asymmetric key $tpmk$. The TPM replies with a random session id sid to protect against replay. It then receives the encrypted command to extend the value n into the PCR and updates the arbitrary state s to become $\#(n, s)$.

The **create key** role does not interact directly with the state. It receives the command with the argument s specifying a state. It then replies with a signed certificate for a freshly created public key k' that binds it to the state value s. The certificate asserts that the corresponding private key k'^{-1} will only be used in the TPM and only when the current value of the state is s. This constraint is leveraged in the **decrypt** role which receives a message m encrypted by k' and a certificate for k' that binds it to a state s. The TPM then consults the state (without changing it) to ensure it is in the correct state before performing the decryption and returning the message m.

Finally, the **quote** role receives the command together with a nonce n. It consults the state and reports the result s in a signed structure that binds the state to the nonce to protect against replay.

Since the **quote** role puts the state s into a message, and the **extend** role puts a message into the state, in our formalization states are the same kind of entity as messages.

We similarly formalize Alice's actions. Her access to the TPM state is entirely mediated via the message-based interface to the TPM, so her role has no state events. It is displayed in Fig. 2.

Alice begins by establishing an encrypted session with the TPM in order to extend a fresh value n into the PCR. She then has the TPM create a fresh key that can only be used when the PCR contains the value $\#(\mathsf{obt}, \#(n, s))$, where s is whatever value was in the PCR immediately before Alice performed her extend command. Upon receiving the certificate for the freshly chosen key, she uses it to encrypt her secret v that gives her destination for the night.

The parents may then either choose to further extend the PCR with the value obt in order to enable the decryption of Alice's secret, or they can choose to extend the PCR with the value ref and get a quote of that new value to prove to Alice that they did not take the other option. The adversary roles displayed in Fig. 3 constrain what the parents can do.

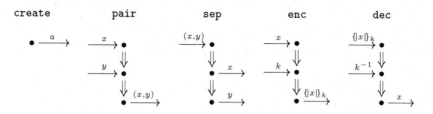

Fig. 3. Adversary roles, where a in the create role must be an atomic message.

It is important to note that, like Alice's role, the adversary roles do not contain any state events. Thus the adversary can only interact with the state via the interface provided by the TPM commands.

We aim to validate a particular security goal of the Envelope Protocol using the enrich-by-need method. The parent should not be able to both learn the secret value v and generate a refusal token.

Security Goal 1. *Consider the following events:*

- *An instance of the Alice role runs to completion, with secret v and nonce n both freshly chosen;*
- *v is observed unencrypted;*
- *the refusal certificate $\{\!|\mathsf{quote}, \#(\mathsf{ref}, \#(n, s)), \{\!|v|\!\}_{k'}|\!\}_{aik}$ is observed unencrypted.*

These events, which we call jointly $\mathbb{A}_0$, are not all present in any execution.

3   State-Respecting Bundles

In this section, we introduce a model of protocol behavior in the presence of global state; it is new in this paper. It enriches the notion of a bundle, which is the longstanding strand space formalization of global behaviors [14,27].

We organize this section as a sequence of refinements, starting from the traditional strand space bundle notion (Definition 1). We then give a direct generalization, *enriched bundles* (Definition 2) to associate states with synchronization events, and to track their propagation. We then introduce informally describe the notion of an *execution*, which explicitly includes both a bundle (as a global record of events and their causal ordering) and a family of state histories, and note that enriched bundles are not restrictive enough to match this notion of execution. This motivates the two axioms of state, leading to our final model of stateful protocol executions, *state-respecting bundles* (Definition 3), which matches the notion of executions. See the extended version of this paper [17] for formal definitions of executions and a proof of our claim that state-respecting bundles and executions match.

Definition 1 (Bundle). *Suppose that Σ is a finite set of strands. Let $\Rightarrow$ be the strand succession relation on $\mathsf{nodes}(\Sigma)$. Let $\rightarrow \subseteq \mathsf{nodes}(\Sigma) \times \mathsf{nodes}(\Sigma)$ be any relation on nodes of Σ such that $n_1 \rightarrow n_2$ implies that n_1 is a transmission event, n_2 is a reception event, and $\mathsf{msg}(n_1) = \mathsf{msg}(n_2)$.*

$\mathcal{B} = (\mathcal{N}, \rightarrow)$ is a bundle over Σ iff $\mathcal{N} \subseteq \mathsf{nodes}(\Sigma)$, and

1. *If $n_2 \in \mathcal{N}$ and n_1 precedes it on the same strand in Σ, then $n_1 \in \mathcal{N}$;*
2. *If n_2 is a reception node, there is exactly one $n_1 \in \mathcal{N}$ such that $n_1 \rightarrow n_2$; and*
3. *The transitive closure $(\Rightarrow \cup \rightarrow)^+$ of the two arrow relations is acyclic.*

$\mathcal{B}$ is a bundle of protocol Π iff *every strand with nodes in $\mathcal{B}$ is either an instance of a role of Π, or else an instance of one of the adversary roles in Fig. 3.*

Any finite behavior should have these properties, since otherwise some participant starts a role of the protocol in the middle, or receives a message no one sent, or else the (looping) pattern of events is causally impossible. By acyclicity, every bundle determines a partial ordering $\preceq_{\mathcal{B}}$ on its nodes, where $n_1 \preceq_{\mathcal{B}} n_2$ means that some path of one or more arrows $\rightarrow, \Rightarrow$ leads from n_1 to n_2 in $\mathcal{B}$.

We incorporate state transition histories directly into the bundles. To do this, we enrich the bundles with a new relation $\rightsquigarrow$ that propagates the current state from one event to another. We do this so that our analysis method can work with a single object that has both message dependencies and state dependencies within it. We also distinguish between *state transitions* and *state observations*. Transitions need to be linearly ordered if they pertain to a single device, but many state observations may occur between a single pair of state transitions. They are like *read* events in parallel computation: There is no need for concurrency control to sequentialize their access to the state, as long as they are properly nested between the right transition events.

This is an advantage of the strand space approach, which focuses on partially ordered execution models. It is important for enrich-by-need analysis, where the exponential number of interleavings must be avoided.

Later in this section, we will introduce a model containing a number of traditional state machines, where we correlate the synchronization nodes with transitions in their state histories. We make this model more rigorous in the extended version of this paper [17], where we prove an exact match between the state respecting behaviors we use here and the more traditional model of state machine histories.

3.1   Enriching Bundles with State

We now enrich the bundles to incorporate states, and to propagate them from node to node, just as transmissions and receptions propagate messages.

The diagrams in Sect. 2 suggest a way to incorporate state into bundles: We enrich them so that each state synchronization event is associated with messages representing states. A transition event is associated with a pair, representing the pre-state before the transition together with the post-state after it. The pre-state must be obtained from an earlier synchronization event. The post-state is produced by the transition, and may thus be passed to later events. We also now distinguish state observation events; these are associated with a single state, which is like a pre-state since it is received from an earlier event that produced it. We also identify initiation events, which initialize a devices state and serve as the beginning of a state computation history.

Initiation nodes $\circ \rightsquigarrow s$ record the event of creating a new state. We use init s to indicate an initiation of state to s.

Observation nodes $s \rightsquigarrow \circ$ record the current state without changing it. We use obsv s to indicate an observation of state s.

Transition nodes $s_0 \rightsquigarrow \circ \rightsquigarrow s_1$ represent the moment at which the state changes from a specific pre-state to a specific post-state. We use tran (s_0, s_1) to indicate a state transition with pre-state s_0 and post-state s_1.

In specifying protocols and their state manipulations, we can use the style illustrated in Fig. 1. There, an observation such as the synchronization node in the quote role, acquires a message on the incoming $\rightsquigarrow$ arrow. In this case, it is a variable s, which is itself a parameter to the role which contributes to the subsequent transmitted message. The decrypt role also has an incoming $\rightsquigarrow$ arrow labeled with s; in this case, the role can proceed to engage in this event only if the value s equals a previously available parameter acquired in the previous reception node. The extend role has a transition node, in which any pre-state s will be updated to a new post-state by hashing in the parameter n.

These pre- and post-state annotations, using parameters that appear elsewhere in the roles, determine subrelations of the transition relation associated with each instance of a role. An instance of the extend role with a particular value n_0 for the parameter n will engage only in state transformations that hash in that value n_0.

Observation events are not strictly necessary; we could model the checking of a state value as a transition $s \rightsquigarrow \circ \rightsquigarrow s$. However, this would require observation events be ordered in a specific sequence. This violates the principled choice that our execution model not include unnecessary ordering.

In the Introduction, we defined a protocol to be a finite set of strands called the *roles* of the protocol. An *enriched protocol* Π^+ will be a protocol Π enriched with a classification of its state synchronization events into init, tran, and obsv nodes, with each of those annotated with messages defining their pre- and post-states. The *regular strands* of Π^+ are all of the substitution instances of the roles of $\varPi^+$, including the instances of the pre- and post-states on the synchronization nodes.

An enriched bundle uses $\rightsquigarrow$ arrows to track the propagation of the state of each device involved in the behavior. This is not a sufficient model for reasoning about state, which requires also the two axioms of Definition 3, but it provides the objects from which we will winnow the state-respecting bundles.

Definition 2 (Enriched bundles). $\mathcal{B}^+ = (\mathcal{N}, \rightarrow, \rightsquigarrow)$ *is an* enriched bundle *iff* $(\mathcal{N}, \rightarrow)$ *is a bundle, and moreover:*

1. $n_1 \rightsquigarrow n_2$ *implies that* n_1 *is an* init *or* tran *event and* n_2 *is an* obsv *or* tran *event, and the post-state of* n_1 *equals the pre-state of* n_2*;*
2. *For each* obsv *or* tran *event* n_2*, there exists a unique* n_1 *such that* $n_1 \rightsquigarrow n_2$*;*
3. *The transitive closure* $(\Rightarrow \cup \rightarrow \cup \rightsquigarrow)^+$ *of the three arrow relations is acyclic. We refer to the partial order it determines as* $\prec_{\mathcal{B}^+}$ *or* $\prec$ *when* $\mathcal{B}^+$ *is clear.*

Enriched bundles are not a sufficient execution model, however, because they do not capture what is essentially different about state as compared to messages: the way that the next transition event consumes a state value, such that it cannot be available again unless a new transition creates it again. We can see this by connecting our current set-up to a state-machine model.

Each enriched protocol $\varPi^+$ determines a type of state machine. Its states (included in the set of messages) are all pre-states and post-states of the synchronization nodes of all instances of the roles of $\varPi^+$. A state machine has a set of initial states. In the state machine determined by $\varPi^+$, the initial states are the states $\sigma(s)$ such that some role $\rho \in \varPi^+$ has an initiation event init s, and σ is a substitution determining an instance of ρ.

The state machine determined by $\varPi^+$ has the state transition relation $\rhd$ consisting of all pairs of states (s_1, s_2) where

$[s_1 \rhd s_2]$ iff there exists a state transition node of $\varPi^+$ with pre-state t_1 and post-state t_2 and a substitution σ, such that $s_1 = \sigma(t_1)$ and $s_2 = \sigma(t_2)$.

A *state history* or *computation* is a finite or infinite sequence of states $s_0, s_1, \dots$ that starts with an initial state s_0, and, for every i, if s_{i+1} is defined then $s_i \rhd s_{i+1}$.

There may be a collection of devices $\{D_i\}_{i \in I}$ that instantiate this type of state machine. A *execution* consists of a bundle $\mathcal{B}$ (Definition 1) together with a state history for each device $\{D_i\}_{i \in I}$, where each transition is caused by a state synchronization node of $\mathcal{B}$.

The enriched bundles are not a sufficient model for reasoning about state, because there are enriched bundles that do not correspond to any execution in this sense. We will illustrate this in Sect. 4.

3.2   Our Axioms of State

The initiation and transition events are meant to describe the sequence of states that a device passes through. The notion of bundle says nothing about the "out-degree" of an event. A message transmission event can satisfy more than one

Fig. 4. State-respecting semantics. (1) State produced (either from a tran or init event) cannot be consumed by two distinct transitions. (2) Observation occurs after the state observed is produced but before that state is consumed by a subsequent transition.

message reception. However, a state event (initiation or transition) can satisfy *at most one* state transition event.

Observations must occur in a constrained place in the sequence of states. They acquire an incoming $\rightsquigarrow$ arrow from a transition or an initiation. Any such observation occurs before a subsequent change in the state.

These two principles—that transitions do not fork, and observations must precede a transition that consumes their state—motivate our execution model. They are illustrated in Fig. 4.

Definition 3 (State-respecting bundle). *Let* $\mathcal{B}^+ = (\mathcal{N}, \rightarrow, \rightsquigarrow)$ *be an enriched bundle with precedence order* $\prec$. $\mathcal{B}^+$ *is* state-respecting *if and only if:*

1. *if* $n \rightsquigarrow n_0$ *and* $n \rightsquigarrow n_1$, *where* n_0 *and* n_1 *are* tran *events, then* $n_1 = n_0$;
2. *Let the relation* $\prec^+$ *be the smallest transitive relation including* $\prec$ *such that whenever* n_0 *is an* obsv *and* n_1 *is a* tran, *then*

$$n \rightsquigarrow n_0 \text{ and } n \rightsquigarrow n_1 \text{ implies } n_0 \prec^+ n_1. \qquad (2)$$

Then $\prec^+$ *is acyclic.*

We call Clause 1 the *No State Split Principle*. Clause 2 is the *Observation Ordering Principle*.

These two axioms are adequate to provide a model of state. In particular, in the extended version of this paper [17], we prove that the executions in the sense we formalize there correspond exactly to the state-respecting bundles of Definition 3. Given a state-respecting $\mathcal{B}^+$, we show how to follow its $\rightsquigarrow$ arrows, thereby generating one state machine computation starting from each initiation node. This process would fail if the state axioms did not hold. Conversely, given a family of computations, we can use its steps to determine what states to assign to each synchronization node, and how to draw$\rightsquigarrow$ arrows between them.

3.3   Enrich-by-need for Stateful Protocols

In order to analyze stateful protocols with respect to state-respecting bundles (Definition 3), we adapted the Cryptographic Protocol Shapes Analyzer (CPSA) which performs automated protocol analysis with respect to (traditional) bundles (Definition 1). CPSA uses the enrich-by-need method as described in the Introduction. That is, it progressively extends an execution fragment $\mathbb{A}$ into a set of execution fragments $\{\mathbb{B}_i\}$. The extending occurs only as needed, namely, when the execution fragment does not contain enough information to fully describe a bundle. For message-only protocols, extending is necessary exactly when a message received at node n cannot be derived by the adversary using previously sent messages as inputs to a web of adversary strands.

We adapted CPSA in several ways to account for the properties of state synchronization nodes in state-respecting bundles. First, we added state synchronization nodes to the internal data structures of the tool. We then augmented the tool to recognize that extending is necessary when a state synchronization node n has pre-state s, but there is no node n_0 with post-state s such that $n_0 \rightsquigarrow n$. Finally, we implemented the corresponding rules for extending execution fragments by adding state synchronization nodes that supply the necessary state. In doing so, we experimented with two versions, one works for enriched bundles that need not satisfy the two axioms from Definition 3, and one which enforces these axioms. This former version allow us to perform analyses that lead to bundles satisfying Definition 2 which do not correspond to any executions of the state-machine model. The latter eliminates these ersatz results.

One advantage to the use of state-respecting bundles is that it allowed us to integrate an analysis of the stateful part of the protocol in a modular fashion. Our current release of CPSA [24] simply adds techniques for state-based reasoning without altering the message passing analysis algorithms. The analysis of protocols that do not contain state synchronization nodes remains unchanged. We thus provide a clean separation of the two distinct aspects of stateful protocols in an integrated whole.

The next section explores several examples that demonstrate the results of these two versions and hopefully provide some intuition about why the two axioms of state are necessary.

4   Analysis of the Envelope Protocol

The two conditions of Definition 3 identify the crucial aspects of state that distinguish state events from message events. They axiomatize necessary properties of state that are not otherwise captured by the properties of enriched bundles. In order to give the reader some intuition for these properties, we present several analyses of the Envelope Protocol in this section. We begin by contrasting two analyses; one is based on enriched bundles that only satisfy Definition 2, while the other is based on state-respecting bundles that also satisfy Definition 3.

Enriched vs. state-respecting bundles. Recall that the Envelope Protocol was designed to satisfy Security Goal 1. That is, there should be no executions in which (1) Alice completes a run with fresh, randomly chosen values for v and n, (2) v is available unencrypted on the network, and (3) the refusal certificate Q is also available on the network. Whether we use enriched bundles or state-respecting bundles as our model of execution, the analysis begins the same way. The relevant fragment of the point at which the two analyses diverges is depicted in Fig. 5. The reader may wish to refer to the figure during the following

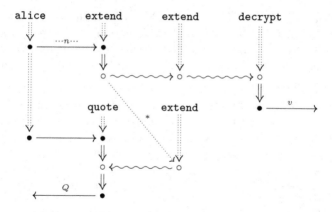

Fig. 5. A crucial moment in the CPSA analysis of the Envelope Protocol, demonstrating the importance of our first axiom of state.

description of the enrich-by-need process. The first three steps describe how we infer the existence of the top row of strands from right to left. The last two steps explain how we infer the strands in the bottom row from left to right.

1. The presence of v in unencrypted form implies the existence of a decrypt strand to reveal it.
2. The decrypt strand requires the current state to be $\#(\mathsf{obt}, \#(n, s))$, so our new principle of state explanation implies the existence of an extend strand with input value obt.
3. This newly inferred extend strand, in turn must have its current state $\#(n, s)$ explained which is done by another extend strand that receives the value n from Alice.
4. The presence of the quoted refusal token Q implies the existence of a quote strand to produce it.
5. The quote strand requires the state to be $\#(\mathsf{ref}, \#(n, s))$, which allows us to infer the third extend strand.

At this point in the analysis, the underlying semantics of bundles begins to matter. Our analysis still must explain how the state became $\#(n, s)$ for this last extend strand. If we use enriched bundles that do not satisfy Definition 3, then we may re-use the extend strand inferred in Step 3 as an explanation. This would cause us to add a ⤳ arrow between these two state events (along the dotted arrow ∗ of Fig. 5) forcing us to "split" the state coming out of the earlist extend strand. Further steps allow us to discover an enriched bundle compatible with our starting point, contrary to Security Goal 1. Importantly, however, all enriched bundles that extend the fragment with the split state are non-state-respecting.

If, on the other hand, we only allow state-respecting bundles, Condition 1 of Definition 3 does not allow us to re-use the extend strand inferred in Step 3 to explain the state found on the strand of Step 5. Instead, we are forced to infer yet another **extend** strand that receives Alice's nonce n. However, since Alice uses an encrypted session that provides replay protection, the adversary has no way to return the TPM state to $\#(n, s)$. Thus, although there are enriched bundles that violate Security Goal 1, there are no state-respecting bundles that do so.

A flawed version. We also performed an analysis of the Envelope Protocol, removing the assumption that Alice's nonce n is fresh, to demonstrate our state-respecting variant's ability to automatically detect attacks. The analysis proceeds similarly; as in the previous analysis we decline to add a $\rightsquigarrow$ arrow along $*$ thanks to our stateful semantics. However, the alternative possibility that a fresh **extend** strand provides the necessary state proves to work out. Because n is not freshly chosen, the parent can engage in a distinct **extend** session with the same n.

Note that our analysis does not specify that $s = \mathsf{s_0}$, where s is the state of the PCR when first extended. For the case where $s = \mathsf{s_0}$, the attack is to reboot the TPM after obtaining one value (either the refuse token or Alice's secret), re-extend the boot state with n, and then obtain the other. More generally, as long as s is a state that the parent can induce, a similar attack is possible.

4.1   The Importance of Observer Ordering

The Envelope Protocol example demonstrates the crucial importance of capturing our first axiom of state correctly. The second axiom, involving the relative order of observations and state transition, is no less crucial to correct understanding of stateful protocols.

Another example protocol, motivated by a well-known issue with PKCS #11 (see, e.g. [9]), illustrates the principle more clearly. Suppose a hardware device is capable of producing keys that are meant to be managed by the device and not learnable externally. If the device has limited memory, it may be necessary to export such a key in an encrypted form so the device can utilize external storage.

Thus, device keys can be used for two distinct purposes: for encryption/decryption of values on request, or for encrypting internal keys for external storage. It is important that the purpose of a given key be carefully tracked, so that the device is not induced to decrypt one of its own encrypted keys.

Suppose that for each key, the device maintains a piece of state, namely, one of three settings:

- A **wrap** key is used only to encrypt internal keys.
- A **decrypt** key may be used to encrypt or decrypt.
- An **initial** key has not yet been assigned to either use.

If a key in the **wrap** state can later be put in the **decrypt** state, a relatively obvious attack becomes possible: while in the wrap state, the device encrypts some internal key, and later, when the key is in the decrypt state, the device decrypts the encrypted internal key.

However, if keys can never exit the wrap state once they enter it, this attack should not be possible. If we were to represent this protocol within CPSA, we would include the following roles:

- A create key role that generates a fresh key and initializes its state to initial
- A set wrap role that transitions a key from initial or decrypt to wrap.
- A set decrypt role that transitions a key from initial to decrypt.
- A wrap role in which a user specifies two keys (by reference), and the device checks (with an observer) that the first is in the wrap state and if so, then encrypts the second key with the first and transmits the result.
- A decrypt role in which a user specifies a key (by reference) and a ciphertext encrypted under that key, and the device checks (with an observer) that the key is in the decrypt state and if so, then decrypts the ciphertext and transmits the resulting plaintext.

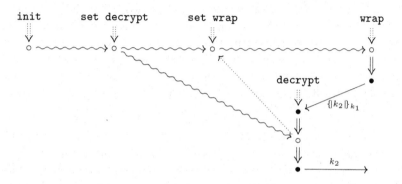

Fig. 6. Observer ordering example

Note that the attack should not be possible. However, the bundle described in Fig. 6 is a valid bundle, and fails to be state-respecting only because of our axiom about observers. Our second axiom induces an ordering so that the observer in the decrypt strand occurs before the following transition event in the set wrap strand. The induced ordering is shown in the figure with a single dotted arrow; note the cycle among state events present with that ordering that is not present without it.

5 Related Work

The problem of reasoning about protocols and state has been an increasing focus over the past several years. Protocols using TPMs and other hardware security modules (HSMs) have provided one of the main motivations for this line of work.

A line of work was motivated by HSMs used in the banking industry [18,28]. This work identified the effects of persistent storage as complicating the security analysis of the devices. There was also a strong focus on the case of PKCS #11 style devices for key management [5,6,12]. These papers, while very informative, exploited specific characteristics of the HSM problem; in particular, the most important mutable state concerns the *attributes* that determine the usage permitted for keys. These attributes should usually be handled in a monotonic way, so that once an attribute has been set, it will not be removed. This justifies using abstractions that are more typical of standard protocol analysis.

In the TPM-oriented line of work, an early example using an automata-based model was by Gürgens et al. [13]. It identified some protocol failures due to the weak binding between a TPM-resident key and an individual person. Datta et al.'s "A Logic of Secure Systems" [8] presents a dynamic logic in the style of PCL [7] that can be used to reason about programs that both manipulate memory and also transmit and receive cryptographically constructed messages. Because it has a very detailed model of execution, it appears to require a level of effort similar to (multithreaded) program verification, unlike the less demanding forms of protocol analysis.

Mödersheim's set-membership abstraction [21] works by identifying all data values (e.g. keys) that have the same properties; a change in properties for a given key K is represented by translating all facts true for K's old abstraction into new facts true of K's new abstraction. The reasoning is still based on monotonic methods (namely Horn clauses). Thus, it seems not to be a strategy for reasoning about TPM usage, for instance in the Envelope Protocol.

Guttman [15] developed a theory for protocols (within strand spaces) as constrained by state transitions, and applied that theory to a fair exchange protocol. It introduced the key notion of *compatibility* between a protocol execution ("bundle") and a state history. This led to work by Ramsdell et al. [23] that used CPSA to draw conclusions in the states-as-messages model. Additional consequences could then be proved using the theorem prover PVS [22], working within a theory of both messages and state organized around compatibility.

A group of papers by Ryan with Delaune, Kremer, and Steel [10,11], and with Arapinis and Ritter [2] aim broadly to adapt ProVerif for protocols that interact with long-term state. ProVerif [1,4] is a Horn-clause based protocol analyzer with a monotonic method: in its normal mode of usage, it tracks the messages that the adversary can obtain, and assumes that these will always remain available. Ryan et al. address the inherent non-monotonicity of adversary's capabilities by using a two-place predicate att(u, m) meaning that the adversary may possess m at some time when the long-term state is u. In [2], the authors provide a compiler from a process algebra with state-manipulating operators to sets of Horn clauses using this primitive. In [11], the authors analyze protocols with specific syntactic properties that help ensure termination of the analysis. In particular, they bound the state values that may be stored in the TPMs. In this way, the authors verify two protocols using the TPM, including the Envelope Protocol.

Meier, Schmidt, Cremers, and Basin's tamarin prover [20] uses multiset rewriting (MSR) as a semantics in which to prove properties of protocols. Since MSR suffices to represent state, it provides a way to prove results about protocols with state. Künnemann studied state-based protocol analysis [19] in a process algebra akin to StatVerif, which he translated into the input language of tamarin to use it as a proof method. Curiously, the main constructs for mutable state and concurrency control (locking) are axiomatized as properties of traces rather than encoded within MSR (see [19, Fig. 10]).

Our work. One distinguishing feature of this work is our extremely simple modification to the plain message passing semantics to obtain a state-respecting model. These are the two Axioms 1–2 in Definition 3. We think it is an attractive characteristic of the strand space framework that state reflects such a clean foundational idea. Moreover, this foundational idea motivated a simple set of alterations to the enrich-by-need tool CPSA.

6 Protocol Security Goals

The enrich-by-need analysis performed in our enhanced version of CPSA is fully compatible with the language of goals found in previous work such as [26]. The goal language is based on two classes of predicates: role-related predicates that relate an event or parameter value to its use within a specific protocol role, and predicates that are protocol-independent and describe important properties of bundles. The latter includes the ordering of events as well as assumptions about freshly chosen values and uncompromised keys. Both classes of predicates apply within state-respecting bundles in a natural way. The role-related predicates are sensitive only to the position of an event in the sequence of events of a role, and to the choice of parameter values in that instance of the role. Indeed, nodes that represent state transitions or observations are handled in exactly the same way, since they have positions in the role and parameter values in just the same way as the message transmission and reception events.

Thus, the state-respecting version of CPSA can verify formulas expressing security goals in exactly the same way as the previous version, and with the same semantic definitions.

7 Conclusion

In this paper, we have argued that CPSA—and possibly other formalized protocol analysis methods—can provide reliable analysis when protocols are standardized, even when those protocols are manipulating devices with long-term state. A core idea of the formalization are the two axioms of Definition 3, which encapsulate the difference between a message-based semantics and the state-respecting semantics.

References

1. Abadi, M., Blanchet, B.: Analyzing security protocols with secrecy types and logic programs. J. ACM **52**(1), 102–146 (2005)
2. Arapinis, M., Ritter, E., Ryan, M.D.: Statverif: verification of stateful processes. In: Computer Security Foundations Symposium (CSF), pp. 33–47. IEEE (2011)

3. Arapinis, M., Ryan, M., Ritter, E.: StatVerif: verification of stateful processes. In: IEEE Symposium on Computer Security Foundations. IEEE CS Press, June 2011
4. Blanchet, B.: An efficient protocol verifier based on Prolog rules. In: 14th Computer Security Foundations Workshop, pp. 82–96. IEEE CS Press, June 2001
5. Cortier, V., Keighren, G., Steel, G.: Automatic analysis of the security of XOR-based key management schemes. In: Grumberg, O., Huth, M. (eds.) TACAS 2007. LNCS, vol. 4424, pp. 538–552. Springer, Heidelberg (2007)
6. Cortier, V., Steel, G.: A generic security API for symmetric key management on cryptographic devices. In: Backes, M., Ning, P. (eds.) ESORICS 2009. LNCS, vol. 5789, pp. 605–620. Springer, Heidelberg (2009)
7. Datta, A., Derek, A., Mitchell, J.C., Pavlovic, D.: A derivation system and compositional logic for security protocols. J. Comput. Secur. **13**(3), 423–482 (2005)
8. Datta, A., Franklin, J., Garg, D., Kaynar, D.: A logic of secure systems and its application to trusted computing. In: 2009 30th IEEE Symposium on Security and Privacy, pp. 221–236. IEEE (2009)
9. Delaune, S., Kremer, S., Ryan, M.D.: Composition of password-based protocols. In: Proceedings of the 21st IEEE Computer Security Foundations Symposium (CSF'08), pp. 239–251. IEEE Computer Society Press, June 2008
10. Delaune, S., Kremer, S., Ryan, M.D., Steel, G.: A formal analysis of authentication in the TPM. In: Degano, P., Etalle, S., Guttman, J. (eds.) FAST 2010. LNCS, vol. 6561, pp. 111–125. Springer, Heidelberg (2011)
11. Delaune, S., Kremer, S., Ryan, M.D., Steel, G.: Formal analysis of protocols based on TPM state registers. In: IEEE Symposium on Computer Security Foundations. IEEE CS Press, June 2011
12. Fröschle, S., Sommer, N.: Reasoning with past to prove PKCS#11 keys secure. In: Degano, P., Etalle, S., Guttman, J. (eds.) FAST 2010. LNCS, vol. 6561, pp. 96–110. Springer, Heidelberg (2011)
13. Gürgens, S., Rudolph, C., Scheuermann, D., Atts, M., Plaga, R.: Security evaluation of scenarios based on the TCG's TPM specification. In: Biskup, J., López, J. (eds.) ESORICS 2007. LNCS, vol. 4734, pp. 438–453. Springer, Heidelberg (2007)
14. Guttman, J.D.: Shapes: surveying crypto protocol runs. In: Cortier, V., Kremer, S. (eds.) Formal Models and Techniques for Analyzing Security Protocols, Cryptology and Information Security Series. IOS Press (2011)
15. Guttman, J.D.: State and progress in strand spaces: proving fair exchange. J. Autom. reasoning **48**(2), 159–195 (2012)
16. Guttman, J.D.: Establishing and preserving protocol security goals. J. Comput. Secur. **22**(2), 201–267 (2014)
17. Guttman, J.D., Liskov, M.D., Ramsdell, J.D., Rowe, P.D.: Formal support for standardizing protocols with state (extended version). Arxiv, September 2015. http://arxiv.org/abs/1509.07552
18. Herzog, J.: Applying protocol analysis to security device interfaces. IEEE Secur. Priv. **4**(4), 84–87 (2006)
19. Kremer, S., Künnemann, R.: Automated analysis of security protocols with global state. In: IEEE Symposium on Security and Privacy, pp. 163–178 (2014)
20. Meier, S., Schmidt, B., Cremers, C., Basin, D.: The TAMARIN prover for the symbolic analysis of security protocols. In: Sharygina, N., Veith, H. (eds.) CAV 2013. LNCS, vol. 8044, pp. 696–701. Springer, Heidelberg (2013)
21. Mödersheim, S.: Abstraction by set-membership: verifying security protocols and web services with databases. In: ACM Conference on Computer and Communications Security, pp. 351–360 (2010)

22. Owre, S., Rushby, J.M., Shankar, N.: PVS: a prototype verification system. In: Kapur, D. (ed.) CADE 1992. LNCS, vol. 607, pp. 748–752. Springer, Heidelberg (1992). http://pvs.csl.sri.com
23. Ramsdell, J.D., Dougherty, D.J., Guttman, J.D., Rowe, P.D.: A hybrid analysis for security protocols with state. In: Albert, E., Sekerinski, E. (eds.) IFM 2014. LNCS, vol. 8739, pp. 272–287. Springer, Heidelberg (2014)
24. Ramsdell, J.D., Guttman, J.D.: CPSA: A cryptographic protocol shapes analyzer (2009). http://hackage.haskell.org/package/cpsa
25. Ramsdell, J.D., Guttman, J.D., Millen, J.K., O'Hanlon, B.: An analysis of the CAVES attestation protocol using CPSA. MITRE Technical report MTR090213, The MITRE Corporation, December 2009. http://arxiv.org/abs/1207.0418
26. Rowe, P.D., Guttman, J.D., Liskov, M.D.: Measuring protocol strength with security goals. Submitted to IJIS in the SSR 2014 special issue, April 2015. http://web.cs.wpi.edu/~guttman/pubs/ijis_measuring-security.pdf
27. Thayer, F.J., Herzog, J.C., Guttman, J.D.: Strand spaces: proving security protocols correct. J. Comput. Secur. **7**(2/3), 191–230 (1999)
28. Youn, P., Adida, B., Bond, M., Clulow, J., Herzog, J., Lin, A., Rivest, R., Anderson, R.: Robbing the bank with a theorem prover. In: Security Protocols Workshop (2007). http://www.cl.cam.ac.uk/techreports/UCAM-CL-TR-644.pdf

Author Index

Bernstein, Daniel J. 109

Chen, Cheng 140
Choi, Daeseon 43
Chou, Tung 109
Chuengsatiansup, Chitchanok 109
Clarke, Dylan 3

Escobar, Santiago 86

Fujiki, Yuri 203

González-Burgueño, Antonio 86
Guttman, Joshua D. 246

Hao, Feng 3, 21
Hokino, Masatoshi 203
Hülsing, Andreas 109
Hwang, Jung Yeon 43

Jin, Seung-Hun 43

Kaneko, Takeaki 203
Kayuni, Mwawi Nyirenda 185
Khan, Mohammed Shafiul Alam 165, 185
Kim, Seung-Hyun 43

Lambooij, Eran 109
Lange, Tanja 109
Li, Wanpeng 185
Liskov, Moses D. 246

McCorry, Patrick 3
Meadows, Catherine 86
Mehrnezhad, Maryam 21
Meseguer, José 86
Milius, Stefan 218
Mitchell, Chris J. 165, 185

Niederhagen, Ruben 109

Onda, Sakura 203

Ramsdell, John D. 246
Rowe, Paul D. 246
Ruland, Karl Christoph 70

Sakimura, Natsuhiko 203
Santiago, Sonia 86
Sassmannshausen, Jochen 70
Sato, Hiroyuki 203
Sekar, Gautham 154
Shahandashti, Siamak F. 3, 21
Sibold, Dieter 218
Song, Boyeon 43

Teichel, Kristof 218

van Vredendaal, Christine 109

Yang, Kang 140
Yau, Po-Wah 185

Zhang, Jiang 140
Zhang, Zhenfeng 140

Printed in the United States
By Bookmasters